INFORMATION PROCESSING IN ANIMALS: Conditioned Inhibition

Edited by
RALPH R. MILLER
NORMAN E. SPEAR
The State University of New York at Binghamton

LAWRENCE ERLBAUM ASSOCIATES, PUBLISHERS
1985 Hillsdale, New Jersey London

Lawrence Erlbaum Associates, Inc., Publishers
365 Broadway
Hillsdale, New Jersey 07642

Library of Congress Cataloging in Publication Data
Main entry under title:

Information processing in animals—conditioned inhibition.

Proceedings of the 1983 Binghamton Symposium on Information Processing in Animals: Conditioned Inhibition, held June 14–16, 1983 at the State University of New York at Binghamton.
Includes bibliographies and indexes.
1. Conditioned response—Congresses. 2. Inhibition—Congresses. 3. Animal psychology—Congresses. I. Miller, Ralph R., 1940– . II. Spear, Norman E. III. Binghamton Symposium on Information Processing in Animals: Conditional Inhibition (1983)
QP416.I54 1985 591.5′1 84-18810

ISBN 0–89859–506–1

Printed in the United States of America
10 9 8 7 6 5 4 3 2 1

Contents

Preface

Interest in conditioned inhibition has repeatedly waxed and waned over the last 70 years. In the early part of this century the concept was central to Pavlov's (1927) theorizing. There followed a period of relative neglect after which Hull (1943) and Konorski (1948) revitalized (and greatly modified) the idea of conditioned inhibition in the 1940s. Next came a period of disinterest in which but a few researchers, most notably Konorski and his colleagues, concerned themselves with the concept.

Conditioned inhibition regained the attention of learning theorists in the late 1960s. This period saw the general acceptance of the Rescorla's (1969) and Hearst's (1972) carefully constructed definition of a conditioned inhibitor as a stimulus that passes both summation and retardation tests (see LoLordo and Fairless' chapter in this volume). The acceptance of this definition gave new vigor to the study of conditioned inhibition, which had previously been a rather amorphous name that was applied to a number of only marginally related concepts and phenomena. The achievement of this period of interest are well summarized in Boakes and Halliday's (1972) *Proceedings of the Sussex Conference on Conditioned Inhibition* (''Inhibition and Learning'').

Concern for conditioned inhibition waned again in the mid and late 1970s, but now in the 1980s it has returned to center stage as the subject of numerous experiments and considerable theorizing. Many of the researchers responsible for this recent resurgence gathered together at the State University of New York at Binghamton on June 14, 15, and 16 of 1983, to exchange thoughts and summarize their current views concerning conditioned inhibition. This volume is the Proceedings of the 1983 Binghamton Symposium on Information Processing in Animals: Conditioned Inhibition. We view it as a successor to the 1972 Boakes and Halliday anthology.

In our capacity as editors, we are delighted with the quality of the individual contributions as well as the comprehensiveness of the collective chapters. Yet, all our organization efforts notwithstanding, there are two notable omissions. One is Anthony Dickinson's motivational view of conditioned inhibition, for which the interested reader is directed to his 1980 monograph. And the other is conceptual developments that have occurred in Eliot Hearst's laboratory, for which the reader is referred to the Kaplan and Hearst (in press) chapter in the continuing Harvard series on the Quantitative Analysis of Behavior, although many of the ideas expressed by Kaplan and Hearst can also be found in some of the chapters in the present volume (e.g., Fowler, Kleiman, & Lysle; Mackintosh & Cotton; and Miller & Schachtman).

Continuing in the tradition of past investigators of conditioned inhibition, animals have been the primary subjects of choice in recent years. Conceptually, things have been far more fluid. At the beginning of this decade, two theories of conditioned inhibition were popular: Konorski's (1967) and Wagner and Rescorla's (1972) view of conditioned inhibition as the manifestation of a negative CS–US association, and Konorski's (1948) and Rescorla's (1979) view of conditioned inhibition as the elevation of the threshold for reactivating the memory trace of a US. Now but a few years later, as can be seen in the present chapters, few researchers identify with either of these positions. Theoretically, many of the present chapters appear to favor one or another form of two relatively new theoretical views of conditioned inhibition: One of these focuses on the relativity of a target stimulus' (X) associative strength with respect to the context or the A prestimulus in a Pavlovian A+/AX-preparation (e.g., Fowler, Kleiman, & Lysle; Mackintosh & Cotton, and Miller & Schachtman), and the other hinges on the similarity of a conditioned inhibition to a conditional discriminative stimulus (e.g., Bolles; Holland; Jenkins; and Rescorla).

One of the great pleasures of the conference was to see the convergence of data. In a science in which embarrassingly many studies fail to replicate, we heard speaker after speaker report similar data. Although some participants and observers found this tendency toward redundancy slightly disturbing, many others were delighted to have a solid data base through which competing interpretations could be compared.

Yet, one major definitional schism did arise that resulted in the creation of two data bases that defied integration. One set of investigators, Bolles, Jenkins, Rescorla, and to some degree Holland, chose to regard the summation test for conditioned inhibition as more fundamental than the retardation test and focused on preparations that left the "conditioned inhibitor" capable of passing the former but not the latter test. In contrast, the other contributors addressed their attention to stimuli that passed summation and the retardation tests. At this time it is not possible to tell which is of greater value—the traditional, more limited definition of conditioned inhibition that possibly refers to a unitary phenomenon, or the newer, broader definition that likely refers to a number of phenomena having perhaps only passage of the summation test in common.

The differences between the two definitions, and consequently the phenomena studied, are nontrivial. Based largely on Holland's penetrating studies, it appears as if "traditional" conditioned inhibitors (those stimuli that will pass both summation and retardation tests) may best be created by having the putative conditioned inhibitor and the conditioned exciter upon which it is based present *simultaneously,* e.g., Pavlov's A+/AX− paradigm or context+/context and X−. Such conditioned inhibitors frequently not only pass both tests for conditioned inhibition, but they also appear to readily transfer their inhibitory powers to other conditioned exciters that have been paired to the same US and to lose their inhibitory characteristics when consistently paired with the US with which they were initially negatively correlated. In contrast, some researchers have chosen to regard negative conditional discriminate stimuli as instances of conditioned inhibition because such stimuli pass summation tests. These investigators have focused largely on *serial* stimulus presentation during conditioning, e.g., A+/X followed by A−. Such putative conditioned inhibitors appear to fail retardation tests, do not readily transfer to other conditioned exciters (unless the subject receives extensive discriminatory training with the new CS), and retain their inhibitory control over A even when subsequently paired with the US to the point of becoming themselves conditioned exciters.

Despite Holland's emphasis of the simultaneous/serial cues distinction, some researchers, e.g., Jenkins, have obtained the "conditional discrimination syndrome" using simultaneous cues. This suggests that the simultaneous/serial cues distinction may not map as neatly onto the traditional conditioned inhibition/conditional discrimination distinction as Holland's data suggest. Although this leaves unresolved the critical factor(s) responsible for producing conditional discriminative stimuli as opposed to conditioned inhibitors, it does not dispute the existence of two distinct types of cues having in common only similar results on summation tests.

Those researchers electing to use only the summation test to define conditioned inhibition do not deny that the more traditional type of conditioned inhibitor also fits their definition, but they have chosen to focus their attention largely on those stimuli that pass the summation test but not the retardation test for conditioned inhibition. Whether or not this new definition of conditioned inhibition ultimately proves to be of value in illuminating the mechanisms responsible for traditional conditioned inhibition, there can be no question that the present analyses of such stimuli by Bolles, Holland, Jenkins, and Rescorla add greatly to our understanding of conditional discrimination.

In light of the sharp distinctions that arise from whether or not passage of a retardation test is assumed to be essential for a stimulus to be called a conditioned inhibitor, we have attempted to organize the chapters in terms of which definition was used. This effort was not totally successful as some of the contributors have focused on factors orthogonal to the one- versus two-test distinction; however, we believe that it is, all things considered, the best organizing principle available.

First, the reader finds LoLordo and Fairless' comprehensive historical theoretical review serving as an introduction to all that follows. Then comes a series of chapters (2–8) that employ the more traditional two-test definitions of a conditioned inhibitor to either analyze the nature of conditioned inhibition or to consider the impact of such conditioned inhibitors upon other phenomena. Next, Holland's chapter (9) serves as an excellent transition and introduction to the new one-test definition of a conditioned inhibitor, as it squarely confronts the empirical differences that arise from the two definitions. Then, there are three chapters (10–12) that use the one-test definition of conditioned inhibition. And finally there is a chapter (13) by Soltysik in which he uses the concept of conditioned inhibition in the framework of the protection-from-extinction phenomenon to illuminate the basis of maintained responding without reinforcement. Soltysik's experimental paradigm is A+/A followed by X−, which does not fit neatly into either the simultaneous or X-first serial groupings aforementioned. However, this is not to imply that his chapter is any the less relevant. In fact, as Solytsik makes clear in his chapter, his paradigm is particularly appropriate in light of his central concern being the protection-from-extinction phenomenon.

It is perhaps noteworthy that today no one appears to use an operational definition of conditioned inhibition, i.e., one that hinges on what the experimenter does. Rather, all the contributors have employed behavioral working definitions of conditioned inhibition that hinge on how the animal behaves in response to the stimulus in question, regardless of how the animal arrived in that state.

The June, 1983 Binghamton Symposium that gave rise to this volume was generously supported by the SUNY Conversations in the Disciplines Program, Lawrence Erlbaum Associates, and the SUNY–Binghamton Offices of the Vice-President, Provost, and Dean of Arts and Sciences. Additional support was provided by the Center for Neurobehavioral Sciences and the Department of Psychology at SUNY–Binghamton. Thanks are also due Todd Schachtman for his preparation of the index, Janice Felter, Rita Sawicki, Teri Tanenhaus, and Roxanne Vavra for their secretarial assistance, and Larry Erlbaum and his entire staff for their willing and able assistance in producing this volume. Finally, a special debt of gratitude is due Joan Miller, who looked after the palates and stomachs of all those who attended the Symposium.

The Symposium was as successful as it was not only because of the high quality of the formal presentations but also due to the extensive exchanges that took place during the formal discussion periods and throughout the days and nights of the Symposium. Many a model had to be rethought in light of these conversations between participants and expert observers who were anything but silent observers. Thus, the editors and contributors are grateful to the distinguished guests who attended the Symposium and made important contributions to the intellectual and social climate. In addition to those in the ranks of the SUNY–Binghamton faculty, these observers included Joe Ayres, Peter Balsam,

Bruce Brown, Ruth Colwill, Helen Daly, Paul DeVito, John Donahoe, Paula Durlach, Charles Flaherty, William Gordon, Douglas Grant, Lynn Hammond, Riley Hinson, Werner Honig, Peter Kaplan, William Pavlik, John Pearce, Alan Randich, David Riccio, Jean Richards, Rick Richardson, Robert Rosellini, Shepard Siegal, Robert St. Claire–Smith, Fred Stollnitz, Roger Tarpy, David A. Thomas, and Arthur Tomie.

R.R.M.
N.E.S.

REFERENCES

Boakes, R. A., & Halliday, M. S. (Eds.) (1972). *Inhibition and learning*. London: Academic Press.

Dickinson, A. (1980). *Contemporary animal learning theory*. Cambridge: UK: Cambridge University Press.

Hearst, E. (1972). Some persistent problems in the analysis of conditioned inhibition. In R. A. Boakes & M. S. Halliday (Eds.), *Inhibition and learning*. London: Academic Press.

Hull, C. L. (1943). *Principles of behavior*. New York: Appleton–Century–Crofts.

Kaplan, P., & Hearst, E. (in press). Trace conditioning, contiguity, and context. In M. Commona, R. Herrnstein, & A. R. Wagner (Eds.), *Harvard Symposium on Quantitative Analysis of Behavior:* (Vol. 3), *Acquisition*. Cambridge, MA: Ballinger.

Konorski, J. (1948). *Conditioned reflexes and neural organization*. Cambridge, UK: Cambridge University Press.

Konorski, J. (1967). *Integrative activity of the brain: An interdisciplinary approach*. Chicago: University of Chicago Press.

Pavlov, I. P. (1927). *Conditioned reflexes*. London: Oxford University Press.

Rescorla, R. A. (1969). Conditioned inhibition of fear resulting from negative CS–US contingencies. *Journal of Comparative and Physiological Psychology, 67*, 504–509.

Rescorla, R. A. (1979). Conditioned inhibition and extinction. In A. Dickinson & R. A. Boakes (Eds.), *Mechanisms of learning and motivation*. Hillsdale, NJ: Erlbaum.

Wagner, A. R., & Rescorla, R. A. (1972). Inhibition in Pavlovian conditioning: Application of a theory. In R. A. Boakes & M. S. Halliday (Eds.), *Inhibition and learning*. London: Academic Press.

1 Pavlovian Conditioned Inhibition: The Literature Since 1969

Vincent M. LoLordo and Jeffrey L. Fairless
Dalhousie University

INTRODUCTION

Our review of the literature on Pavlovian conditioned inhibition will include two major topics. First, we discuss the various procedures that have yielded conditioned inhibition and attempt to analyze the factors that give rise to the inhibition. Rescorla reviewed the literature on this question in 1969. We focus on research conducted since then; thus we do not say much about the seminal empirical work of Pavlov (1927) and Konorski (1948, 1967). Second, we discuss the question "What is learned when inhibitory conditioning occurs?" Before we can deal with these matters, however, we must discuss the definition of inhibition, and its assessment.

Rescorla (1969a) suggested that a stimulus should be: "called a conditioned inhibitor, if, as a result of experience of the organism with some operation relating that stimulus to the US, the stimulus comes to control a tendency opposite to that of the conditioned excitor" (p. 78). For Rescorla, conditioned inhibition was UCS-specific; if, for example, as a result of being placed in a certain relation to an electric shock UCS, a CS acquired the ability to oppose excitatory CRs to signals for the shock, it would be called a conditioned inhibitor. If it also came to oppose excitatory CRs to signals for very different UCSs, e.g., food or water, it would be assumed that the putative inhibitor was controlling some other process instead of, or in addition to, conditioned inhibition. Rescorla's 1969 definition of inhibition has been widely accepted, and we use it in this chapter. However, it becomes evident that the authors of several of the chapters in this volume (Holland, Jenkins, Rescorla) now believe that at least some stimuli that deserve the name ''conditioned inhibitors'' do not satisfy this definition. Discussion of this matter is best left to those later chapters.

Measures of Conditioned Inhibition: Summation and resistance to reinforcement

In our discussion of procedures that produce conditioned inhibition we rely on Rescorla's definition in order to focus the inquiry, not meaning to rule out the possibility that appetitive excitors necessarily act as aversive inhibitors and vice versa (e.g., Dickinson & Dearing, 1979).

How can we measure the ''opposition'' referred to in the definition of inhibition? The conventional wisdom, following Rescorla (1969a), is that we have strong evidence that some stimulus is a conditioned inhibitor if it both (1) attenuates excitatory CRs to some signal for the relevant UCS in a stimulus summation test, and (2) is harder to transform into an excitatory CS for that UCS in a resistance to reinforcement (retardation) test than is some control stimulus. Rescorla maintained that either of the two tests alone is inadequate, because either can be ''passed'' by stimuli other than conditioned inhibitors. This is particularly clear in the case of the resistance to reinforcement test, which is well known to respond positively to a CS that has lost associability as a result of being presented repeatedly in the absence of a UCS (e.g., Reiss & Wagner, 1972; Rescorla, 1971; see the chapter by Hall, Kaye, & Pearce in this volume for a discussion of procedures in which a CS loses associability). Unfortunately, only a minority of the studies of inhibition conducted since 1970 have used both tests.

Other assessment procedures have sometimes been used in place of summation and resistance to reinforcement tests. These tests include the ''superconditioning'' procedure, in which a stimulus is said to be an inhibitor if it promotes superconditioning to a novel CS that is compounded with it and followed by the appropriate UCS, and tests in which the putative inhibitor is presented contingent on an instrumental response, in order to assess its conditioned reinforcing or punishing effect. These tests, like the disinhibition and spontaneous recovery test procedures devised by Pavlov (1927), are not as straightforward as the summation and resistance to reinforcement tests, because additional assumptions are required before inhibition can be inferred from their oucomes.

Withdrawal from a Localized CS

One measure that appears frequently in these pages is withdrawal from a localized visual CS that signals a decreased probability of a food UCS (Wasserman, Franklin, & Hearst, 1974). Withdrawal has become a popular measure of inhibition largely because of a unique feature that it has. Unlike other inhibitory

CRs, which are inferred from effects a putative inhibitor has upon excitatory CRs in transfer tests, withdrawal emerges as a new response to the inhibitory CS during inhibitory conditioning. The approach–withdrawal scores obtained from such a procedure yield a session-by-session measure of the acquisition of conditioned inhibition (Hearst, Bottjer, & Walker, 1980; Hearst & Franklin, 1977). An acquisition curve for inhibition would be tedious and expensive to obtain using the more conventional transfer tests.

Withdrawal reflects the opposition referred to in the definition of inhibition in the sense that it is directionally opposite the approach responses that pigeons direct at localized (excitatory) signals for food. Moreover, performance of the withdrawal response depends on the presence of excitation, just as do other inhibitory CRs. This was demonstrated by Kaplan and Hearst (in press), who observed a withdrawal response to a key light CS that was nonreinforced in a context in which intertrial food (UCS) was often presented. After the excitatory conditioning of context was extinguished by repeatedly placing the pigeons in the conditioning chamber in the absence of food, presentations of the key light failed to evoke withdrawal. Subsequent intertrial feedings in that context reinstated the withdrawal response.

If the results of experiments using the approach–withdrawal measure are to be integrated with other research on inhibition, it would be useful to know how strongly withdrawal correlates with performance in resistance to reinforcement and summation tests. There have been no studies that compared withdrawal from a key light CS with its subsequent ability to inhibit approach to, and pecking at, another visual stimulus in a summation test. However, in several studies withdrawal has been compared with subsequent resistance to reinforcement when the key light was subsequently paired with food. Generally, the two measures correlate fairly well, but there have been some discrepancies. For example, Hearst and Franklin (1977) observed similar levels of withdrawal in groups that received two different magnitudes of negative contingency between the CS and food, but in a subsequent resistance to reinforcement test the group that had received the larger negative contingency was slower to acquire an autoshaped key peck. On the other hand, Kaplan (1984) found no differences in retardation of conditioning of an approach response to a key-light signal for food among groups of pigeons that had previously shown strong, moderate, or no withdrawal from that CS when it had been placed early, in the middle, or late in an interfood interval, respectively. More comparisons of the outcomes of transfer tests with approach–withdrawal scores must be made before we can say very much about the extent to which the tests reveal the same processes.

Batson and Best (1981; also see Best, 1975) have recently suggested a new measure of conditioned inhibition in the conditioned taste-aversion (CTA) paradigm. They presented rats with a conditioned inhibition (A+, AX−) paradigm in which consumption of coffee was followed by administration of poison (LiCl) on some days, whereas on others it was followed by the opportunity to consume vinegar without being poisoned. After several cycles of this procedure, on the test day the rats had to choose between the putative inhibitor, vinegar, and water in their home cages, an environment quite different from the one in which they had received conditioning trials. Three conditioned inhibition groups, which received different intervals between coffee and vinegar on AX− days, drank a higher proportion of vinegar in the test than a control group that had previously received vinegar 24 hrs after each unreinforced presentation of coffee. This outcome is provocative because there was apparently no excitation present during the test. Thus this experiment is one of the few instances of a conditioned inhibitor evoking a change in behavior in the absence of excitation. Because such an outcome may teach us something about the operation of inhibitors, this test should be repeated, and its results compared with those of the standard summation and resistance to reinforcement tests.

ANALYSIS OF SOME PROCEDURES THAT PRODUCE CONDITIONED INHIBITION

In this section we analyze the experimental conditions that have been thought to produce Pavlovian conditioned inhibition.

The UCS–CS Relationship in Backward Conditioning

Several theorists have proposed accounts of inhibitory conditioning that specify that inhibition may result from an association formed between a backward CS and an excitation-opposing process evoked either by the termination of the UCS or by a secondary response to that UCS (Denny, 1971; Konorski, 1948; Mowrer, 1960; Schull, 1979; Solomon, 1980; Solomon & Corbit, 1974). As a result of this conditioning the backward CS should come to behave like a conditioned inhibitor in summation and resistance to reinforcement tests, using the same UCS that had been used during the conditioning phase.

We first describe the theories proposed by Denny (1971) and Solomon and Corbit (1974), emphasizing in each case the mechanism said to be responsible for the inhibitory effect of backward conditioning. We then review certain studies that demonstrate the inhibitory impact of backward CSs. In so doing we attempt to assess the roles of the UCS–CS relation and other potential causes of inhibition within the backward conditioning paradigm.

According to the opponent process theory of acquired motivation proposed by Solomon and Corbit (Solomon, 1980; Solomon & Corbit, 1974; see also Schull, 1979), onset of the UCS unconditionally elicits a primary affective process, called the *a* process. The *a* process activates an opponent affective process, the *b* process. The net affective state of the organism at any time is given by the algebraic sum of these two processes, which have opposite valences. When the *a* process is larger than the *b* process, the organism is in the A state; when the *b* process is larger, it is in the B state. At the beginning of training the *a* process predominates early in the UCS, creating an A state, whereas the *b* process predominates when the UCS has ended, creating a B state. The latter persists for some seconds following offset of an electric shock UCS, and backward CSs appropriately placed in the postshock period should become conditioned elicitors of the B state. According to Solomon and Corbit, the *b* process will be strengthened through repeated presentations of the UCS, whereas the *a* process is relatively stable. Thus after repeated, massed presentations of the UCS, it will evoke an attenuated A state, followed by a larger and longer lasting B state. The implications of this postulate for inhibitory conditioning are taken up later.

Denny (1971) has proposed a related view that has been described in most detail for the aversive case. In this account, the termination of a UCS like electric shock is considered a secondary elicitor, evoking a state called relief, whose characteristic response components are in opposition to those associated with the primary fear state elicited by shock onset. On the basis of several experiments concerned with the acquisition and extinction of shock avoidance in rats (reviewed in Denny, 1971), Denny argued that relief begins approximately 5 sec after shock offset and persists for 10–15 sec thereafter. Denny further concluded that a second, opponent state, called relaxation, begins about 30 sec after shock offset and persists for approximately 120 sec. Unlike relief, relaxation will not occur unless the minimum interval between shocks is more than 150 sec. Because relaxation is a long-latency process, it should not be active at the time when a backward CS is presented. However, the postulated time course of relief should result in contiguity of the CS with relief in most backward conditioning experiments.

The Role of a Safety Signal in Backward Conditioning. In addition to the UCS–CS relationship, other relations that might themselves produce inhibition are imbedded within the backward conditioning paradigm. The first of these relations is found in an experiment by Moscovitch and LoLordo (1968). They noted that in most backward conditioning experiments the UCS–CS relation is accompanied by a long minimum interval between the backward CS and the next UCS. Thus, when the UCS is electric shock the CS might become an inhibitor of fear by virtue of its value as a "safety signal," rather than because it is contiguous with the occurrence of unconditioned relief. Moscovitch and LoLordo reasoned that this safety signal effect should occur, regardless of the duration of the interval between shock offset and CS onset, but should be eliminated when shocks are allowed to occur following unpredictable and occasionally very brief post-CS intervals. They addressed this possibility by exposing three groups of dogs to different Pavlovian conditioning treatments, which are illustrated in Fig. 1.1. Each treatment involved presentation of a 5-sec tone CS shortly after termi-

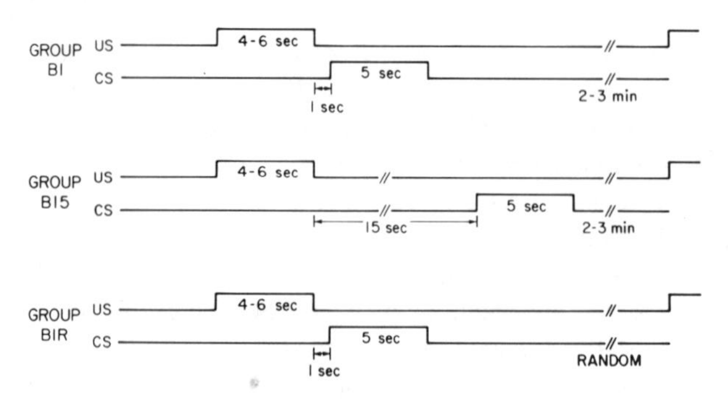

FIG. 1.1. Time relations among events for groups B1, B15, and B1R in the experiment by Moscovitch and LoLordo (1968). All groups had the same intertrial interval of 2.5 min. (© 1968 by the American Psychological Association. Reprinted by permission.)

nation of a 3-mA grid shock. For the two safety signal groups, B1 and B15, this CS also preceded shock-free intertrial intervals (ITIs) that were 2, 2.5, or 3 min long. For the third group, Group B1R, subsequent shock-CS trials were equally likely to begin during each sec of the intertrial interval, with the mean intertrial interval equal to that of the other two groups. Subsequently all dogs received a summation test in which the tone CS was presented while they performed an unsignaled, free-operant avoidance response (Sidman, 1953) in a two-way shuttle box.

Figure 1.2 shows that the tone suppressed avoidance responding in the two safety-signal groups but had no impact on responding in the random ITI group. It seems, then, that the tone must precede a long minimum shock-free interval in order to become an inhibitor of fear (see Dunham & Carr, 1976 for a related result). Moreover, the equivalent test performance of Groups B1 and B15, which experienced 1-sec and 15-sec intervals, respectively, between the end of shock and the beginning of the CS, suggests that it was the reliable prediction of a long shock-free period, rather than contiguity with an opponent process, that gave the CS its inhibitory effect.

Several years later Maier, Rapaport, and Wheatley (1976) suggested that both UCS–CS intervals studied by Moscovitch and LoLordo should have been good ones for the conditioning of relief, so that a comparison of the two values would not have provided a good test of the necessity of a conditioned opponent for backward inhibitory conditioning. In order to evaluate this possibility, Maier et al. used Denny's (1971) data as a rough guideline for optimal placement of the CS. One group of rats received the 10-sec tone CS 3 sec after the offset of a 5-sec, 1-mA grid shock in a procedure designed to produce conditioned relief. The other group received the tone 30 sec after shock termination, when relief had already ended or was waning. Each of these groups received the same intertrial

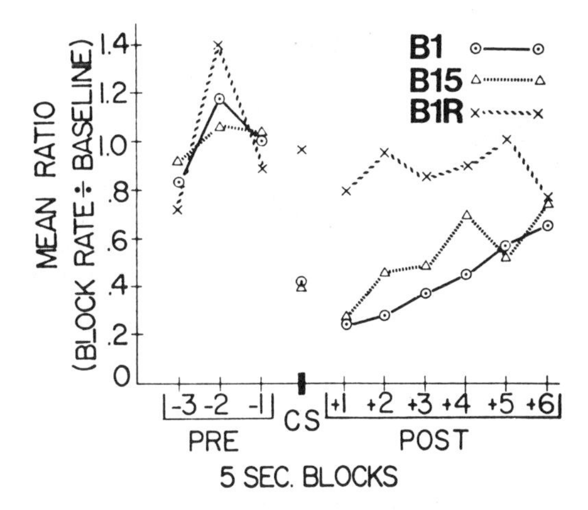

FIG. 1.2. Mean response ratios for successive 5-sec periods before, during, and after test CS presentations for groups B1, B15, and B1R in the experiment by Moscovitch and LoLordo (1968). The response ratio equals the response rate during a 5-sec period divided by the mean rate for the 15-sec pre-CS period. (© 1968 by the American Psychological Association. Reprinted by permission.)

intervals, 2, 3, or 4 min; thus potential safety-signal effects were equated for the two groups.

The tone had different effects on the shuttle-avoidance performance of the two groups in a subsequent test (Fig. 1.3). It suppressed avoidance in the 3-sec backward group but did not affect responding in either the 30-sec backward group or a group that had experienced a zero contingency between tone and shock during conditioning. This finding is consistent with the notion that the 3-sec CS was contiguous with an opponent process, whereas the 30-sec CS was not.

Plotkin and Oakley (1975) obtained similar results in a study of backward conditioning of the rabbit's nictitating membrane response. They first subjected different groups of rabbits to various Pavlovian conditioning treatments. For two backward conditioning groups the 200-msec tone CS followed the offset of the brief extraorbital shock UCS by either 200 or 500 msec. The interval between the offset of the CS and the next shock was 55 sec for both groups. A third group received explicitly unpaired (EUP) presentations of the CS and shock; the CS occurred at the midpoint of the 110-sec interval between successive shocks. The effects of these treatments upon the CS were evaluated in a subsequent resistance to reinforcement test.

The backward groups conditioned more slowly than the explicitly unpaired group and did not differ from one another. Although the explicitly unpaired CS

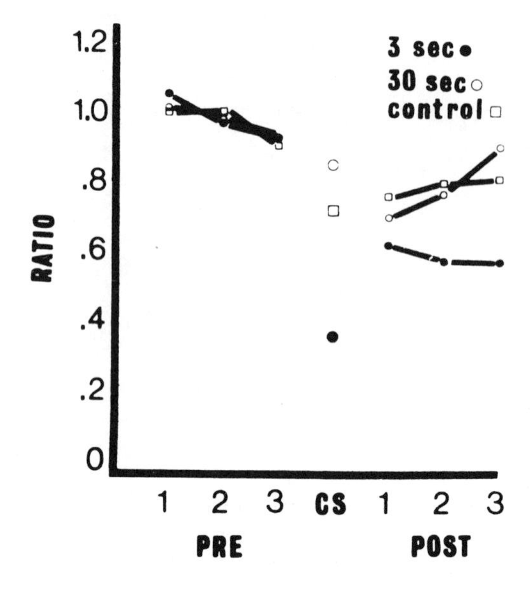

FIG. 1.3. Mean response ratios for successive 10-sec periods before, during, and after test CS presentations for groups that received a 3-sec interval between UCS and CS (3 sec), a 30-sec interval between UCS and CS (30 sec), and a zero-contingency control treatment (control) in the experiment by Maier, Rapaport, and Wheatley (1976). The response ratio equals the response rate during a 10-sec period divided by the mean rate for the 30-sec pre-CS period. (Reproduced by permission of The Psychonomic Society, Austin, Texas and the authors.)

predicted a long shock-free interval, test performance of Group EUP did not differ from that of a naive control group. Presumably, in both backward groups the CS was contiguous with relief, whereas this was not the case in Group EUP.

The data of Maier et al. and Plotkin and Oakley suggest that even when a long minimum UCS-free period following CS does not itself produce inhibition, the UCS–CS relationship will combine with that factor to do so. Moreover, in combination with the data of Moscovitch and LoLordo, these data suggest that neither factor is sufficient by itself.

In retrospect, it is difficult to determine whether the observed effect of randomizing the intertrial intervals is compatible with opponent-process models, primarily because the models do not make sufficiently quantitative predictions. The models would predict that when one shock follows another by a short interval in this procedure, the fear (*a* process) evoked by shock would summate with, and might overwhelm, the relief from the previous trial, perhaps resulting in fear conditioning to the backward CS on that trial. Thus, at a minimum, inclusion of short intershock intervals should markedly reduce the rate at which the CS comes to conditionally elicit the opponent state. Whether the models can predict the total failure of inhibition to develop under such a procedure remains unclear.

Solomon and Corbit's view does make an interesting prediction about the effects of backward conditioning procedures with randomized intertrial intervals. Extensive prior exposure to the shock UCS, which causes the *b*-process to begin sooner, reach a larger value, and continue longer than it did initially, will result in a small net A state and a large B state at the start of backward conditioning.

Because the A state is the reinforcer for fear conditioning, shocks that occur soon after the backward CS will not condition fear to that CS, and the CS will become inhibitory, even if it would not have become inhibitory in the absence of prior exposure to the UCS.

Relatively little research has been devoted to the possible inhibitory effects of a related procedure called cessation conditioning, in which the onset of the CS reliably precedes the termination of the UCS. Moscovitch and LoLordo compared the effectiveness of the CS from their backward procedure with that of a CS that preceded the cessation of each shock by 1 sec. In a summation test avoidance responding was not suppressed during presentations of the cessation CS, although some suppression did occur in the period immediately following CS offset.

In a subsequent doctoral dissertation Moscovitch (1972) demonstrated that cessation conditioning could produce powerful inhibitory effects when the shock duration was allowed to vary from 10–30 sec, so that shock onset would no longer be a good predictor of shock termination (Fig. 1.4, Group C). In this study a 5-sec CS preceded the termination of each shock by 5 sec, and the intertrial interval varied from 2–3 min. When cessation conditioning was repeated, but with intertrial intervals ranging from 0.5 sec to 15 min, the inhibition was markedly reduced but not eliminated (Group CR).

Data reported by Segundo, Galeano, Sommer–Smith, and Roig (1961) support Moscovitch's findings. Cats received electric shocks varying in duration from 30 sec to 30 min. These shocks were interrupted 2–5 sec after onset of a 4–10 sec-long tone CS. After 3–10 sessions the cats reliably relaxed, i.e., stopped crouching, mewing, and licking the shocked limb, lifted their heads and opened their eyes when the tone came on, even though shock was still being administered. Responses to tone looked like the responses to shock termination. For some cats, tone offset was followed by a long shock-free period; for others, it

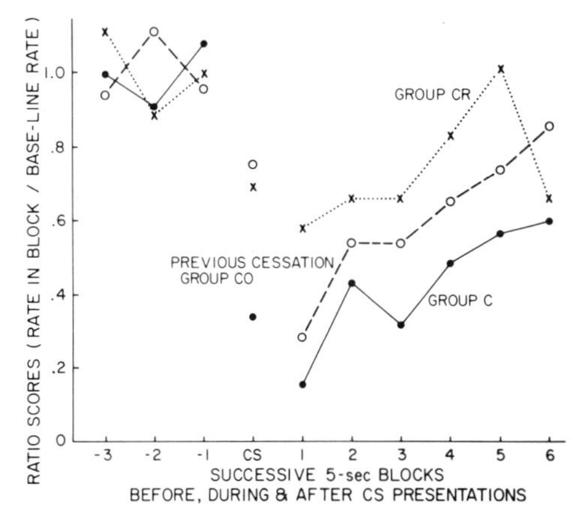

FIG. 1.4. Mean response ratios for successive 5-sec periods before, during, and after test CS presentations for groups C and CR in the experiment by Moscovitch (1972), and for group CO in the experiment by Moscovitch and LoLordo (1968). The response ratio equals the response rate during a 5-sec period divided by the mean rate for the 15-sec pre-CS period. (Reproduced by permission of the authors.)

was followed by shock within 5 sec. Segundo et al. reported that this variable had no effect on the outcome; however, they did not present the data.

The studies reviewed thus far suggest that inclusion of very short shock-free periods after the CS appears to have a greater disruptive effect upon inhibitory conditioning when CS onset follows UCS termination than when it precedes UCS termination. However, this may not be generally true. Moscovitch and Segundo et al. presented much longer shocks than those presented by Moscovitch and LoLordo; perhaps the balance between the *a* and *b* processes tends to favor the *b* process as shocks get longer, so that conditioning of the B state will be less disrupted by subsequent shocks when shocks are long. If this is so, backward inhibitory conditioning would also be resistant to disruption by subsequent shocks when those shocks are long.

The Role of Concurrent Sources of Conditioned Excitation in Backward Conditioning. The backward-conditioning studies discussed thus far include another potential cause of inhibition. In all of them the putative inhibitor was nonreinforced in the presence of other cues—in these cases, contextual cues—that had themselves been reinforced. This case is the most common instance of the negative discrepancy that is the basis of the Rescorla–Wagner (1972) model's account of losses in associative strength. The explicitly unpaired control groups in the studies of Maier et al. and Plotkin and Oakley also received reinforced presentations of contextual cues. However, according to the Rescorla–Wagner model these groups should not have had as much conditioned excitation present at the moment of CS presentation as their backward conditioning counterparts. This is so because contextual excitation produced by the previous UCS should have had more time to extinguish prior to CS presentation in the EUP groups than in the backward conditioning groups. Consequently, comparisons between these two sorts of group confound differences in the contiguity of CS and unconditioned relief with differences in contextual excitation present when the CS occurs, especially when the UCS–CS interval is long in the EUP group and nonreinforcement is thought to have relatively large effects.

In order to assess the contribution of contextual excitation to backward inhibitory conditioning, the association between the context and the UCS has to be manipulated experimentally. This objective could be accomplished in any one of at least three ways. First, one could rely on the finding that the amount of excitatory conditioning of context is markedly reduced when the UCS is preceded by a discrete CS (Durlach, 1983; Odling–Smee, 1975a, b, 1978a, b; but see Jenkins, Barnes, & Barrera, 1981) and arrange a signal for the UCS in a backward conditioning study. Second, one could reduce the associability of context with the UCS by giving the animals extended preexposure to the experimental setting prior to backward conditioning in that setting (Balaz, Capra, Kasprow, & Miller, 1982; Tomie, 1981). Finally, and perhaps most directly, one could move the animal to a different setting between the termination of the UCS and the onset of the backward CS. Relatively little research has been done using these procedures.

LeClerc (1983; also see LeClerc & Reberg, 1980) has studied the effects of signaling shocks on the development of inhibition of fear in a backward conditioning procedure. The backward CS was the availability of a raised platform beginning 10 sec after shock termination and lasting for 60 sec. Rats showed an increasing tendency to jump up on and remain on this platform as backward conditioning proceeded. In one experiment the rate of acquisition of this approach response was unaffected by signaling the UCS on each trial. However, as LeClerc noted, if context conditioning occurred very rapidly, this manipulation might not have ruled it out. Thus in a second experiment he gave two groups of rats extensive presentations of signaled versus unsignaled shock prior to the start of backward conditioning with signaled versus unsignaled shock UCSs. According to the Wagner–Rescorla model, in the signaled shock group context should have been overshadowed by the signal during the first phase, ensuring that context excitation would be weaker in the signaled shock group than in the unsignaled shock group at the start of backward conditioning.

This experiment yielded ambiguous data. Although the signaled shock group appeared to acquire the platform approach response more slowly, there was considerable variance in this group and the difference between groups was not statistically reliable.

Overmier and Patterson (1983) have also compared backward conditioning procedures using signaled versus unsignaled UCSs. Dogs restrained in a harness received 30 brief but intense shocks, each followed by a 10-sec tone. For half the dogs each shock was signaled by a 10-sec click; for the other half it was unsignaled. Effects of the tone were assessed in a summation test involving repeated presentation of the tone while the dogs were jumping a hurdle in the shuttle box to avoid shock. The backward CS suppressed avoidance responding only for the dogs that had received unsignaled shocks. This outcome would not have been predicted by Solomon and Corbit's (1974) version of opponent-process theory and directly contradicts Schull's (1979) modification of the theory, which predicts stronger backward inhibitory conditioning when the UCS is signaled. However, it is readily explained by an account that bases inhibitory conditioning upon the nonreinforcement of the inhibitor in the presence of conditioned excitation. Presumably the forward CS overshadowed the context, so that there was less contextual excitation present in the signaled shock group than in the unsignaled shock group at the time of tone presentation. The only concern we have about this experiment is that both the forward and backward CSs were auditory stimuli, so that stimulus generalization could have diminished the inhibitory effect of the backward CS in the signaled shock group. This is a minor concern, because the authors have replicated this effect using a larger number of conditioning trials, which would reduce generalization between the stimuli.

As far as we know, no one has looked at the effects of nonreinforced preexposure to context upon subsequent acquisition of inhibition in a backward conditioning procedure. However, there has been one study in which a context shift occurred between presentation of the UCS and the backward CS. Grelle and James (1981) conducted a backward conditioning experiment explicitly designed to assess the necessity of an excitatory context by removing rats from one setting immediately after the administration of a 2-sec grid shock and placing them in a distinctively different retaining box in which shock never occurred. Two values of the intertrial interval, 40 sec and 6 min, alternated randomly, and a 3-min auditory CS occurred 15 sec after the rat was placed in the retaining box during long, but not during short, ITIs (see Fig. 1.5). The authors reasoned that the backward CS should precede the peak phase of the unconditioned relief response evoked by shock termination and thus should come to evoke conditioned relief. Moreover, because the rat was moved to a new context immediately after shock, there should have been no source of conditioned excitation present when the CS was presented. The control group received the same number of events during conditioning as the experimental group, but all its CS presentations occurred *en masse* beginning 6 min after the last shock of the session.

The effects of the putative inhibitor were assessed in two tests involving escape from fear. Half the rats in each group received a summation test, in which they were placed in the grid box in which they had been shocked. Ten sec after the CS was turned on, a guillotine door was raised, and the rats were permitted to escape into a novel chamber. The other rats in each group were placed in the grid chamber, and 15 sec later the guillotine door was raised. If the rat moved into the novel chamber, the CS was presented and remained on for the 10-sec period of confinement in that chamber. The CS did function as a conditioned reinforcer in this test; rats in the experimental group learned to escape more quickly than controls. However, when the CS was presented in the former shock box, it failed to slow escape learning in the experimental group, relative to the controls. Thus, although the less direct measure indicated that the CS had become an inhibitor of fear, the outcome of the more direct summation test did not support this conclusion.

In a second experiment Grelle and James took a slightly different approach to the same goal. An auditory CS was repeatedly paired with shock in a black

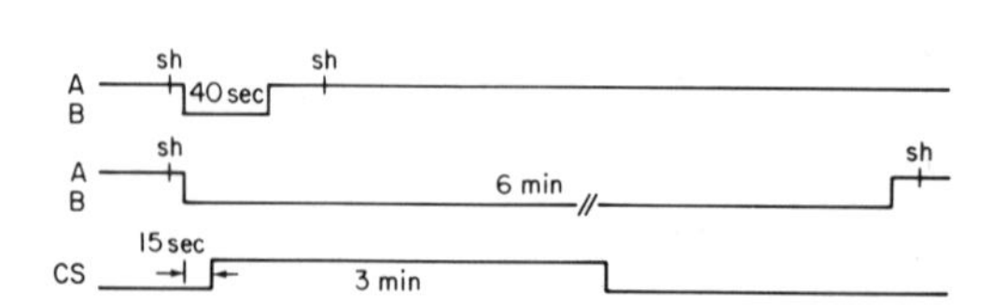

FIG. 1.5. Time relations among events for the experimental group in the first experiment by Grelle and James (1981). After receiving a shock in context A, the rat is moved to context B for either 40 sec (top line), or 6 min (middle line). The 3-min CS is presented 15 sec after the rat is placed in B during long intertrial intervals only.

chamber, and on other trials, in a white chamber, a visual stimulus was presented 0.5 sec after the termination of the auditory stimulus. Thus the visual stimulus was said to be paired with the relief evoked by termination of a conditioned aversive stimulus. Because shock and light never occurred in the same environment, there should have been no first-order conditioned excitation present when light was presented in the white chamber. However, it is possible that repeated presentations of the auditory CS in the white chamber would result in the conditioning of second-order excitation to cues of that chamber. To assess the possibility that any inhibition of fear conditioned to the light would be a consequence of such second-order contextual conditioning, a control group was treated just like the experimental group except that light followed the termination of white noise by 120.5 sec instead of 0.5 sec (see Fig. 1.6). Effects of the visual CSs were assessed in a summation procedure, using the lick suppression technique. Half the subjects in the experimental group received test presentations of the white noise CS+ alone; the other half received a compound of the CS+ and the visual CS−. White noise alone produced marked suppression of licking; the compound produced very little, suggesting that light had become an inhibitor of fear. Rats in the control group showed strong suppression to the compound, indicating that the brief interval between white noise termination and the onset of light was responsible for the inhibition observed in the experimental group.

On the basis of these studies Grelle and James (1981) argued that conditioning of relief is the optimal way of producing an inhibitor of fear, because inhibition was produced after only 20 inhibitory conditioning trials in each study, as compared with the much larger number typically used with more traditional procedures. However, though they are provocative, the data of Grelle's and James' first study are equivocal, and their work does not definitively rule out the presence of conditioned excitation at the time of CS presentation as a necessary condition of backward inhibitory conditioning. Moreover, their first experiment included a feature that is not required by theories based on conditioning of the *b*-process or of relief. The interval between the end of one shock and the beginning of the next could be either short or long, and the backward CS occurred only in the long intervals. It remains to be seen whether this feature was a necessary condition of the observed effect.

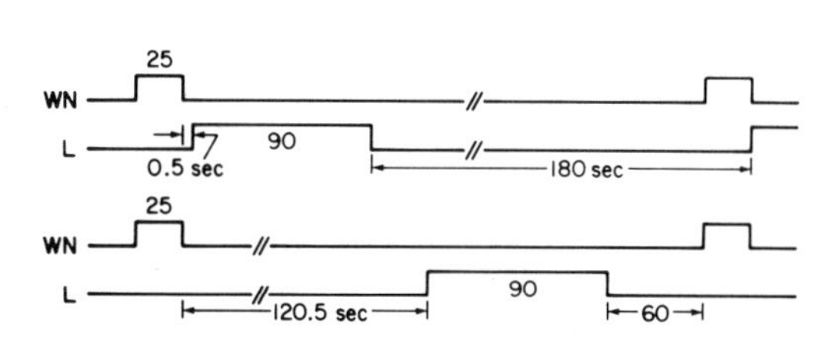

FIG. 1.6. Time relations among events for groups WN-L and WN-L Delay in Experiment 2 of Grelle and James (1981). Both groups received white noise (WN)-shock pairings in one context (not shown). In a different context group WN-L received backward conditioning trials on which light followed white noise by 0.5 sec. The control group, WN-L Delay, experienced a 120.5-sec interval between white noise and light.

To summarize the research on backward conditioning, repeated UCS–CS pairings will result in the CS becoming a conditioned inhibitor, if there is a fairly long minimum interval between the CS and the next UCS, and conditioned excitation is present when the CS is present. However, the causal role of the UCS–CS relation in producing inhibition remains unclear, primarily because it has not been shown conclusively that the backward relationship can operate independently of the presence of conditioned excitation at the time of backward CS presentation.

The Negative Correlation between CS and UCS

When the CS is reliably followed by a relatively long UCS-free interval in a backward conditioning experiment, CS and UCS can be said to be negatively correlated. Such a negative correlation is also a part of other procedures that have been shown to produce conditioned inhibition, e.g., Pavlov's conditioned inhibition or A+, AX− procedure, differential conditioning, and the explicitly unpaired (EUP) procedure. In most of these procedures the negative correlation is confounded with another potential cause of inhibition, the presence on nonreinforced trials of some other stimulus that has otherwise been reinforced.

In a recent review Damianopoulos (1982) suggested that we can assess the sufficiency of the negative correlation between CS and UCS for inhibition by showing that this correlation yields inhibition by itself, and that the inhibition disappears when the correlation is removed. Specifically, he asserted that it must be shown that the negative correlation yields inhibition even when the CS is presented in the absence of any concurrent excitation. Damianopoulos argued that a 1977 experiment by Baker provided the best test of the sufficiency of the negative correlation for conditioned inhibition. Rats first received several sessions of tone–shock pairings, designed to allow background cues to lose any associative strength they might have acquired initially. Then they received a between-days, discriminative conditioning procedure; on some days tone–shock pairings occurred, and on other days only the CS−, a clicker, was presented.

In subsequent summation and retardation tests, using the CER procedure, performance in response to the CS− was equivalent to performance of a CS-only control group. Figure 1.7 illustrates the results of the summation test, which yielded no evidence that a negative correlation between CS and UCS was sufficient for the conditioning of inhibition (group Disc. in Fig. 1.7). Subsequent research in which the CS+ and the CS− were from different sensory modalities confirmed these results, suggesting that the failure of the discriminative conditioning procedure to produce inhibition was not simply a consequence of stimulus generalization from tone to clicker.

Baker also showed that removal of the CS+ from the discriminative conditioning procedure, so that a group of rats received UCSs on some days and CS−s on other days, did result in a strong inhibitory effect (see group Neg. Corr. in Fig. 1.7; alse see Nieto, 1984). This procedure retains the negative correlation

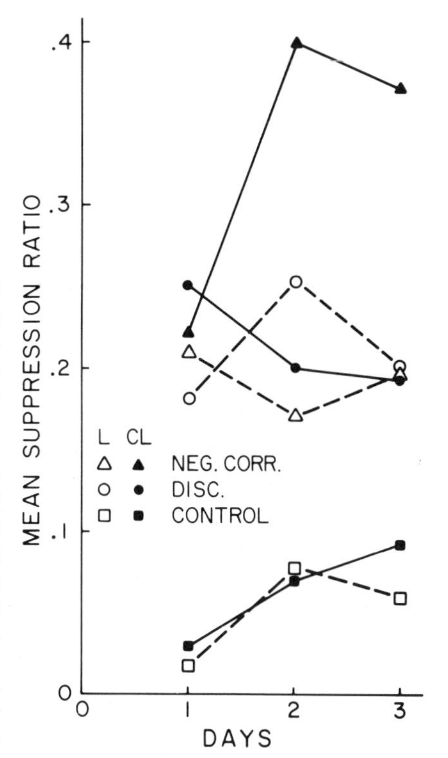

FIG. 1.7. Suppression to the light (L) and clicker–light compound (CL) in the summation test of an experiment (Exp. 4) by Baker (1977). The groups were preexposed to a between-days negative correlation between the clicker and shocks (Neg. Corr.), a between-days negative correlation but with the shocks signaled by a tone (Disc.), or to only the clicker (Control). The suppression ratio equals the number of responses during the CS divided by the sum of that number and the number of responses in a comparable pre-CS period. (© 1977 by the American Psychological Association. Reprinted by permission of publisher and author.)

between CS and UCS and also allows contextual cues to become strongly excitatory, so that CS− is nonreinforced in the presence of those excitatory cues.

Except for its between-sessions character, and the consequent alternation of blocks of pairings of CS+ and UCS with blocks of CS− trials, Baker's discriminative conditioning procedure is like the standard discriminative conditioning procedure introduced by Pavlov (1927), and there is a fairly substantial literature on the effects of this procedure. We review this literature, insofar as it bears on the question of the sufficiency of the negative correlation between CS and UCS for conditioned inhibition.

The Role of Excitatory Conditioning of Context in Differential Conditioning Procedures. In the autoshaping procedure differential conditioning has consistently yielded an inhibitory CS−, with such diverse assessment procedures as summation and retardation of acquisition tests (Wessells, 1973), withdrawal from a localized CS− (Hearst & Franklin, 1977; Kaplan & Hearst, in press; Wasserman, Franklin, & Hearst, 1974), and the production of incremental generalization gradients around CS− (Hearst & Franklin, 1977; Tomie & Kruse, 1980). However, none of these results constitutes strong support for the sufficiency of the negative correlation between CS and UCS, because none of the studies assessed the amount of excitatory conditioning of context. Thus we do not know whether CS− was nonreinforced in the presence of a strong excitor.

This is not a trivial point, because findings from several studies suggest that preceding food by a discrete CS+ may be insufficient to prevent excitatory acquisition to the context. Balsam (1984) has shown that acquisition of autoshaped pecking at a key-light signal for food was equally retarded when the pigeons had previously received unsignaled food in that context, or when they had received food signaled by an auditory CS+. In a related study Jenkins et al. (1981, p. 273) gave three groups of pigeons repeated pairings of a white noise stimulus and grain. Then the birds received pairings of a key-light stimulus and grain. For two of the groups grain was also presented at the same rate during the intervals between key light–grain pairings; one group received unsignaled grain, but for the other grain was reliably preceded by the white noise. The presentation of intertrial grain resulted in retarded acquisition of the autoshaped key peck to the key light CS+. Furthermore, signaling the grain did not significantly reduce the magnitude of the attenuation. A similar effect held in a second study in which intertrial food was preceded by another key-light stimulus. These outcomes suggest that there may have been strong excitatory conditioning of context in the differential autoshaping studies. However, Durlach (1983) obtained the opposite result in a similar study. Three groups of pigeons received food after 25% of the presentations of a key-light CS. Two of these groups received enough intertrial food to produce a zero contingency between the key light and food. Intertrial food was unsignaled for one of these groups, whereas it was preceded by a previously conditioned tone for the other. Free food attenuated autoshaping of a peck at the key-light CS, but the effect was markedly reduced if the free food was signaled. The discrepancy between this outcome and the one observed by Jenkins et al. may arise from differences in the amount of conditioning in the two studies. Durlach administered many more trials than Jenkins et al., and the nonsignificant divergence of the two groups at the end of the Jenkins et al. study might have become significant if more trials had been administered. Nonetheless, the results of the aforementioned studies do suggest that there may be strong contextual excitation present during at least the initial sessions of a differential autoshaping procedure. Rescorla, Durlach, and Grau (in press) have pointed out that at least some of the contextual excitation arising from such a procedure may be second-order excitation resulting from pairings of context and the first-order excitor.

Rescorla (1982) used an autoshaping procedure to make a within-subjects comparison of the inhibitory effects of a differential CS− with a CS− from a Pavlovian A+, AX− conditioned inhibition procedure. During the conditioning phase stimuli A and B were reinforced, whereas the AC compound and stimulus D were not. In a subsequent summation test the effects of C versus D in suppressing responding to A and B were assessed. Stimulus C, the Pavlovian conditioned inhibitor, was clearly more inhibitory than D, but D did markedly suppress responding to A and may have been an inhibitor. Other autoshaping studies had failed to find marked differences between the effects of CS-s from discriminative conditioning versus or explicitly unpaired procedures (Hearst & Franklin, 1977; Wasserman et al., 1974).

The CS− from differential conditioning has also become inhibitory in several studies that used aversive UCSs and evaluated the putative inhibitor by presenting it while the animal was engaging in avoidance behavior in a form of summation test (Bull & Overmier, 1968; LoLordo, 1967; Rescorla & LoLordo, 1965; Weisman & Litner, 1969a, 1971; also see Hendersen, 1973 for some methodological concerns). In studies that compared the procedures, the magnitude of the suppression of avoidance responding produced by the differential CS− was equivalent to the amount evoked by CS−s from the explicitly unpaired and Pavlovian conditioned inhibition procedures (Rescorla & LoLordo, 1965; Weisman & Litner, 1969a). In these studies avoidance training preceded Pavlovian conditioning, so it is likely that some aspect of the conditioning contexts had acquired generalized excitatory strength by the start of the differential conditioning treatments, and the experiments would not be good tests of the sufficiency of a negative correlation between CS and UCS for the conditioning of inhibition.

Differential conditioning has also typically resulted in an inhibitory CS− in the Conditioned Emotional Response (CER) procedure, at least when CS+ and CS− are from different sensory modalities (Cappell, Herring, & Webster, 1970; Hammond, 1966, 1967, 1968; Reberg & Black, 1969). The clearest demonstration of this outcome is in a pair of papers by Hammond (1967, 1968), who compared the effects of a differential CS− with the effects of a stimulus that was independent of CS–UCS pairings. The CS− was more effective than the random control CS in alleviating suppression to a CS+. Moreover, when paired with shock, the former CS− was slower to evoke conditioned suppression than was the former random CS. This effect tended to disappear if the rats had been subjected to very extensive discrimination training, a result that would have been predicted on the basis of extinction of the original context conditioning as discriminative conditioning progressed, followed by extinction of the inhibitory effect of CS− (Wagner & Rescorla, 1972).

There are some data that indicate that at least transitory excitatory conditioning of context occurred in these studies. Hammond (1966) observed suppression of the baseline rate of food-reinforced lever pressing for roughly 10 sessions, a large number for the CER procedure. Moreover, Bouton and King (1983) have observed that rats that have received CS-shock pairings in some context tended to escape from that context if given the opportunity to move into a smaller side chamber. This measure of context excitation, which was more sensitive than the baseline rate of lever pressing, would probably reveal that context is excitatory in differential, as well as in simple, conditioning.

On the other hand, Yadin and Thomas (1981) have conducted a fear conditioning study that suggests that a differential CS− may acquire an inhibitory effect even in the absence of any context excitation. They recorded activity of the lateral septal nucleus during intermodal differential aversive conditioning of rats. Figure 1.8 illustrates that the CS+ evoked a decrease in activity of this area,

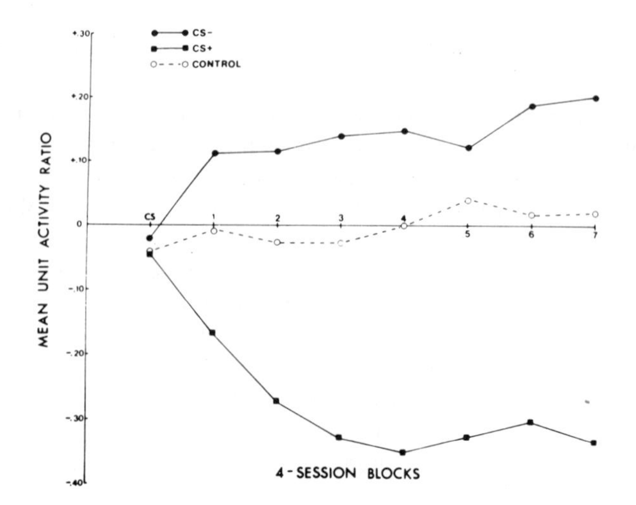

FIG. 1.8. Differential effects of CS+ and CS− upon activity in the lateral septal nucleus during discriminative fear-conditioning sessions in the study by Yadin and Thomas (1981). For the control group, CSs and shock were uncorrelated. The point marked 'CS' indicates the level of activity elicited by the stimuli prior to discriminative conditioning. (© 1981 by the American Psychological Association. Reprinted by permission of publisher and author.)

whereas the CS− evoked a reliable increase in activity. These effects were associative, because the stimuli had no effect upon the septal activity of other rats that had experienced them independent of shock. In the latter group, the pre-CS level of septal activity declined over sessions, suggesting that the context was becoming excitatory (see Fig. 1.9). However, there was no such decline in the pre-CS level of septal activity of rats in the differential conditioning group. These results are provocative, but more data are needed before they imply that a negative correlation between CS and UCS is sufficient for inhibition. First, the CS− will have to pass summation and resistance to reinforcement tests. Second, a sensitive behavioral test like that used by Bouton and King (1983) will have to support the septal activity measure in failing to reveal an excitatory effect of the differential conditioning context.

Intramodal versus Intermodal Differential Conditioning. There have been failures to demonstrate that the CS− from discriminative conditioning is inhibitory. Using the CER procedure, Rescorla (1976) found that a CS− failed to become an inhibitor of fear, as measured by its inability to condition inhibition to a second-order CS that was paired with it. Rescorla and Holland (1977) failed to obtain evidence that a CS− from an intramodal Pavlovian appetitive discrimination becomes inhibitory. Such a stimulus had very little effect on responding to

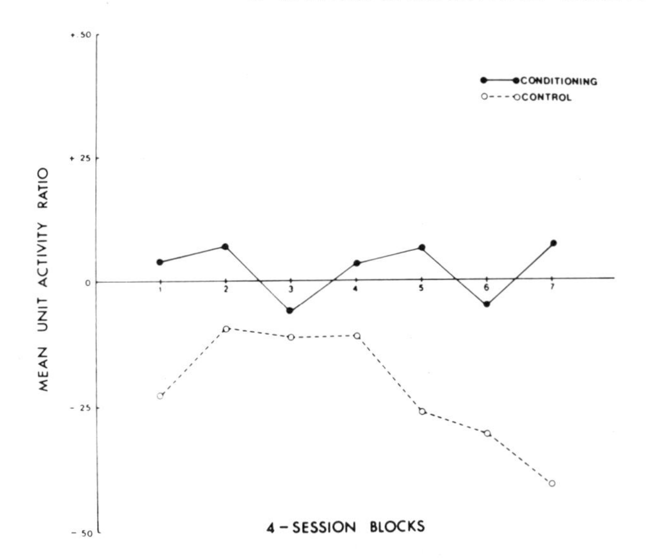

FIG. 1.9. "Spontaneous" or pre-CS activity in the lateral septal nucleus in the study by Yadin and Thomas (1981). Activity ratios are based on a comparison of activity during the pre-CS period in a conditioning session with activity in the absence of the CS in the initial habituation session. (© 1981 by the American Psychological Association. Reprinted by permission of publisher and author.)

auditory or visual signals for food in a summation test. In both of these studies the same stimulus produced marked inhibitory effects when it had been used as the X element in an A+, AX− procedure. Marchant and Moore (1974), who studied the rabbit's nictitating membrane response, also found that the CS− from discriminative conditioning failed to become inhibitory, whereas the X stimulus from the A+, AX− design did become an inhibitor.

These three studies entailed intramodal discriminations. Thomas (1972) has shown that intramodal discriminative conditioning may fail to result in inhibition even in situations in which a CS− from an intermodal discrimination becomes inhibitory. Cats received randomly alternated CS+s and CS−s, with the last two sec of the 5-sec CS+ accompanied by electrical stimulation of the brain. The unconditioned response was fear or rage, and the conditioned response, a change in activity. CS+ produced an increase in activity. So did CS− in the intramodal case, but this response extinguished completely as discriminative conditioning progressed. In the intermodal case, CS− did not evoke a change in activity. In the test the cats were presented with an unsignaled threshold UCS, or with the same UCS following either CS+ or CS−. CS+ facilitated responding to the threshold UCS in both groups, as did CS− in the intramodal group. However, the CS− was inhibitory for the intermodal group, reducing responding to the threshold UCS. Although these results suggest that, all other parameters being equal, intermodal CS−s are more likely to become inhibitory than intramodal ones, they do not permit us to identify the critical determinant of the effect of the CS− from differential conditioning with any confidence, especially because some intramodal studies have resulted in inhibition (Hearst & Franklin, 1977; Hendersen, 1973; Rescorla & LoLordo, 1965; Wessells, 1973).

Additional Comments on the Role of Context in Differential Conditioning. The literature on differential conditioning can be summarized simply. The differential CS− has become inhibitory in most of the studies, but because of the absence of measures of context conditioning there is very little evidence to suggest that a negative correlation between CS and UCS is sufficient for inhibition in the absence of concurrent excitation.

Parenthetically, there is one finding from the literature on discriminative conditioning that is hard to reconcile with an account based on nonreinforcement of the CS− in the presence of conditioned excitation. Weisman and Litner (1971) did intermodal discriminative aversive conditioning, with four groups of rats receiving intertrial intervals of 0.5, 1.0, 2.0, and 5.0 min. Effects of the CSs were assessed in a summation test in which they were presented while the rat performed a wheel-turn avoidance response. CS− produced clear suppression of avoidance in the 2- and 5- min groups but had no effect in the 0.5- and 1-min groups. Although it is likely that context was excitatory for all groups, contextual excitation resulting from presentation of the Pavlovian UCS should have been greater in the short-ITI groups, which had the least opportunity for context extinction. If the magnitude of excitation concurrent with CS− were the critical determinant of inhibition, then CS− should have had the biggest inhibitory effect in the 0.5-min ITI group. Alternatively, because the ITI was fixed for each group in this study, it could be asserted that such fixed temporal cues, and not context per se, provided the additional source of conditioned excitation in this study. On this analysis each group received an AX+, BX− discrimination, with A and B the nominal CSs and X the temporal cue correlated with the end of the ITI. This analysis seems to predict no difference between the groups, rather than the observed direct effect of ITI length upon the magnitude of inhibition.

This result implies that the effect of the length of the shock-free period following CS− upon the magnitude of inhibition is not mediated by the amount of conditioned excitation present when CS− is nonreinforced. Unfortunately, this experiment has not been replicated or extended to other procedures, although Hearst and Franklin (1977) conducted an intramodal discriminative autoshaping experiment in which the ITI following CS−, but not CS+, was varied across groups, and found no between-groups differences. Consequently, it is difficult to know how important the Weisman and Litner (1971) result really is.

Comparing Several Sorts of Inhibitory Conditioning Procedures. Comparisons of the inhibitory effects of the discriminative conditioning procedure with the effects of procedures in which CS− is presented along with other stimuli that have been reinforced separately, e.g., the conditioned inhibition (A+, AX−) and explicitly unpaired (+, X−) procedures, have been made in a number of studies, several of which have been cited. The Rescorla–Wagner model makes clear predictions about the results of these comparisons. The conditioned inhibition procedure should produce the strongest contextual excitation, and thus the most inhibitory CS−, the explicitly unpaired (EUP) procedure, the next strongest, and discriminative conditioning should produce little or no inhibition. The existence of within-compound associations between CS+ and CS− or between the excitatory context and CS− (Cunningham, 1981; Marlin, 1982; Rescorla & Freberg, 1978) would be expected to mask inhibition conditioned to CS−, but such effects could be reduced by separate presentations of the excitor following the inhibitory conditioning treatments (Rescorla, 1982).

We have already noted that the conditioned inhibition procedure has produced stronger inhibition than discriminative conditioning in some studies (Hoffman & Fitzgerald, 1982; Marchant & Moore, 1974; Rescorla, 1976, 1982; Rescorla & Holland, 1977), but that CS−s from the two procedures have been equally inhibitory in others (Mahoney, Kwaterski, & Moore, 1975; Rescorla & LoLordo, 1965; Weisman & Litner, 1969a). There have not been many comparisons of the effects of the explicity unpaired and discriminative conditioning procedures. Most of these comparisons have revealed that the two procedures have equivalent, inhibitory effects (Hearst & Franklin, 1977; Rescorla & LoLordo, 1965; Wasserman, Franklin, & Hearst, 1974; Weisman & Litner, 1969), although a between-sessions study by Baker (1977), cited earlier, yielded a strong inhibitory effect of the EUP procedure, but no effect of the CS− from discriminative conditioning (also see Hoffman & Fitzgerald, 1982). It is hard to evaluate any of these comparisons in the absence of measures of the amount of excitatory conditioning of context. For example, an answer to the question "does the A+, AX− procedure result in stronger inhibition than the A+, X− procedure in a given experiment because stronger excitation accompanies X in the former?" requires a measure of context conditioning.

Nonreinforcement of CS in the presence of conditioned excitation

The A+, AX− Procedure. The inhibitory conditioning procedures that have received the most attention recently are those in which a CS− is nonreinforced in the presence of some other stimulus that has been reinforced separately. The most straightforward example of this procedure is Pavlov's conditioned inhibition procedure (A+, AX−). This procedure has been used frequently in recent years and has nearly always resulted in an inhibitory effect (e.g., Cotton, Goodall, & Mackintosh, 1982; Mahoney, Kwaterski, & Moore, 1975; Marchant & Moore, 1974; Pearce, Montgomery, & Dickinson, 1981; Rescorla, 1973, 1976, 1982; Rescorla & Holland, 1977; Rescorla & LoLordo, 1965; Weisman & Litner, 1969a; Zimmer–Hart & Rescorla, 1974). In Pavlov's laboratory the CS− usually began before the CS+ on nonreinforced trials, but Pavlov asserted that this temporal order was not essential for the development of inhibition. However, Pavlov maintained that inhibition would develop most readily when CS+ and CS− overlapped in time. These aspects of the conditioned inhibition procedure have not received much attention in Western research. In all but two of the aforementioned studies CS+ and CS− were presented simultaneously. In one of the exceptions Rescorla and LoLordo (1965) observed roughly equal inhibition of avoidance behavior in dogs when CS− ended as CS+ began, and when CS− followed presentation of CS+ and a variable trace period (also see Best, 1975; Gokey & Collins, 1980). In the other exception, Rescorla (1973) compared a procedure in which a 30-sec CS− terminated at the start of the CS+ with one in which the CS− remained on during the CS+. In a subsequent summation test only the latter procedure yielded clear evidence of inhibition. Later in this volume Holland reports systematic research comparing the effects of an A+, AX− procedure, in which A and X are presented simultaneously with one in which X precedes A. The two procedures have dramatically different effects, and one may wonder whether it is appropriate to call both "conditioned inhibition" procedures.

The Explicitly Unpaired Procedure. Next, we consider experiments in which context plays the role of the A stimulus in the A+, AX− procedure. First, consider the so-called explicitly unpaired (EUP) procedure, in which the CS− occurs on some trials, and the UCS on others, with the intertrial intervals relatively fixed. In discussing backward conditioning, we described two experiments, one using the nictitating membrane response of the rabbit (Plotkin & Oakley, 1975), and the other, aversive conditioning in rats (Maier et al., 1976), that failed to find an inhibitory effect of the EUP procedure. Some experiments on conditioning of the human galvanic skin response have also failed to find an inhibitory effect of the EUP procedure (see Furedy, Poulos, & Schiffman, 1975). All the other studies using the EUP that we have examined yielded inhibitory effects. These include a number of studies of aversive conditioning of rats, with various response measures and procedures for assessing inhibition (Baker, 1977; Fowler, Fago, Domber, & Hochhauser, 1973; Fowler, Goodman, & Zanich, 1977; Hammond & Daniel, 1970; Hoffman & Fitzgerald, 1982; Hyde, 1976a; Kremer, 1971; Suiter &LoLordo, 1971; Weisman & Litner, 1969a), one study of the rabbit's nictitating membrane response (Bromage & Scavio, 1978), a study of aversive conditioning of dogs (Rescorla & LoLordo, 1965), as well as several experiments on appetitive conditioning of pigeons (Bottjer, 1982; Gaffan &

Hart, 1981; Hearst & Franklin, 1977; Kaplan & Hearst, in press; Wasserman et al., 1974; Wasserman & Molina, 1975).

With the exception of the study of Hoffman and Fitzgerald, all the studies that have yielded inhibition included randomly alternated CS− trials and UCS trials, so that at the end of the intertrial interval either CS− or the UCS might occur. Thus, temporal cues correlated with the end of the intertrial interval could have become excitatory, so that there would have been a source of conditioned excitation present whenever CS− was presented and nonreinforced. On the other hand, the two studies that failed to produce an inhibitory effect of the EUP procedure presented CS+ and CS− in single alternation. If the animals learned that these two events alternated singly, CS− would never have occurred in the presence of an excitatory temporal cue, leaving contextual excitation resulting from the previous presentation of the UCS as the only source of excitation concurrent with CS−. If the passage of time between that UCS presentation and presentation of the CS− permitted extinction of that contextual conditioning, the failure of the single alternation EUP procedures of Plotkin and Oakley and Maier *et al.* to produce inhibition supports the necessity of concurrent excitation for the development of inhibition.

However, the results of two recent studies temper our enthusiasm for this analysis. Hoffman and Fitzgerald (1982), studying heart rate conditioning in rats with a shock UCS, found that a CS− from an EUP procedure in which CS− and shock alternated singly was slower to evoke an excitatory CR in a resistance to reinforcement test than was a CS− from discriminative conditioning. This outcome suggests that concurrent excitation was unnecessary for inhibition. However, performance of the two groups did not differ in a summation test. Kaplan (1984) found that pigeons came to withdraw from a key-light CS that always occurred in the middle of an 87-sec interreinforcement interval and suggested that the CS had become a conditioned inhibitor. Although Kaplan's suggestion is certainly reasonable, in a subsequent resistance to reinforcement test these birds came to acquire excitatory, approach CRs to this CS as rapidly as pigeons that had previously received it later in the interreinforcement interval and had been indifferent to it. Thus, although these two studies cast doubt on the necessity of concurrent conditioned excitation for the development of inhibition in the EUP procedure, they are not conclusive.

Trace conditioning procedures can be considered EUP procedures in which the CS− occurs relatively near the end of the inter-UCS interval. Although trace conditioning is often excitatory (e.g., Ellison, 1964; Kamin, 1961; Newlin & LoLordo, 1976; Smith, Coleman, & Gormezano, 1969), a trace CS has been shown to be inhibitory in several recent studies. Rescorla (1968) found that a 5-sec CS that preceded an electric shock UCS by 25 sec, and thus occurred roughly 80% of the way through the intershock interval, subsequently markedly suppressed the shock-avoidance behavior of dogs. Roughly comparable effects occurred during the early part of a long-delay CS. Both of these effects appear to be examples of what Pavlov (1927) called inhibition of delay.

More recently, Hinson and Siegel (1980), studying eyelid conditioning in rabbits, found that a brief, trace CS that occurred 10 sec before the shock UCS, and thus nearly 3 min after the previous shock, had an inhibitory effect in a summation test re CSs from no-treatment and zero-contingency control groups. Furthermore, in a resistance to reinforcement test, the trace CS was slower to evoke an eyeblink CR than a novel CS, but not slower than the CS from a zero-contingency control group. Putting this result together with Plotkin and Oakley's, it seems to be the case that, for at least some preparations, placing a trace CS near either end of the inter-UCS interval, but not in the middle, will cause it to become inhibitory. In a subsequent chapter in this volume, Fowler describes some data from aversive conditioning of rats that fit this pattern and suggests an account of those data (but see Baker, 1977). In any case, if a trace CS becomes less inhibitory when placed in the middle of the inter-UCS interval than when placed at the end, some mechanism other than context conditioning arising from presentation of the prior UCS would have to be responsible for the outcome, because that conditioning would extinguish as the interval elapsed.

Kaplan (1984) has recently reported some data that do not fit this pattern. In one experiment he presented pigeons with repeated trials consisting of 12-sec illumination of a key, followed by a 12-sec trace period during which intertrial conditions were in effect, the trial ending in 3-sec access to grain. Five groups received these trials at mean intervals of 15, 30, 60, 120, and 240 sec. Birds in the 240-sec ITI group acquired a strong tendency to approach the CS, birds in the 15-sec group came to withdraw from it, and birds in the other groups showed no conditioned tendencies. In a second experiment the inter-UCS interval was held constant across groups, as was the duration of the key-light CS, but the duration of the trace interval varied. Trace interval durations that put the CS early or in the middle of the interreinforcement interval resulted in withdrawal from the CS, whereas the pigeons approached the CS when it occurred near the end of the inter-UCS interval. If we accept withdrawal as reflecting inhibition, this pattern of results is discrepant from the one described in the preceding paragraph; that is, Kaplan failed to find inhibitory effects of CS placements on both sides of that placement that resulted in a neutral CS. The accounts of the effects of backward conditioning, trace conditioning, and EUP treatments that have been offered by Kaplan (1984), Fowler (this volume), and Balsam (1984) will be challenged by this discrepancy (also see Wagner, 1981).

Negative Contingency Procedures. A procedure related to the EUP treatment arose from the contingency view of conditioning (Rescorla, 1967). In this negative contingency procedure UCSs are presented on a probabilistic basis so that p(UCS/CS) < p(UCS/noCS). In 1966 Rescorla showed that when dogs had received electric shock UCSs at a rate of .4/min except during presentation of a 5-sec CS or the 25 sec immediately following CS, when they never were shocked, the CS markedly suppressed the rate of avoidance (see also Grossen & Bolles, 1968; Libby, 1976). Patterson and Overmier (1981) recently showed that the conditioning context becomes excitatory in such a negative contingency treatment.

If the strength of conditioned inhibition varies directly with the magnitude of conditioned excitation present when CS− is nonreinforced, when p(UCS/CS) = 0, increases in p(UCS/noCS) should result in increases in the magnitude of conditioned inhibition. Rescorla (1969b) collected data bearing on this hypothesis, using the Conditioned Emotional Response (CER) procedure with rat subjects. In one study two groups received no shocks during 2-min auditory CS presentations or the 2 min immediately after each CS. Group 0-.1 received shocks at other times at a rate of .1/2 min, whereas Group 0-.4 received them at a rate of .4/2 min. In a subsequent resistance to reinforcement test, both groups acquired conditioned suppression to the auditory CS more slowly than controls that had received independent presentations of CSs and shocks in a zero-contingency procedure or controls that had received negative contingencies between a different, visual CS and shock. Moreover, Group 0-.4 acquired conditioned suppression more slowly than Group 0-.1. In a second experiment, which included a Group 0-.8 and used a summation test procedure, the magnitude of the difference between the amount of conditioned suppression evoked by CS+ and the amount evoked by a compound of CS+ and the CS− was directly related to the shock frequency in the absence of CS− during conditioning, confirming the results of the first experiment.

In a subsequent experiment, also using the CER procedure, Witcher and Ayres (1980) failed to fully confirm this result. In their first experiment both Group 0-.1 and Group 0-.8 showed equal retardation of acquisition, re naive controls. However, in a second experiment, a summation test revealed an inhibitory effect only in Group 0-.8. It is difficult to reconcile these two outcomes with each other, because the first suggests that inhibition was as strong in Group 0-.1 as in Group 0-.8, whereas the second suggests that the treatment of Group 0-.1 failed to make the CS at all inhibitory.

Hearst and Franklin (1977) exposed pigeons to a key-light CS that was negatively correlated with the presentation of grain. Three groups received no grain during the 20-sec CS. In the absence of the CS Group 37 received brief access to grain at a rate of 0.37 UCSs/min, Group 74, at 0.74/min, and Group 148, at 1.48/min. As conditioning progressed, all these groups came to withdraw from the key-light CS, and the magnitude of the conditioned withdrawal response was greater in Groups 148 and 74 than in Group 37. In this experiment, unlike those of Rescorla and Witcher and Ayres, UCSs could occur immediately after CS termination. Thus the direct, inhibitory effect of the negative contingency between CS and UCS overcame any excitatory effect of "chance pairings" of CS and UCS, because the frequency of such pairings would have increased with p(UCS/noCS).

Relatively few experiments have looked at the strength of inhibitory conditioning resulting from procedures in which there is a negative contingency between CS and UCS, but p(UCS/CS) > 0 (but see Miller & Schachtman, this volume). In the autoshaping situation Hearst, Bottjer, and Walker (1980) presented three groups of pigeons with brief access to grain at rates of 0.4/12-sec period in the absence of a 12-sec key-light CS. In the presence of the CS the rates of UCS presentation were 0, 0.2, and 0.4/12 sec. The strength of the tendency to withdraw from the CS during conditioning and the magnitude of resistance to reinforcement when the key light was subsequently paired with food both increased with increases in the negative contingency. Hearst and Franklin (1977) compared a group of pigeons that received food at a rate of 0.1/12 sec during CS and 0.2/12 sec in its absence with a group that received UCSs at rates of 0 and .3/12 sec during CS and its absence. In this experiment the CS was present for half the session, so that the groups were equated for total number of food reinforcements. During conditioning both groups withdrew from the CS and did not differ in the strength of withdrawal (also see LoLordo, Fairless, & Stanhope, in press). However, when the CS was paired with food in a subsequent resistance to reinforcement test, the group that had received the stronger negative contingency was slower to acquire approach to the CS.

Negative contingency treatments in which p(UCS/CS) > 0 have also been studied by Ayres and his colleagues in several studies using the CER procedure in rats. Ayres, Benedict, and Witcher (1975) reported excitatory effects of slightly negative contingencies. More recently, Witcher and Ayres (1980) varied the probability of shock during a 2-min CS (and the 2 min immediately after CS) from 0–0.6/2 min across groups, with the probability of shock at 0.8/2 min at other times. This manipulation produced orderly results in both summation and resistance to reinforcement tests. However, the CS appeared to be associatively neutral for a group that received shock with probabilities 0.2 and 0.8 in the presence and absence of the CS, respectively, and it was excitatory for a group that received probabilities of 0.6 and 0.8. Neither we nor the authors can explain this effect.

To summarize this section, most of the data are compatible with the claim that the development of conditioned inhibition depends on the nonreinforcement, or less frequent reinforcement, of the CS in the presence of some excitatory stimulus, and that, with other parameters held constant, the amount of inhibition conditioned is a direct function of the amount of excitation that is present when the CS is nonreinforced. However, if trace CSs reliably produce inhibition, even at long inter-UCS intervals, which would be expected to produce considerable extinction of contextual cues, it will be difficult to account for such inhibition in terms of concurrent conditioned excitation.

Some other Procedures for Producing Inhibition

Several other procedures have been shown to generate inhibition in at least one experiment but have not been studied very much.

Damianopoulos (1982) has suggested that one can determine whether nonreinforcement of a putative inhibitor in the presence of concurrent conditioned excitation is sufficient for conditioned inhibition by ensuring that this condition occurs in the absence of a negative correlation between the putative inhibitor and the UCS. He argued that this could be accomplished by presenting A+ trials in some initial sessions, and then only AX− trials in the following sessions. A control group, which should experience neither concurrent excitation nor the negative correlation, would receive A+ trials in phase one, followed by independent A− trials and X− trials in the second phase. If X becomes inhibitory in the experimental group, relative to the control, Damianopoulos argues that concurrent excitation is sufficient for the conditioning of inhibition.

Rescorla (1979a, p. 105–107) has presented some data that suggest that the A+, then AX− procedure produces inhibition. Rats that had received nonreinforced presentations of a flashing-light CS in compound with a previously reinforced tone CS were slower than controls to conditionally suppress food-getting behavior when the light was subsequently paired with shock, and they showed more inhibitory summation than controls when the light was added to another CS+ in a summation test. In this study, one of the control groups received A− rather than AX− trials during the second, extinction phase of the experiment, whereas the other control group received nonreinforced presentations of X in compound with a neutral stimulus. Cunningham (1981) has confirmed this result for the summation test. Thus concurrent excitation in the presence of a nonreinforced CS seems to be sufficient for inhibition, and a negative correlation does not seem to be a necessary part of a complex of factors that includes concurrent excitation. Of course all of this depends on using a definition of negative correlation that requires that at least some UCSs occur after the putative inhibitor, but outside the range of effective temporal contiguity for the response system in question. If "negative correlation" is equated with "negative contingency," then Rescorla's procedure entails a negative correlation and doesn't bear on the question at issue.

The second-order conditioned inhibition procedure, in which, following the standard A+, AX− procedure, a novel stimulus B is paired with X in the absence of reinforcement, is theoretically interesting for several reasons. Not only does such a procedure exclude a negative correlation between B and the UCS; it also nonreinforces B in the absence of any excitatory stimulus. An inhibitory outcome of this procedure would suggest that the first-order inhibitor is active in the absence of excitation, and that this inhibition can be conditioned to a second-order CS in the absence of excitation. Unfortunately, there has been relatively little systematic research on second-order conditioned inhibition. The effect was obtained frequently in Pavlov's and other Russian laboratories, with the most detailed report of Russian work in the Western literature being by Lindberg (1933). More recently Rescorla (1976), using a CER procedure with rats, demonstrated that it was more difficult to transform the B stimulus from the A+, AX−, then BX− procedure into a second-order excitor than it was to transform a control stimulus, suggesting that the B stimulus was inhibitory. The generality of this effect, which would not be predicted by the Rescorla–Wagner model, remains to be determined.

Finally, Kremer (1978) paired each of two stimuli, A and B, with an electric shock UCS, and then presented a novel CS, X, in compound with A and B, and followed this compound by the same UCS (thus, ABX+). Interspersed among compound trials were A+ and B+ trials designed to maintain the associative strengths of A and B at their asymptotic values. The Wagner–Rescorla model predicts that X should become inhibitory as a result of this procedure, and the data supported this prediction. The X stimulus strongly facilitated food-reinforced lever pressing on the first trial of a resistance to reinforcement test. Moreover, acquisition of conditioned suppression was significantly retarded in this stimulus, re controls. These results are particularly striking because they contradict any account of inhibition that requires a negative correlation or negative contingency between a CS and the UCS, or, for that matter, any theory that requires differential reinforcement of various stimuli. Because of its theoretical importance this procedure should be replicated and extended to procedures other than the CER (see Rescorla, 1979b).

What is Learned in Conditioned Inhibition?

The question of what is learned by animals subjected to Pavlovian inhibitory conditioning procedures can be divided into two relatively independent subtopics. The first includes discussions of the specific mechanisms, associative or otherwise, whereby conditioned inhibitors attain their impact on behavior. The second involves attempts to detail qualities of the CS-elicited memorial representations that are assumed by many to be a result of exposure to inhibitory contingencies. Although one's position on mechanism most likely entails implications regarding memory and vice versa, we have found it useful, for the moment, to approach these topics separately.

Associative Linkages

The issues involved in the first subtopic are perhaps best illustrated by contrasting Pavlov's (1927) nonassociative approach with Konorski's (1948, 1967) strong claim for the UCS specificity of inhibitory action. It was Pavlov's view that inhibitory processes radiate diffusely to affect cortical regions surrounding the inhibitory focal point in the CS center. This irradiation results, Pavlov believed, in attenuated responding to all conditioned excitors. Thus, an X derived from an A+, AX− procedure should attenuate excitation elicited by B, a stimulus not involved in the inhibitory training procedure. Furthermore, this transfer of the behavioral impact of X should occur regardless of the nature of the responding elicited by B, or of the UCS that had been paired with B. For example, Pavlov cites data that indicate that extinction of appetitive responding to a metronome will attenuate conditioned responding both to other appetitive CSs (homogenous CRs) and to aversive CSs (heterogenous CRs). For Pavlov, transfer of inhibitory effects depended only on the spatial proximity of the cortical centers involved and their relative "physiological strengths." The nature of the UCS centers with which the various excitors had been linked by training was, in this respect, irrelevant.

It was Konorski's (1948) view that the apparent nonspecificity of inhibitory transfer observed by Pavlov was an artifact of his reliance on a single response system, salivation, as the model for both aversive and appetitive conditioning. Konorski's own view was that a conditioned inhibitor's impact would transfer only to excitors based on the same UCS used in the original inhibition training. This prediction was derived from the view that such training results in the formation of an inhibitory link between the CS− and the UCS center. Subsequent presentation of the CS− still produces excitation in the corresponding CS cortical center, but the result of this excitation is to activate the inhibitory CS–UCS link. The net result is an increase in the activation threshold of the UCS center, so that a previously effective CS+ will be rendered ineffective in a summation test. Konorski's (1967) later view complicated the mechanics of the inhibitory linkage and seems to make more specific predictions concerning inhibitory transfer within a UCS class (i.e., aversive to aversive or appetitive to appetitive). Nonetheless, we agree with Rescorla's (1979a) assertion that the early and later versions of Konorski's model are difficult to distinguish in this regard. Later, however, we argue that the 1967 theory does have unique implications relevant to the second aspect of the "what is learned" question to be treated here.

Contemporary considerations of this issue clearly have much in common with Konorski's UCS-specific inhibitory mechanism. Nonetheless, it has been only quite recently (Rescorla & Holland, 1977) that the precise nature of the presumed associative linkages has received comprehensive empirical analysis. Working within the context of both aversive and appetitive preparations with rats, Rescorla and Holland considered a number of potential loci of the inhibitory action of a CS− derived from an A+, AX− procedure. The first possibility, that the impact of X is specific to the A stimulus used in training, is discounted by the by-now common demonstration of transfer to an alternative excitor (e.g., Cotton et al., 1982; Wagner, Mazur, Donegan, & Pfautz, 1980). Rescorla and Holland provided another demonstration of such transfer; their shock-based inhibitory X attenuated responding to an additional aversive excitor B, and did so even for a group in which responding to the training excitor A had been extinguished. Of course, this demonstration that X's inhibitory strength is independent of A does not contradict Pavlov's prediction of very diffuse transfer effects. What is not consistent with Pavlov's nonassociative view is the subsequent demonstration by Rescorla and Holland that if B had been transformed into an appetitive excitor by being paired with food, X no longer attenuated conditioned responding to B. Thus, the UCS-specific nature of a conditioned inhibitor was confirmed.

Another possible locus of inhibitory action considered by Rescorla and Holland (1977) is upon the Pavlovian CR itself; that is, the impact of an inhibitor may be upon the peripheral response system involved, rather than upon any central representation of the UCS, which may itself remain covertly active in the presence of X. To address this issue Rescorla and Holland took advantage of the fact that light and clicker CSs, although both readily conditionable as food-based excitors, support CRs having different topographies (Holland, 1977). All rats were given clicker–food and light–food pairings as well as nonreinforced compound trials involving a tone (X) and, for different groups, either the light excitor or the clicker excitor. The observation that in a subsequent summation test (Fig. 1.10) X not only attenuated the CR characteristic of the training excitor (e.g., rearing in response to light) but subsequently also attenuated the topographically different response to the test excitor (e.g., startle in response to clicker) strongly suggests that the inhibitor acts upon a locus more central than that responsible for the CS-specific response topographies. This result, in combination with the observation of UCS-specificity described previously, led Rescorla and Holland to conclude that a third possible locus for inhibitory action, the UCS center, is the most viable one; that is, an inhibitory X attenuates excitatory conditioned responding by suppressing the activity of a central representation corresponding to the UCS (or UCS class) employed in training.

It should be noted that Rescorla and Holland also considered the possibility that the inhibitory effect of X arises as a result of its impact on the excitatory A-UCS linkage. They rejected this alternative for the A+, AX− case on the basis of the CS transfer effects already noted. However, in separate chapters of the present volume both Rescorla and Holland present evidence that just such an inhibitory relation between the CS- and the A-UCS link can result from conditioned inhibition procedures in which CS− precedes CS+ on nonreinforced trials. These findings seem to validate their earlier warning (Rescorla, 1975; Rescorla & Holland, 1977) that different training procedures may yield inhibitors that, although behaviorally similar in some respects, nontheless accomplish their impact via distinctly different mechanisms. If this proves to be the case generally, it will be necessary to qualify the generic term inhibitor.

The Details of the Representation

Concern with the details of the memorial representations presumed to underlie the behavioral effects of conditioned inhibitors has become more prominent in

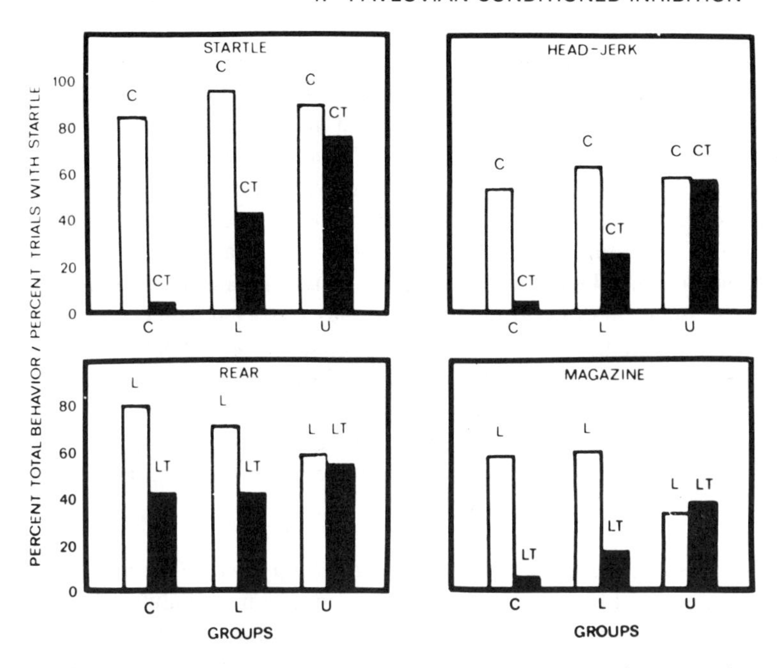

FIG. 1.10. The effects of an inhibitory tone upon the relative frequency of appetitive CRs characteristic of a clicker CS+ (top panels) or a light CS+ (bottom panels) in the inhibitory transfer study by Rescorla and Holland (1977). During training, Group C experienced the clicker as the excitatory element on nonreinforced compound trials involving the tone, whereas the light served this function in Group L. Group U experienced the tone alone on nonreinforced training trials. Both the clicker and the light were excitatory for all groups. (Reproduced by permission of Academic Press, San Francisco, and the author.)

recent years and has developed concurrently with the increasing application of information-processing models to problems in animal learning (e.g., Dickinson, 1980; Wagner, 1981). Of course the roots of these contemporary considerations can, as with most issues in the area, be identified in the early writings of Pavlov. However for Pavlov the question of the precise details of the cortical processes responsible for conditioned behavior was an important one primarily with respect to conditioned excitors. His position on this point affirmed what later became known as the principle of stimulus substitution; that is, he thought that CSs activated a process whose characteristics mimicked those elicited by the UCS itself sufficiently to produce adaptive, preparatory responding. Perhaps the most striking modern support for this kind of view comes from the autoshaping experiments of Jenkins and Moore (1973; Moore, 1973), in which pigeons were observed to "eat" key-light stimuli paired with food and to "drink" key lights paired with water. These selective differences in response topography can at least be taken as evidence that the memories elicited by the respective stimuli involve details sufficient to support reinforcer-appropriate behavior.

It is not surprising that Pavlov failed to identify an analogous stimulus substitution principle for the case of conditioned inhibitors. His account of inhibitory action pointed to suppressed activity focused in the affected CS center. Nothing in this account would allow an inhibitor to be at all UCS-specific. But if animals can act adaptively in the presence of informative excitors, they may do so also in the presence of inhibitors, and this action might have a form appropriate for circumstances correlated with the absence of a particular UCS. Konorski's early (1948) theory shared with Pavlov's a lack of interest in this possibility. Although in that account the effect of an inhibitor was restricted to the particular UCS employed in training, this specificity did not require the attribution to animals of "knowing" that this specific UCS was not forthcoming. The question of what Konorski (1967) would later term the epicritic or sensory components of conditioned inhibitors is simply not likely to arise from a perspective that defines inhibitors exclusively with respect to their excitation-opposing properties.

The Opposing Affect View of Conditioned Inhibition. One approach to the informational qualities of states elicited by conditioned inhibitors is consistent with the standard excitatory-opposition view of inhibition. This is the assertion that inhibitors elicit an affective state opposite to that elicited unconditionally by the reinforcer. We have already discussed this notion in some detail with respect to the opponent-process interpretations of backward conditioning. The idea is clearly a pervasive one, and findings from a number of experimental approaches have encouraged the view that inhibitors based on one affective class of UCS share functional attributes with excitors from the opposite affective class. These include the positive conditioned reinforcing effects of shock-based inhibitors (Grelle & James, 1981; Weisman & Litner, 1969b), and the conditioned punishing effects of food-based inhibitors (Hyde, 1976b; see LoLordo, 1969, for a review of early demonstrations of this effect), as well as the apparent tendency of pigeons to avoid stimuli associated with the nonavailability of food (Terrace, 1971; Wasserman et al., 1974).

This opposing-affect view can be readily taken as consistent with the standard resistance-to-reinforcement test. One need only assume that affect appropriate to a UCS is a prerequisite of excitatory responding and that, because inhibitors based on that UCS lie on the opposite side of neutrality from this required affective threshold, they will condition more slowly than novel cues. Of course one can retain the notion of opposition and an emphasis on the resistance-to-reinforcement effect while remaining mute about affective states; the formal equations of the Rescorla–Wagner model are the most familiar case in point. But a simple identification of the associative strength dimension basic to the Rescorla–Wagner model with a bipolar affective dimension makes that model a statement about the interaction of opposing affective states (e.g., Dickinson & Dearing, 1979; Fowler, 1978).

Aversive-Appetitive and Appetitive-Aversive Transfer Studies. The opposing-affect view leads quite naturally to predictions concerning the ease with which inhibitors from one affective domain will become excitors within the opposite affective domain. For example, the presumed aversiveness of a food-based inhibitor, considered the cause of its retarded acquisition when paired with food, should by the same logic cause the stimulus to acquire excitatory aversive properties with relative ease. Dickinson and Pearce (1977) have provided a review of some studies relevant to this issue, although from a broader perspective that includes a presumed mutually inhibitory relation between aversive and appetitive excitors. Our purpose here is to focus on the relatively few published attempts to transform inhibitors of one affective class into excitors of the other, most of which have appeared since the Dickinson and Pearce survey.

Scavio (1974) subjected rabbits to an initial aversive conditioning treatment in which different groups experienced a 400-msec tone CS either paired or explicitly unpaired (semirandom alternation of CS and UCS) with a 50-msec, 4.0-mA eyelid shock. An additional group received neither stimulus in this phase. After a robust nictitating membrane (NM) response had developed in the paired group, all animals were given an appetitive conditioning treatment in which the same tone CS was paired with water injected into the oral cavity. Animals that had experienced an excitatory CS-shock relation in phase 1 were slowest to acquire the appetitive jaw-movement response in phase 2. However, the performance of rabbits from the unpaired condition was indistinguishable from that of the naive controls, and the opposing-affect view was in this respect not supported (but see Krank, 1982). A subsequent study by Bromage and Scavio (1978) employed very similar aversive NM conditioning parameters and did find some evidence that a CS− for shock can produce an accelerated appetitive acquisition curve. In this study two levels of water deprivation were combined factorally with the contingency treatments (i.e., CS and UCS paired, explicitly unpaired, or no stimulus). Previous exposure to the unpaired CS and shock resulted in facilitated acquisition of the jaw-movement CR for the low-deprivation condition but not for the high-deprivation condition (see Fig. 1.11). Bromage and Scavio (1978) suggested that this interaction may be due to a ceiling effect operative in the high deprivation animals. Presumably this high-deprivation level translates into a high degree of appetitive system activation, with the result that the positive affect associated with the shock-based CS− from phase 1 could contribute little additional appetitive activation. Because water deprivation in the earlier Scavio study was intermediate to the two levels employed by Bromage and Scavio, this same argument might apply to the failure in the former experiment to demonstrate facilitated acquisition in the unpaired group.

Although these aversive-to-appetitive transfer studies offer some support for the opposing-affect view of inhibition, recent reports concerning appetitive-to-aversive transfer seem to contradict it. Scavio and Gormezano (1980) initially

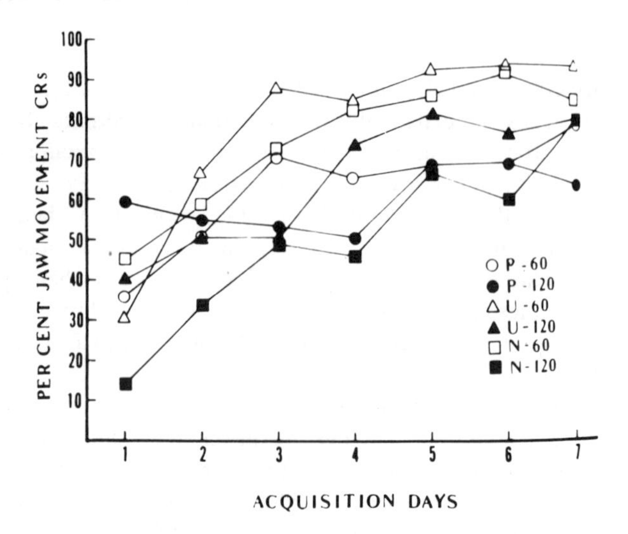

FIG. 1.11. Mean percentage appetitive jaw-movement CRs over conditioning days during phase 2 of the Bromage and Scavio (1978) study. The letters P (paired), U (unpaired), or N (no training) refer to the phase 1 aversive conditioning treatment. The numbers refer to water deprivation level (60 ml/day or 120 ml/day). (Reproduced by permission of the Psychonomic Society, Austin, Texas and the authors.)

exposed different groups of rabbits to either paired or explicitly unpaired tone–water conditioning treatments. In the subsequent transfer phase all *S*'s received pairings of tone and paraorbital shock. The paired group, which had acquired appetitive jaw-movement CRs in phase 1, acquired the nictitating membrane CR more rapidly than did a naive control, whereas the unpaired group was slowest in acquiring the response. The performance of both groups contradicts the opposing-affect view, although Scavio and Gormezano suggest that the retardation of aversive CR acquisition observed in the unpaired tone–water group was due to a loss in salience or associability of the CS (cf. Mackintosh, 1975). Indeed, the absence of a CS-alone treatment group in the appetitive phase of this study leaves open the possibility that the slower acquisition observed in the unpaired group was due to CS preexposure per se and was independent of any putative inhibitory conditioning.

This latent inhibition interpretation of the Scavio and Gormezano (1980) results is made less viable by a set of experiments reported by DeVito and Fowler (1982). Rats were employed in a lick-suppression paradigm to demonstrate both excitatory and inhibitory appetitive-to-aversive transfer. In the two excitatory-transfer studies, which differed only in the shock intensities used, rats first received paired (group CS+), unpaired (group CS−), or random (group CSo) presentations of a flashing light and response-independent food. A fourth group (group USa) received only food during this phase. All rats were subsequently given pairings of the flashing light and shock during sessions in which they could lick a tube for water reinforcement. This licking response was soon suppressed during light presentations for all groups. During subsequent presentations of the light in extinction, group differences appeared; suppression extinguished more slowly in group CS+ than in either group CSo or group USa, which did not differ from one another. Suppression in group CS−, on the other hand, extinguished the most rapidly, suggesting that the least amount of aversive conditioning had accrued to the stimulus in this group, which is an outcome inconsistent with the opposing-affect viewpoint.

The third experiment reported by DeVito and Fowler addressed the possibility that the attenuated aversive conditioning observed in the transfer phase for group CS− was due to a decrement in stimulus salience. They reasoned that such a loss of salience should also render an appetitive CS− less effective when it is subsequently used as the added element in an A+, AX− aversive conditioning procedure. Groups of rats were first given differential appetitive treatments identical to those just described. All rats then received on-baseline tone–shock pairings intermixed with nonreinforced presentations of this tone compounded with the flashing-light CS from phase 1. Rats from the appetitive CS− treatment came to discriminate this nonreinforced compound from the excitatory element more

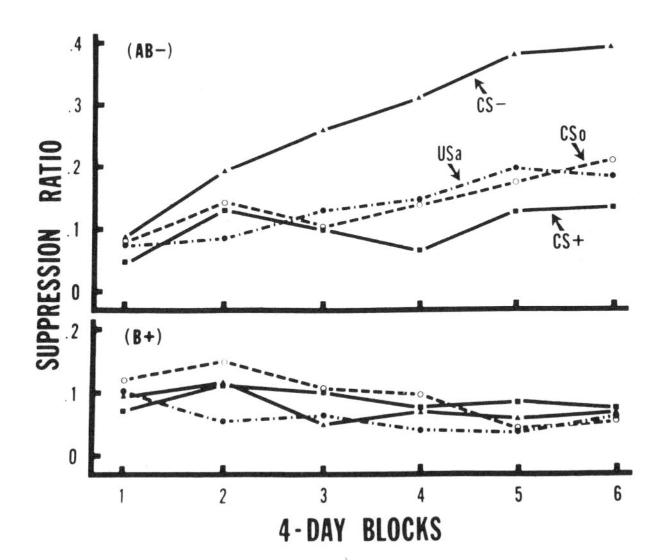

FIG. 1.12. Mean suppression to the excitatory B element (lower panel) and the nonreinforced AB compound (upper panel) in the appetitive-to-aversive inhibitory transfer study of DeVito and Fowler (1982). The A element had been paired (CS+), explicitly unpaired (CS−), or uncorrelated with food (CSo) during phase 1. For group USa, A was novel at the outset of aversive conditioning. (Reproduced by permission of Academic Press, San Francisco and the authors.)

rapidly than did the other treatment groups (see Fig. 1.12). This outcome discounts a loss of salience interpretation of the excitatory transfer studies, and together these experiments suggest that the simple characterization of appetitive inhibitors as aversive is inappropriate for predicting the outcome of Pavlovian appetitive-to-aversive transfer. DeVito and Fowler interpret their results by making a distinction between the affective and the signaling characteristics of conditional stimuli. By this account, an appetitive inhibitor is associated with the affective result of nonreinforcement (presumably aversive), as well as with a negative signal value, which develops because the inhibitor reliably predicts the absence of a biologically important event. It is this latter component, DeVito and Fowler argue, that attenuates acquisition of *positive* signaling value when the stimulus is paired with shock and facilitates acquisition of *negative* signal value when the stimulus is used as the X element in an A+, AX− procedure involving shock.

Division of an inhibitor's impact into two components has a precedent in Konorski's 1967 theory. Konorski proposed that a cue presented in the absence of otherwise available reinforcement results in the formation of two distinct associative connections. One connection is between the CS center and an antidrive center. This antidrive center has an innate, reciprocally inhibitory link with the drive center appropriate to the UCS. In the case of food inhibitors, CS− presentation activates the antihunger center and thereby suppresses hunger motivation. The second link proposed by Konorski connects the CS center with no-UCS units. In the case of food, these units are the antagonistic counterparts of neural centers selectively responsive to such sensory attributes as taste and texture.

This sensory component of inhibitors as proposed by Konorski (1967) involves considerably more representational detail than is implied by the signaling notion put forward by Fowler and his associates (DeVito & Fowler, 1982; Fowler et al., 1973; Goodman & Fowler, 1976). More importantly, the two views differ with respect to the ease with which the behavioral influence of the other component, CS-elicited affect, can be transformed by counterconditioning. Konorski considered the antidrive (and drive) connection the most inertial, and his early work on the topic (Konorski & Szwejkowska, 1956) yielded outcomes in appetitive-to-aversive transfer opposite to those of DeVito and Fowler. DeVito and Fowler explained their observation of retarded aversive acquisition by an appetitive inhibitor by asserting the opposite claim; that it is the more cognitive, signal value of appetitive conditional stimuli that is their most durable aspect. DeVito and Fowler reconciled this view with the otherwise contradictory aversive-to-appetitive transfer results reviewed earlier (Bromage & Scavio, 1978; Scavio, 1974) by suggesting that the affective impact of aversive cues is biologically more significant, and consequently more inertial, than appetitive affect. Thus, aversive inhibitors may show an accelerated appetitive acquisition curve, when paired with food, although this entails changing the signal value of the stimulus from negative to positive. Clearly, more extensive empirical treatment of this issue is required.

Stimulus Compounding Studies. An alternative method for assessing the interaction of inhibitors of one affective class with excitors of the other has been proposed by Dickinson and his associates (Dickinson, 1977; Dickinson & Dearing, 1979). This technique involves an extension of the logic employed by Kamin (1969) in investigating the tendency of a previously conditioned aversive excitor to block conditioning of an added neutral element on subsequent reinforced compound trials. Kamin's original interpretation of this effect was expressed in terms of redundancy; because the initial excitor already predicts the outcome of compound trials (i.e., shock reinforcement), the added element is redundant and so will not become excitatory. If both shock and the unavailability of food are aversive events, and if similarity of affect is an important aspect of redundancy, an appetitive inhibitor may block aversive conditioning to an added element during shock reinforced compound trials. Dickinson (1977) tested this notion by first exposing rats to various contingencies between a light and free food during sessions in which food was also available on a VI 2-min instrumental bar-press schedule. For group P the last 30 sec of the 1-min light coincided with the response-independent delivery of food pellets. Group U experienced an A+, AX− treatment; free food occurred in the presence of a clicker excitor but did not occur when the light and clicker were compounded. Group R received a combination of both the P and the U schedules, whereas Group L received presentations of the light alone. The light was later presented to all animals in compound with a tone and reinforced with foot shock. When tested alone later the tone controlled greater suppression in group P than in the other three groups. For group U, the tone appeared to have no effect at all on instrumental responding. Because this was true also of group R, the apparent absence of aversive conditioning to the tone in group U cannot be attributed to blocking by the appetitive inhibitor. However, Dickinson and Dearing reviewed an unpublished experiment by Dickinson in which such a blocking effect was observed.

Goodman and Fowler (1983) applied similar logic in an experiment that employed the suppression of water-tube licking as the measure of differential aversive conditioning. Rats were exposed to either a paired or an explicitly unpaired (random alternation) relation between a 5-sec flashing light and response-independent food. During subsequent lick sessions this light was combined with the last 5 sec of a 20-sec clicker, and the termination of the compound was contiguous with a 0.5-mA foot shock. Suppression to the 15-sec clicker-alone portion of the serial compound was measured during both acquisition trials and subsequent extinction trials. Group differences emerged only during extinction; suppression to the clicker was slower to extinguish for the appetitive CS+ group than for either the CS− group or a group that had received food alone during phase 1. The extinction of suppression in the CS− group was numerically, but not statistically, slower than that observed in the rats that had received food alone. Goodman and Fowler speculated that the inability to demonstrate a clear aversive blocking effect by appetitive inhibitors in their study and in the one Dickinson (1977) study might be due to the relatively low shock intensities employed. This notion gained support from the second Goodman and Fowler experiment, in which a 1.0-mA shock was employed during the serial compound phase. In this case the tone suppression extinguished most rapidly in group CS−, suggesting that the appetitive inhibitor had blocked the acquisition of aversive conditioning.

Although the findings of these appetitive-to-aversive blocking studies are somewhat tenuous, they are consistent with the extensive set of Pavlovian aversive-to-instrumental appetitive transfer studies conducted by Fowler and his colleagues (Fowler, Fago, Domber, & Hochhauser, 1973; Fowler, Goodman, & Zanich, 1977; Ghiselli & Fowler, 1976; Goodman & Fowler, 1976). In his review of these studies, Fowler (1978) attributed the attenuation of appetitive discrimination learning by the response contingent presentation of an aversive inhibitor to a blocking-like effect.

The problem remains of reconciling the results of these blocking studies with the appetitive-to-aversive transfer results reviewed earlier. The transformation of an appetitive inhibitor into an aversive excitor proceeds relatively slowly, an outcome inconsistent with an opposing-affect interpretation of an inhibitor's ability to disrupt appetitive excitation. On the other hand, some evidence exists that appetitive inhibitors attenuate the aversive conditioning of an added element, an outcome that *is* congruent with an opposing-affect view of an inhibitor's action. Goodman and Fowler (1983) suggest that both results can be reconciled through Kamin's notion of redundancy. Thus, it is conditioning of the added element in the Dickinson (1977; Dickinson & Dearing, 1979) and Goodman and Fowler (1983) studies that suffers from the redundancy entailed in preceding shock by a predictor of an aversive consequence, the absence of food. In the transfer studies, they argue, the appetitive CS− itself fails to be readily conditioned because of its own redundancy. It might also be the case that the signal characteristics of conditional stimuli proposed by DeVito and Fowler (1982) play a predominant role in single-cue transfer tests, whereas affective contrast and consequent redundancy or surprise prevail in compound acquisition tests. In any event, any attempt at a general account of these phenomena is complicated considerably by the contrasting results of appetitive-to-aversive and aversive-to-appetitive transfer tests.

Inhibitory Stimuli and Sensory Characteristics of the UCS. Earlier we outlined Konorski's (1967) account of the two associative linkages engendered by inhibitory conditioning, one between the CS− and an antidrive center and the other between the CS− and a no-UCS center. This latter center, when activated, inhibits those components of the UCS gnostic unit in which are encoded the specific sensory attributes of the UCS, e.g., its taste, color, or location of application. It is not clear from Konorski's writing how activation by an inhibitor of no-UCS units should affect the representations of sensory attributes of the omitted UCS. Would he claim, for instance, that the activation of no-UCS units tells the animal precisely what it is that will not be available, or does it simply prevent the representations of sensory attributes from being evoked? Konorski unfortunately said very little about the behavioral impact of this epicritic aspect of inhibitors.

The question of whether inhibitors somehow provide information about the sensory attributes of omitted UCSs is beginning to receive some experimental attention (for an earlier, abortive effort, see LoLordo, 1967). We describe three experiments bearing on this question, which is also discussed in the chapter by Mackintosh and Cotton in this volume.

Pearce, Montgomery, and Dickinson (1982) administered an A+, AX− treatment to rabbits in an eyelid conditioning experiment, with the UCS being shock to one eye. Controls received only the A+ trials. Then half the animals in each group were given a resistance to reinforcement test in which X was reliably followed by shock to the same eye, whereas the other half received X followed by shock to the contralateral eye. In the test each experimental group showed marked retardation of acquisition of an eye blink CR, relative to its control. Indeed, as Fig. 1.13 illustrates, the magnitude of the inhibitory effect was as great for the rabbits with the changed shock locus as it was for those with the constant shock locus. If there had instead been incomplete transfer of inhibition when shock locus was changed, there would have been grounds for asserting that, in addition to relief (or the inhibition of fear), the inhibitor also informed the animals that there would be no shock to the originally shocked eye. The complete transfer observed by Pearce et al. does not permit this inference.

Nieto (1984) conducted a formally similar study, using rat subjects and the conditioned emotional response procedure. One group of rats received a noise negatively correlated with electric shock in Baker's between-days procedure, whereas another group received the CS negatively correlated with a qualitatively different UCS, a loud blast from a horn. Then both groups received pairings of light and shock, followed by a summation test in which the suppressive effects of light alone and the light-noise compound were compared. The two groups showed marked and equivalent inhibitory effects in the summation test. Thus as in the study by Pearce et al., there was no evidence that the rats had associated the inhibitor with specific sensory qualities of the omitted reinforcer.

Kruse, Overmier, Konz and Rokke (1983) employed a differential choice task to investigate the specificity of the expectancies associated with Pavlovian inhibitors. In the choice task rats were trained to select one of two levers which were inserted simultaneously in the presence of alternative discriminative stimuli.

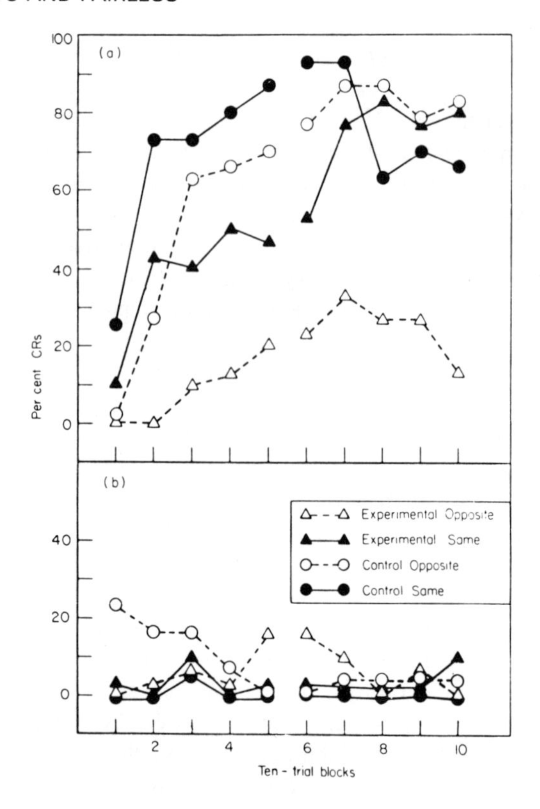

FIG. 1.13. Acquisition of a defensive eyeblink CR as a function of a change (opposite) or no change (same) in shock locus between inhibition training and acquisition test in the Pearce et al. (1981) study. For groups labeled "control" the CS was novel at the outset of test. The upper panel illustrates responding of the reinforced eye; the lower one, of the nonreinforced eye. (Reproduced by permission of Academic Press, London and the author.)

Presence of a clicker S+ indicated that the left lever was correct, and its selection resulted in the delivery of a food pellet. A tone S+ indicated that the right lever was correct, and correct choices of this lever resulted in delivery of a sugar water solution. Incorrect choices were always terminated without reinforcement. When *S*s were performing this discrimination with a high degree of accuracy, Kruse et al. instituted an explicitly unpaired Pavlovian conditioning procedure designed to render the CS−, interpretation of white noise, inhibitory with respect to either the food or the sugar solution for different groups. This CS− was later presented during a session of differential choice trials, both in the presence of each S+ (compound trials) or alone and in place of the S+ (CS− alone trials). The purpose of the CS− alone trials was to determine whether a CS− for one reinforcer would direct the rat's choices to the lever associated with the alternative reinforcer. Such an effect upon the direction of behavior would provide strong evidence for the claim that the inhibitor informs the organism about some sensory attribute of the omitted UCS. However, this choice-directing effect did not occur. Nonetheless, the compound test trials did provide some evidence of CS− specificity. The presence of the inhibitor reliably increased choice latencies, but only when compounded with the S+ appropriate to the reinforcer employed during the Pavlovian inhibitory conditioning phase. Since explicitly unpaired training occurred in the absence of both S+s, this result cannot be attributed to any configural learning, and implies that Pavlovian inhibitors carry with them some specificity regarding the reinforcer whose absence they predict. At the very least, this outcome suggests that a more fine grained breakdown of motivational states than the common appetitive-aversive division is appropriate. In any event, the approach to the issue taken by Kruse et al. is a promising one, and may be more appropriate than summation and transfer tests for determining whether the inhibitor conveys any information about the sensory qualities of the omitted UCS.

A number of recently developed theoretical approaches provide a straightforward answer to the question of an inhibitor's information value. According to these views, the learning engendered by Pavlovian procedures is described through reference to a single, positive dimension (Balsam, 1984; Gibbon & Balsam, 1981; Jenkins, Barnes & Barrera, 1981; Miller and Schachtman, this volume). Here we shall briefly discuss some implications that we think follow from the Scalar Expectancy Theory (SET) of autoshaping developed by Gibbon and Balsam (1981; Gibbon, 1981).

Stated very simply, SET proposes that pigeons exposed to an autoshaping situation acquire two separate expectancies of reinforcement, one that reflects the rate of food delivery during a discrete cue, and another that reflects the overall food delivery rate (see Jenkins, Barnes, & Barrera, 1981, for a similar view). These two expectancies interact to determine performance in response to a CS. Specifically, SET's performance rule, the ratio rule, asserts that an excitatory CR, pecking at the CS, will emerge when the expectancy ratio C/T reaches approximately two, where C is the average time between reinforcements during the experimental session and T is the average time that the trial stimulus is present between reinforcements.

Although both the theoretical and empirical work of Gibbon and Balsam have dealt primarily with excitatory conditioning as reflected by autoshaped keypecking, there has been some discussion of SET's approach to inhibitory effects as well (e.g., Balsam, 1984; Gibbon, 1981; Kaplan, 1984). Balsam has extended SET's analyses to the trace conditioning and EUP procedures, which produce inhibition under some circumstances. In these procedures, at a given interreinforcement interval, the value of the expectancy ratio decreases as either the trace interval or the overall CS-UCS interval increases. When this formulation was applied to Kaplan's (1984) data, withdrawal from the CS emerged at values of the ratio less than 1.5.

One important implication of SET arises from its assertion that the two expectancies that enter into this ratio rule are acquired and retained independently. It should thus be possible to alter a pigeon's response to a key light CS through an appropriate alteration of the overall food expectancy. Thus, although the key light associated expectancy remains constant, the behavioral output should be altered by such a manipulation. The observation by Kaplan and Hearst (in press) that pigeons will no longer withdraw from an explicitly unpaired key light after extended nonreinforced exposure to the experimental setting is consistent with this view. Even more striking is their finding that an initially excitatory CS, although subsequently rendered inhibitory by being used in an EUP procedure, will again elicit approach and pecking as a consequence of the extinction of context excitation.

In an experiment recently conducted in our laboratory in collaboration with Kelly Stanhope this line of reasoning was carried a step further. Consider an experiment in which pigeons receive food at a moderate rate during a key light CS, but at a high rate in its absence. SET correctly predicts that the pigeons will come to withdraw from the CS, but also asserts that the CS will come to evoke an expectancy of food appropriate to the moderate reinforcement rate in its presence. This expectancy should be manifested in approach and pecking when the CS is presented in a discriminably different context previously associated with a low rate of food reinforcement. In our experiment, two groups of pigeons were subjected to a negative contingency between a key light and food. For one group (P=0), food never occurred in the presence of the key light. For the other group (P>0), food occurred at a low but nonzero rate during the CS. Food presentations occurred at the same high rate for both groups in the absence of the key light. In addition to these withdrawal training sessions, all pigeons received food alone sessions in a discriminably different setting. During these sessions, the key light was never presented, and food presentations occurred at a rate lower than that associated with either the key light or the withdrawal training context. Eventually, in the high-reinforcement rate context birds in both groups came to withdraw from the key light, and there was no between groups difference in the magnitude of this effect. A within subjects resistance-to-reinforcement test was then conducted in the low food rate setting. Both the withdrawal-eliciting key light and a novel key light were presented and reinforced an equal number of times during each test session. Reinforcement occurred only in the presence of the CSs during test. None of the pigeons pecked either CS during the early part of the test session, contrary to SET's prediction that birds in group P>0 would peck the former CS− at this time. However, the relative speed with which key pecking was acquired to the two stimuli was a function of the withdrawal training condition. Although most birds in group P=0 acquired pecking at the novel key light first, the reverse was true for group P>0, in which most birds acquired a response to the former CS− first. This savings effect suggested that the animals in group P>0 had come to appreciate the nonzero food rate associated with the key light during withdrawal training. However, a second experiment using a between-groups design failed to confirm the savings effect. Thus the weight of the evidence from these two experiments is against SET.

Miller and Schachtman (this volume) have also argued that a single, excitatory dimension is adequate for explaining behavioral effects commonly considered inhibitory. We will leave further discussion of that view to their contribution. It is interesting to note, however, that treatment of Pavlovian conditioned inhibition has in some sense come full circle since Pavlov's introduction of the issue, from Skinner's (1938) early assertion that the inhibition concept is superfluous, through Konorski's detailed, cortical models of the phenomenon, and again now to the consideration of views that question the necessity of the construct. What is certain is that the issues involved will remain a rich source of empirical enquiry and theoretical concern.

ACKNOWLEDGMENTS

Preparation of this chapter was supported by a research grant from NSERC of Canada to V. M. L., and by postgraduate scholarships from NSERC and the I. W. Killam Foundation to J. L. F. The authors would like to thank A. Delamater, A. Droungas, and J. Kruse for their comments on an earlier draft of this chapter.

REFERENCES

Ayres, J. J. B., Benedict, J. O., & Witcher, E. S. (1975). Systematic manipulation of individual events in a truly random control in rats. *Journal of Comparative and Physiological Psychology, 88,* 97–103.

Baker, A. G. (1977). Conditioned inhibition arising from a between-sessions negative correlation. *Journal of Experimental Psychology: Animal Behavior Processes, 3,* 144–155.

Balaz, M. A., Capra, S., Kasprow, W. J., & Miller, R. R. (1982). Latent inhibition of the conditioning context: Further evidence of contextual potentiation of retrieval in the absence of context-US associations. *Animal Learning and Behavior, 10,* 242–248.

Balsam, P. (1984). Relative time in trace conditioning. In J. Gibbon & L. Allan (Eds.), *Timing and time perception.* New York: New York Academy of Sciences.

Batson, J. D., & Best, M. R. (1981). Single-element assessment of conditioned inhibition. *Bulletin of the Psychonomic Society, 18,* 328–330.

Best, M. R. (1975). Conditioned and latent inhibition in taste aversion learning: Clarifying the role of learned safety. *Journal of Experimental Psychology: Animal Behavior Processes, 1,* 97–113.

Bottjer, S. W. (1982). Conditioned approach and withdrawal behavior in pigeons; effects of a novel extraneous stimulus during acquisition and extinction. *Learning and Motivation, 13,* 44–67.

Bouton, M. E. & King, D. A. (1983). Contextual control of the extinction of conditioned fear: Tests for the associative value of the context. *Journal of Experimental Psychology: Animal Behavior Processes, 9,* 248–265.

Bromage, B. K., & Scavio, M. J. Jr. (1978). Effects of an aversive CS+ and CS− under deprivation upon successive classical appetitive and aversive conditioning. *Animal Learning and Behavior, 6,* 57–65.

Bull, J. A. III, & Overmier, J. B. (1968). Additive and subtractive properties of excitation and inhibition. *Journal of Comparative and Physiological Psychology, 66,* 511–514.

Cappell, H. D., Herring, B., & Webster, C. D. (1970). Discriminated conditioned suppression: Further effects of stimulus compounding. *Psychonomic Science, 19,* 147–148.

Cotton, M. M., Goodall, G., & Mackintosh, N. J. (1982). Inhibitory conditioning resulting from a reduction in the magnitude of reinforcement. *Quarterly Journal of Experimental Psychology, 34B,* 163–180.

Cunningham, C. L. (1981). Association between the elements of a bivalent compound stimulus. *Journal of Experimental Psychology: Animal Behavior Processes, 7,* 425–436.

Damianopoulos, E. N. (1982). Necessary and sufficient factors in classical conditioning. *Pavlovian Journal of Biological Science, 17,* 215–229.

Denny, M. R. (1971). Relaxation theory and experiments. In F. R. Brush (Ed.), *Aversive conditioning and learning* (pp. 235–295). New York: Academic Press.

DeVito, P. L., & Fowler, H. (1982). Transfer of conditioned appetitive stimuli to conditioned aversive excitatory and inhibitory stimuli. *Learning and Motivation, 13,* 135–154.

Dickinson, A. (1977). Appetitive-aversive interactions: Superconditioning of fear by an appetitive CS. *Quarterly Journal of Experimental Psychology, 29,* 71–83.

Dickinson, A. (1980). *Contemporary animal learning theory.* Cambridge: Cambridge University Press.

Dickinson, A., & Dearing, M. F. (1979). Appetitive-aversive interactions and inhibitory processes. In A. Dickinson & R. A. Boakes (Eds.), *Mechanisms of learning and motivation: A memorial volume to Jerzy Konorski* (pp. 203–231). Hillsdale, NJ: Lawrence Erlbaum Associates.

Dickinson, A., & Pearce, J. (1977). Inhibitory interactions between appetitive and aversive stimuli. *Psychological Bulletin, 84,* 690–711.

Dunham, P. J., & Carr, A. (1976). Pain-elicited aggression in the squirrel monkey: An implicit avoidance contingency. *Animal Learning and Behavior, 4,* 89–95.

Durlach, P. (1983). Pavlovian learning and performance when CS and US are uncorrelated. In M. Commons, R. Herrnstein, & A. Wagner (Eds.), *Quantitative analyses of behavior:* (*Vol III*), *Acquisition.* New York: Ballinger.

Ellison, G. D. (1964). Differential salivary conditioning to traces. *Journal of Comparative and Physiological Psychology, 57,* 373–380.

Fowler, H. (1978). Cognitive associations as evident in the blocking effects of response-contingent CSs. In S. H. Hulse, H. Fowler, & W. K. Honig (Eds.), *Cognitive processes in animal behavior* (pp. 109–153). Hillsdale, NJ: Lawrence Erlbaum Associates.

Fowler, H., Fago, G. C., Domber, E. A., & Hochhauser, M. (1973). Signaling and affective functions in Pavlovian conditioning. *Animal Learning and Behavior, 1,* 81–89.

Fowler, H., Goodman, J. H., & Zanich, M. L. (1977). Pavlovian aversive to instrumental appetitive transfer: Evidence for across-reinforcement blocking effects. *Animal Learning and Behavior, 5,* 129–134.

Furedy, J. J., Poulos, C. X., & Schiffman, K. (1975). Contingency theory and classical autonomic excitatory and inhibitory conditioning. Some problems of assessment and interpretation. *Psychophysiology, 12,* 98–105.

Gaffan, E. A., & Hart, M. M. (1981). Pigeons' withdrawal from an appetitive conditioned inhibitor under two training procedures. *Quarterly Journal of Experimental Psychology, 33B,* 77–94.

Ghiselli, W. B., & Fowler, H. (1976). Signaling and affective functions of conditioned aversive stimuli in an appetitive choice discrimination: US intensity effects. *Learning and Motivation, 7,* 1–16.

Gibbon, J. (1981). The contingency problem in autoshaping. In C. M. Locurto, H. S. Terrace, & J. G. Gibbon (Eds.), *Autoshaping and Conditioning Theory* (pp. 285–308). New York: Academic Press.

Gibbon, J., & Balsam, P. D. (1981). The spread of association in time. In C. M. Locurto, H. S. Terrace, & J. G. Gibbon (Eds.), *Autoshaping and conditioning theory* (pp. 219 253). New York: Academic Press.

Gokey, D. S., & Collins, R. L. (1980). Conditioned inhibition in feature negative discrimination learning with pigeons. *Animal Learning and Behavior, 8,* 231–236.

Goodman, J. H., & Fowler, H. (1976). Transfer of the signaling properties of aversive CSs to an instrumental appetitive discrimination. *Learning and Motivation, 7,* 446–457.

Goodman, J. H. & Fowler, H. (1983). Blocking and enhancement of fear conditioning by appetitive CSs. *Animal Learning and Behavior, 11,* 75–82.

Grelle, M. J., & James, J. H. (1981). Conditioned inhibition of fear: Evidence for a competing response mechanism. *Learning and Motivation, 12,* 300–320.

Grossen, N. E., & Bolles, R. C. (1968). Effects of a classical conditioned 'fear signal' and 'safety signal' on non-discriminated avoidance behavior. *Bulletin of the Psychonomic Society, 11,* 321–322.

Hammond, L. J. (1966). Increased responding to CS− in differential CER. *Psychonomic Science, 5,* 337–338.

Hammond, L. J. (1967). A traditional demonstration of the active properties of Pavlovian inhibition using differential CER. *Psychonomic Science, 9,* 65–66.

Hammond, L. J. (1968). Retardation of fear acquisition when the CS has previously been inhibitory. *Journal of Comparative and Physiological Psychology, 66,* 756–759.

Hammond, L. J., & Daniel, R. (1970). Negative contingency discrimination: Differentiation by rats between safe and random stimuli. *Journal of Comparative and Physiological Psychology, 72,* 486–491.

Hearst, E., Bottjer, S. W., & Walker, E. (1980). Conditioned approach-withdrawal behavior and some signal-food relations in pigeons. *Bulletin of the Psychonomic Society, 16,* 183–186.

Hearst, E., & Franklin, S. R. (1977). Positive and negative relations between a signal and food: Approach-withdrawal behavior to the signal. *Journal of Experimental Psychology: Animal Behavior Processes, 3,* 37–52.

Hendersen, R. W. (1973). Conditioned and unconditioned fear inhibition in rats. *Journal of Comparative and Physiological Psychology, 84,* 554–561.

Hinson, R. E., & Siegel, S. (1980). Trace conditioning as an inhibitory process. *Animal Learning and Behavior, 8,* 60–66.

Hoffman, J. W., & Fitzgerald, R. D. (1982). Bidirectional heart rate responses in rats associated with excitatory and inhibitory stimuli. *Animal Learning and Behavior, 10,* 77–82.

Holland, P. C. (1977). Conditioned stimulus as a determinant of the form of the Pavlovian conditioned response. *Journal of Experimental Psychology: Animal Behavior Processes, 3,* 77–104.

Hyde, T. S. (1976a). The effects of Pavlovian CSs on two food-reinforced baselines—with and without noncontingent shock. *Animal Learning and Behavior, 4,* 293–298.

Hyde, T. S. (1976b). The effect of Pavlovian stimuli on the acquisition of a new response. *Learning and Motivation, 7,* 223–229.

Jenkins, H. M., Barnes, R. A., & Barrera, F. J. (1981). Why autoshaping depends on trial spacing. In C. M. Locurto, H. S. Terrace, & John Gibbon (Eds.), *Autoshaping and conditioning theory* (pp. 255–284). New York: Academic Press.

Jenkins, H. M., & Moore, B. R. (1973). The form of the auto-shaped response with food or water reinforcers. *Journal of the Experimental Analysis of Behavior, 20,* 163–181.

Kamin, L. J. (1961). Trace conditioning of the conditioned emotional response. *Journal of Comparative and Physiological Psychology, 54,* 149–153.

Kamin, L. J. (1969). Predictability, surprise, attention and conditioning. In B. A. Campbell & R. M. Church (Eds.), *Punishment and aversive behavior* (pp. 279–296). New York: Appleton-Century-Crofts.

Kaplan, P. (1984). The importance of relative temporal parameters in trace autoshaping: From excitation to inhibition. *Journal of Experimental Psychology: Animal Behavior Processes, 10,* 113–126.

Kaplan, P. S., & Hearst, E. (in press). Contextual control and excitatory vs. inhibitory learning: Studies of extinction, reinstatement, and interference. In P. Balsam & A. Tomie (Eds.), *Context and learning.* Hillsdale, NJ: Lawrence Erlbaum Associates.

Konorski, J. (1948). *Conditioned reflexes and neuron organization.* Cambridge: Cambridge University Press.

Konorski, J. (1967). *Integrative activity of the brain: An interdisciplinary approach.* Illnois: University of Chicago Press.

Konorski, J., & Szwejkowska, G. (1956). Reciprocal transformations of heterogeneous conditioned reflexes. *Acta Biologiae Experimentalis, 17,* 141–165.

Krank, M. D. (1982). *Motivational interactions: The role of inhibition in Pavlovian aversive to appetitive transfer.* Ph.D. thesis, McMaster University.

Kremer, E. F. (1971). Truly random and traditional control procedures in CER conditioning in the rat. *Journal of Comparative and Physiological Psychology, 76,* 441–448.

Kremer, E. F. (1978). The Rescorla–Wagner model: Losses in associative strength in compound conditioned stimuli. *Journal of Experimental Psychology: Animal Behavior Processes, 4,* 22–36.

Kruse, J. M., Overmier, J. B., Konz, W. A., & Rokke, E. (1983). Pavlovian conditioned stimulus effects upon instrumental choice behavior are reinforcer specific. *Learning and Motivation, 14,* 165–181.

LeClerc, R. (1983). *Analysis of the approach component of sign-tracking in aversive conditioning.* Unpublished manuscript.

LeClerc, R., & Reberg, D. (1980). Sign tracking in aversive conditioning. *Learning and Motivation, 11,* 302–317.

Libby, M. E. (1976). The effects of aversive conditioned stimuli on the timing of unsignalled avoidance responding in rats. *Learning and Motivation, 7,* 117–131.

Lindberg, A. A. (1933). The formation of negative conditioned reflexes by coincidence in time with the process of differential inhibition. *Journal of General Psychology, 8,* 392–419.

LoLordo, V. M. (1967). Similarity of conditioned fear responses based on different aversive events. *Journal of Comparative and Physiological Psychology, 64,* 154–158.

LoLordo, V. M. (1969). Positive conditioned reinforcement from aversive situations. *Psychological Bulletin, 72,* 193–203.

LoLordo, V. M., Fairless J. L., & Stanhope, K. (in press). Effect of context shift upon responses to conditioned inhibitors. In F. R. Brush & J. B. Overmier (Eds.), *Affect, conditioning, and cognition: Essays on the determinants of behavior.* Hillsdale, NJ: Lawrence Erlbaum Associates.

Mackintosh, N. J. (1975). A theory of attention: Variations in the associability of stimuli with reinforcement. *Psychological Review, 82,* 276–298.

Mahoney, W. J., Kwaterski, S. E., & Moore, J. W.(1975). Conditioned inhibition of the rabbit nictitating membrane response as a function of CS–US interval. *Bulletin of the Psychonomic Society, 5,* 177–179.

Maier, S. F., Rapaport, P., & Wheatley, K. L. (1976). Conditioned inhibition and the UCS–CS interval. *Animal Learning and Behavior, 4,* 217–220.

Marchant, H. G. III, & Moore, J. W. (1974). Below-zero conditioned inhibition of the rabbit's nictitating membrane response. *Journal of Experimental Psychology, 102,* 350–352.

Marlin, N. A. (1982). Within-compound associations between the context and the conditioned stimulus. *Learning and Motivation, 13,* 526–541.

Moore, B. R. (1973). The role of directed Pavlovian reactions in simple instrumental learning in the pigeon. In R. A. Hinde & J. Stevenson-Hinde (Eds.), *Constraints on learning* (pp. 159–188). London: Academic Press.

Moscovitch, A. (1972). *Pavlovian cessation conditioning*. Unpublished Ph.D. dissertation, University of Pennsylvania.

Moscovitch, A., & LoLordo, V. M. (1968). Role of safety in the Pavlovian backward fear conditioning procedure. *Journal of Comparative and Physiological Psychology, 66,* 673–678.

Mowrer, O. H. (1960). *Learning theory and behavior*. New York: Wiley.

Newlin, R. J., & LoLordo, V. M. (1976). A comparison of pecking generated by serial, delay, and trace autoshaping procedures. *Journal of the Experimental Analysis of Behavior, 25,* 227–241.

Nieto, J. (1984). Transfer of conditioned inhibition across different aversive reinforcers in the rat. *Learning and Motivation, 15,* 37–57.

Odling–Smee, F. J. (1975a). The role of background stimuli during Pavlovian conditioning. *Quarterly Journal of Experimental Psychology, 27,* 201–209.

Odling–Smee, F. J. (1975b). Background stimuli and the inter-stimulus interval during Pavlovian conditioning. *Quarterly Journal of Experimental Psychology, 27,* 387–392.

Odling–Smee, F. J. (1978a). The overshadowing of background stimuli: Some effects of varying amounts of training and UCS intensity. *Quarterly Journal of Experimental Psychology, 30,* 737–746.

Odling–Smee, F. J. (1978b). The overshadowing of background stimuli by an informative CS in aversive Pavlovian conditioning with rats. *Animal Learning and Behavior, 6,* 43–51.

Overmier, J. B., & Patterson, J. (May, 1983). *A transfer of control assessment of Solomon's opponent-process model*. Paper presented at the meeting of the Midwestern Psychological Association, Chicago.

Patterson, J., & Overmier, J. B. (1981). A transfer of control test for contextual associations. *Animal Learning and Behavior, 9,* 316–321.

Pavlov, I. P. (1927). *Conditioned reflexes*. (Translated by G. V. Anrep) London: Oxford University Press.

Pearce, J. M., Montgomery, A., & Dickinson, A. (1981). Contralateral transfer of inhibitory and excitatory eyelid conditioning in the rabbit. *Quarterly Journal of Experimental Psychology, 33B,* 45–61.

Plotkin, H. C., & Oakley, D. A. (1975). Backward conditioning in the rabbit (*Oryctolagus cuniculus*). *Journal of Comparative and Physiological Psychology, 88,* 586–590.

Reberg, D., & Black, A. H. (1969). Compound testing of individually conditioned stimuli as an index of excitatory and inhibitory properties. *Psychonomic Science, 17,* 30–31.

Reiss, S., & Wagner, A. R. (1972). CS habituation produces a "latent inhibition effect" but no active "conditioned inhibition." *Learning and Motivation, 3,* 237–245.

Rescorla, R. A. (1967). Inhibition of delay in Pavlovian fear conditioning. *Journal of Comparative and Physiological Psychology, 64,* 114–120.

Rescorla, R. A.(1968). Pavlovian conditioned fear in Sidman avoidance learning. *Journal of Comparative and Physiological Psychology, 66,* 1–5.

Rescorla, R. A. (1969a). Pavlovian conditioned inhibition. *Psychological Bulletin, 72,* 77–94.

Rescorla, R. A. (1969b). Conditioned inhibition of fear resulting from negative CS–US contingencies. *Journal of Comparative and Physiological Psychology, 67,* 504–509.

Rescorla, R. A. (1971). Summation and retardation tests of latent inhibition. *Journal of Comparative and Physiological Psychology, 75,* 77–81.

Rescorla, R. A. (1973). Second-order conditioning: Implications for theories of learning. In F. J. McGuigan & D. B. Lumsden (Eds.), *Contemporary approaches to conditioning and learning* (pp. 127–150). New York: Wiley.

Rescorla, R. A. (1975). Pavlovian excitatory and inhibitory conditioning. In W. K. Estes (Ed.), *Handbook of learning and cognitive processes: (Volume 2), Conditioning and behavior theory* (pp. 7–35). Hillsdale, NJ: Lawrence Erlbaum Associates.

Rescorla, R. A. (1976). Second-order conditioning of Pavlovian conditioned inhibition. *Learning and Motivation, 7,* 161–172.

Rescorla, R. A. (1979a). Conditioned inhibition and excitation. In A. Dickinson & R. A. Boakes (Eds.), *Mechanisms of learning and motivation: A memorial volume for Jerzy Konorski* (pp. 83–110). Hillsdale, NJ: Lawrence Erlbaum Associates.

Rescorla, R. A. (1979b). Some comments on a model of conditioning. (Unpublished manuscript).

Rescorla, R. A. (1982). Some consequences of associations between the excitor and the inhibitor in a conditioned inhibition paradigm. *Journal of Experimental Psychology: Animal Behavior Processes, 8,* 288–298.

Rescorla, R. A., Durlach, P. J., & Grau, J. W. (in press). Contextual learning in Pavlovian conditioning. In P. Balsam & A. Tomie (Eds.), *Context and learning.* Hillsdale, NJ: Lawrence Erlbaum Associates.

Rescorla, R. A., & Freberg, L. (1978). The extinction of within-compound flavor associations. *Learning and Motivation. 9,* 411–427.

Rescorla, R. A., & Holland, P. C. (1977). Associations in Pavlovian conditioned inhibition. *Learning and Motivation, 8,* 429–447.

Rescorla, R. A., & LoLordo, V. M. (1965). Inhibition of avoidance behavior. *Journal of Comparative and Physiological Psychology, 59,* 406–412.

Rescorla, R. A., & Wagner, A. R. (1972). A theory of Pavlovian conditioning: Variations in the effectiveness of reinforcement and non-reinforcement. In A. H. Black & W. F. Prokasy (Eds.), *Classical conditioning II. Theory and research* (pp. 64–99). New York: Appleton-Century-Crofts.

Scavio, M. J. (1974). Classical-classical transfer: Effects of prior aversive conditioning upon appetitive conditioning in rabbits. *Journal of Comparative and Physiological Psychology, 86,* 107–115.

Scavio, M. J., & Gormezano, I. (1980). Classical-classical transfer: Effects of prior appetitive conditioning upon aversive conditioning in rabbits. *Animal Learning and Behavior, 8,* 218–224.

Schull, J. (1979). A conditioned opponent theory of Pavlovian conditioning and habituation. *Psychology of Learning and Motivation, 13,* 57–90.

Segundo, J. P., Galeano, C., Sommer-Smith, J. A., & Roig, J. A. (1961). Behavioral and EEG effects of tones 'reinforced' by cessation of painful stimuli. In J. F. Delafresnaye (Ed.), *Brain mechanisms and learning* (pp. 265–291). Oxford: Blackwells Scientific Publications.

Sidman, M. (1953). Avoidance conditioning with brief shock and no exteroceptive warning signal. *Science, 118,* 157–158.

Skinner, B. F. (1938). *The behavior of organisms.* New York: Appleton-Century-Crofts.

Smith, M. C., Coleman, S. R., & Gormezano, I. (1969). Classical conditioning of the rabbit's nictitating membrane response at backward, simultaneous, and forward CS–US intervals. *Journal of Comparative and Physiological Psychology, 69,* 226–231.

Solomon, R. L. (1980). The opponent-process theory of acquired motivation: The costs of pleasure and the benefits of pain. *American Psychologist, 35,* 691–712.

Solomon, R. L., & Corbit, J. O. (1974). An opponent-process theory of motivation: I. Temporal dynamics of affect. *Psychological Review, 81,* 119–145.

Suiter, R. D., & LoLordo, V. M. (1971). Blocking of inhibitory Pavlovian conditioning in the conditioned emotional response procedure. *Journal of Comparative and Physiological Psychology, 76,* 137–144.

Terrace, H. S. (1971). Escape from S−. *Learning and Motivation, 2,* 148–163.

Thomas, E. (1972). Excitatory and inhibitory processes in hypothalamic conditioning. In R. A. Boakes & M. S. Halliday (Eds.), *Inhibition and learning.* London: Academic Press.

Tomie, A. (1981). Effects of unpredictable food on the subsequent acquisition of autoshaping: Analysis of the context-blocking hypothesis. In C. M. Locurto, H. S. Terrace, & John Gibbon (Eds.), *Autoshaping and conditioning theory* (pp. 181–215). New York: Academic Press.

Tomie, A., & Kruse, J. (1980). Retardation tests of inhibition following discriminative autoshaping. *Animal Learning and Behavior, 8*, 401–409.
Wagner, A. R. (1981). SOP: A model of automatic memory processes in animal behavior. In N. E. Spear & R. R. Miller (Eds.), *Information processing in animals: Memory mechanisms*. Hillsdale, NJ: Lawrence Erlbaum Associates.
Wagner, A. R., Mazur, J. E., Donegan, N. H. & Pfautz, P. L. (1980). Evaluation of blocking and conditioned inhibition to a CS signaling a decrease in US intensity. *Journal of Experimental Psychology: Animal Behavior Processes, 6*, 376–385.
Wagner, A. R., & Rescorla, R. A. (1972). Inhibition in Pavlovian conditioning: Application of a theory. In R. A. Boakes & M. S. Halliday (Eds.), *Inhibition and learning*. (pp. 301–336). London: Academic Press.
Wasserman, E. A., Franklin, S. R. & Hearst, E. (1974). Pavlovian appetitive contingencies and approach vs. withdrawal to conditioned stimuli in pigeons. *Journal of Comparative and Physiological Psychology, 86*, 616–627.
Wasserman, E. A., & Molina, E. (1975). Explicitly unpaired key light and food presentations: Interference with subsequent auto-shaped key-pecking in pigeons. *Journal of Experimental Psychology: Animal Behavior Processes, 104*, 30–38.
Weisman, R. G., & Litner, J. S. (1969a). The course of Pavlovian excitation and inhibition of fear in rats. *Journal of Comparative and Physiological Psychology, 69*, 667–672.
Weisman, R. G., & Litner, J. S. (1969b). Positive conditioned reinforcement of Sidman avoidance behavior in rats. *Journal of Comparative and Physiological Psychology, 68*, 597–603.
Weisman, R. G., & Litner, J. S. (1971). Role of the intertrial interval in Pavlovian differential conditioning of fear in rats. *Journal of Comparative and Physiological Psychology, 74*, 211–218.
Wessells, M. G. (1973). Autoshaping, errorless discrimination, and conditioned inhibition. *Science, 182*, 941–943.
Witcher, E. S., & Ayres, J. J. B. (1980). Systematic manipulation of CS-US pairings in negative CS-US correlation procedures in rats. *Animal Learning and Behavior, 8*, 67–74.
Yadin, E., & Thomas, E. (1981). Septal correlates of conditioned inhibition and excitation in rats. *Journal of Comparative and Physiological Psychology, 95*, 331–340.
Zimmer-Hart, C. L., & Rescorla, R. A. (1974). Extinction of Pavlovian conditioned inhibition. *Journal of Comparative and Physiological Psychology, 86*, 837–845.

2 Conditioning Context as an Associative Baseline: Implications for Response Generation and the Nature of Conditioned Inhibition

Ralph R. Miller and Todd R. Schachtman
State University of New York at Binghamton

Conditioned inhibition (CI) is ordinarily regarded as the consequence of an animal learning that some event will not occur soon after some other event. This apparitional quality of CI is reminiscent of the nursery rhyme:

> Yesterday upon a stair,
> I met a man who was not there.
> He was not there again today,
> I wish to heck he'd go away.

Conditioned inhibition, being a reflection of something that does not happen, has proven to be an elusive concept. Moreover, the study of CI is further complicated by the difficulty in measuring it. For reasons that LoLordo and Fairless (this volume) review, the most common method of detection relies on a combination of summation and retardation tests (Hearst, 1972; Rescorla, 1969b). The alternative approach–withdrawal measure developed in Hearst's laboratory appears to afford a more direct means of observing CI (Hearst & Franklin, 1977; Wasserman, Franklin, & Hearst, 1974); however, this kind of index may not be sensitive to quite the same factors and variables as are summation and retardation tests. We address the measurement problem in more detail later.

All early conceptualizations of CI incorporated theoretical constructs beyond those required to explain conditioned excitation (CE). For example, Konorski (1948) proposed that CI increased the threshold for reactivation of the US trace, a notion recently resurrected by Rescorla (1979). However, by 1967, Konorski had switched to a view of CI in which associations between the target stimulus (X) and "no-US centers" subtracted from simultaneously existing excitatory associations between X and US centers, a position functionally similar to those of Pavlov (1927) and Hull (1943). The Rescorla–Wagner model (1972; Wagner & Rescorla, 1972) offered us a view of CI with relatively little excess theoretical baggage beyond that necessary for excitation. It postulated a single associative variable that could assume both positive and negative values. Unfortunately, this view encountered many empirical problems, perhaps the most serious being a failure to obtain the predicted extinction of a conditioned inhibitor. Specifically, the Rescorla–Wagner model predicts that CS-alone presentations should reduce the inhibitory value of a CS; however, in practice no such decrement has been observed (Zimmer–Hart & Rescorla, 1974).

The Comparator Hypothesis

In contrast to the Rescorla–Wagner model and the earlier conceptualizations of Pavlov, Hull, and Konorski, in 1969(a) Rescorla proposed a contingency model for CI that assumed conditioned inhibitory associations to a CS are formed when the probability of the US in the presence of the CS, P(US/CS), is less than the probability of the US in the absence of the CS, P(US/$\overline{\text{CS}}$); i.e., the CS predicts a decrease in the likelihood of the US relative to the absence of the CS. Similarly, excitatory associations were assumed to arise when the CS announces an increase in the likelihood of the US, i.e., P(US/CS) $>$ p (US/$\overline{\text{CS}}$). Essentially Rescorla was suggesting that associations depend on a comparison between P(US/CS) and P(US/$\overline{\text{CS}}$). We are now proposing a small but important modification of Rescorla's position. Specifically, rather than have associations depend upon this comparison, we are proposing that P(US/CS) and P(US/$\overline{\text{CS}}$) each result in their own independent associations and that it is a comparison of these two associations that determine the subject's response to the CS. Because P(US/CS) and P(US/$\overline{\text{CS}}$) are necessarily positive numbers (or zero at the very least), associations based on these probabilities are going to be positive, i.e., excitatory. Consequently, the suggestion that conditioned inhibitory behavior results from P(US/CS) $<$ P(US/$\overline{\text{CS}}$) is equivalent to viewing inhibitory responding as a consequence of a particular relationship of exclusively positive associations. Associations involving negative associative strengths or CS-no US relationships are not assumed. Behavior indicative of what is ordinarily regarded as CI can be explained without recourse to theoretical constructs such as negative associations, associations to the absence of the US, or elevation of US trace reactivation thresholds by the CS. The probability of the US in the absence of the CS is taken to reflect the associative strength of the background, i.e., contextual cues. Thus, CI is viewed as the *behavioral consequence* of the positive association between a CS and US being weaker than the positive association between the context and US. This comparator hypothesis may superficially appear similar to the elevated reactivation threshold viewpoint, but the latter attributes the CI-depressed behavior exclusively to associations with the CS, whereas the comparator hypothesis attributes it to differences between associations of the CS and associations of the comparator stimulus. As such, the comparator hypothesis (with its dependence on the animal's being an intuitive statistician) is more cognitive in orientation than many of the models that have preceded it.

Although Rescorla's (1968, 1969a) data were perfectly consistent with the comparator hypothesis (both his version and ours), he put aside the model in order to focus his attention on the interaction between elements of compound stimuli, an endeavor that soon gave rise to the highly influential Rescorla–Wagner (1972) model. Other than Gibbon, Berryman, and Thompson's (1974) elegant restatement of Rescorla's earlier model in more formal contingency terminology, and Fantino's (1969) independently derived version of it applicable to instrumental tasks, the model sat largely dormant until Gibbon and Balsam (1981) and Jenkins, Barnes, and Barrera (1981) applied it to autoshaping in pigeons. However, since Rescorla's work in the late 1960s, no one has focused on the comparator hypothesis' implications for CI.

We were attracted to the comparator model both by its conceptual parsimony and by the unusually sharp distinction that the model makes between learning (of associations) and performance (based on comparisons between associations). Although our working hypothesis is somewhat similar to that of Gibbon and Balsam (1981) and Jenkins et al. (1981), we have intentionally avoided proposing any specific mathematical form of the assumed comparison, e.g., differences or ratios, because we feel that the pursuit of quantitative relationships before qualitative ones have been determined would be premature. Rather we are asking, is the comparator hypothesis illuminating with respect to CI (and CE), and, if so, *to exactly what* is the association arising from P(US/CS) compared? For example, assuming that separate contexts are used for conditioning and testing, both the associative value of the conditioning context and that of the test context are plausible comparators. However, neither Rescorla, Gibbon and Balsam, nor Jenkins et al. (but see Balsam, in press) changed contexts between conditioning and testing to determine whether the conditioning context or the test context served as the baseline US predictor to which the associative strength of the discrete CS was compared.

Evidence Supporting the Comparator Hypothesis

To answer this and related questions, we recently performed a series of lick suppression studies with water-deprived rats. Our basic procedure consisted of initially exposing all animals to each of three radically different contexts, for 20 min/day for 4 days. The contexts, designated A, B, and C, differed in respect to shape, building material, illumination, and floor construction. (For example, one context had a small diameter grid floor with narrow spacing, a second context had a large diameter grid floor with wide spacing, and the last context had a floor composed of two long, narrow, parallel steel plates.) The only feature that

Contexts A and B clearly had in common was a water-filled lick tube that passed through the center of one wall of each context. Context C had no lick tube. The order of context exposure during lick training was alternated each day between ACB and BCA. On these first 4 days, all animals learned the locations of the lick tubes and soon initiated protracted drinking in both the A and B contexts. The use of these two lick contexts for target conditioning was counterbalanced between animals in each treatment group. Henceforth, we refer to the location in which the target CS was presented during conditioning as Context A. The distinctiveness of the three contexts as perceived by our subjects was evident both from current and past research demonstrating the occurrence of associations between USs and the specific context in which the USs were delivered, and from the beneficial effect of testing CS–US associations within the context in which conditioning had occurred (Balaz, Capra, Hartl, & Miller, 1981; Balaz, Capra, Kasprow, & Miller, 1982).

Figure 2.1 summarizes the procedure of a prototypical study. The intent of this initial experiment was simply to demonstrate CE. After the establishment of licking behavior, target conditioning occurred on Days 5, 6, 10, and 11 with a white noise as the CS and footshock as the US. On these days, animals were exposed only to Context A and the daily session length was 20 min. Each rat was presented with the CS six times during each conditioning session. The reinforcement probability with respect to the CS was .33 for the experimental group, and 0 for the explicitly unpaired control group. When footshock was scheduled, it occurred at the offset of the CS. The CS duration was 30 sec, and for analytic purposes the non-CS periods were also divided into 30-sec intervals. Because subsequent studies would include summation tests, between the second and third days of conditioning, all subjects received 3 days (Days 7–9) of exposure to Context C during which conditioning to a nontarget CS took place. Sessions on these nontarget training days were 1 hour long. On the first day no stimuli were

ASSOCIATIVE EXCITATION

GROUP	PHASE 1	PHASE 2	TEST
	Days 5, 6, 10, 11	Days 8–9	Day 13
EX	P (FS/Noise) = $2/6^A$ P (FS/$\overline{\text{Noise}}$) = $0/34^A$	P (FS/Click) = $4/4^C$	Noise^A
EU	P (FS/Noise) = $0/6^A$ P (FS/$\overline{\text{Noise}}$) = $2/34^A$	P (FS/Click) = $4/4^C$	Noise^A

EX = Excitation Group
EU = Explicitly Unpaired Group
P = Probability

FS = Footshock
A = in Context A
C = in Context C

FIG. 2.1. Paradigm to detect excitatory associations between the white noise and footshock using conditioned lick suppression as a behavioral index. See text for a parallel experiment using a different control group.

presented, and on the last 2 of these 3 days, each subject received 4 pairings of a click train with footshock. Prior research had established that the clicks and white noise did not generalize to each other. Regardless of whether or not a summation test was used, the click–shock pairings occurred in this and subsequent studies in order to maintain the equivalency of experimental treatments. Following conditioning of the clicks and white noise, there was one more day of lick experience identical to Days 1–4. This brief lick recovery session in Contexts A and B prevented extended lick suppression prior to the presentation of the CS on the next (test) day in those studies in which testing occurred in contexts where shock had been given; because the test stimulus on the test day was presented upon completion of 25 licks, such differential suppression would have resulted in between-group differences in pre-CS contextual exposure on the test day. (Other of our studies had found that this reshaping procedure provided a smoother behavioral baseline; yet, it was too short to appreciably reduce the associative strength of the context.) On the test day, presentation of the test stimulus upon completion of the first 25 licks insured that all subjects were drinking at the onset of the test stimulus. Differences in context exposure prior to onset of the test stimulus were negligible, the range being 4–10 sec. Our critical measure was a flooding latency, i.e., time to complete 25 additional licks in the presence of the discrete test stimulus. A 20-min maximum test latency assured that ceiling effects were avoided. In this particular experiment, the test stimulus was the white noise alone. Not surprisingly, this procedure yielded evidence of significant associative excitation relative to the explicitly unpaired control group (see Table 2.1).

Noting that the explicitly unpaired control group in this study might have been exhibiting CI to the white noise rather than, or in addition to, the paired group displaying CE, we considered including a truly random (TR) control group. But we decided that, with only 2 footshocks/day for only 4 days, the number of USs was so small that a TR group would be inappropriate. However, a further study was performed in which the control animals, instead of receiving explicitly

TABLE 2.1
Outcome of Excitation Test in Conditioned Excitation Study

Group	*Mean Test Latency*[a] *to Emit Licks 0–25 (± Standard Error)*	*Mean Test Latency*[a,b] *to Emit Licks 25–50 (± Standard Error)*
EX	.54 (±.02)	1.57[c] (± .18)
EU	.64 (± .09)	.82 (± .11)

[a]Log sec.
[b]White noise present during second 25 licks.
[c]Significantly greater than Group EU, $p < .05$.

unpaired presentations of the white noise and footshock, experienced during the 4 conditioning days 6 click trains of which 2 terminated with shock. Consequently during the test phase the white noise was novel for these animals, thereby eliminating any possible conditioned excitatory or inhibitory effects that might have been present in the prior study's control group. As before, testing consisted of a white noise presentation. It should be noted that the novelty of the noise for the control animals in this study might have resulted in a startle response that would *reduce* the difference between the excitation and control groups. Nevertheless, when the white noise was presented during the second 25 licks of this experiment, the excitatory group yielded a mean latency of 2.03 log sec, whereas the control group mean latency was 1.29 log sec. Thus, this latter study, like the first one, supported the conclusion that our 33% reinforcement density yielded an excitatory CS provided $P(US/\overline{CS}) = 0$.

We next attempted to demonstrate that the same CS, still reinforced 33% of the time, could function as a conditioned inhibitor if sufficient unsignaled shocks were administered on the conditioning days. Whereas in our first study, $P(US/CS) > P(US/\overline{CS})$, we now made $P(US/CS) < P(US/\overline{CS})$. The procedure, which is illustrated in Fig. 2.2, was similar to that used for Group EX in our first study except 67% of the unsignaled intervals during the noise–footshock conditioning sessions were reinforced and a summation test was used on Day 13. As before, the unsignaled intervals were equal in duration to the CS presentations, i.e., 30 sec. When unsignaled shock was scheduled for a CS-free interval, it occurred 25 sec into the 30-sec interval so as to avoid the possibility of shock offset being immediately followed by CS onset for the next trial. Note that in this study the only difference between the summation group and baseline control group occurred on the test day. Although we did not include a direct test of the assumption that the unsignaled footshocks resulted in context-shock associations, evidence supporting this assumption under similar circumstances is provided by Patterson and

SUMMATION TEST FOR CONDITIONED INHIBITION

GROUP	PHASE 1	PHASE 2	TEST
	Days 5, 6, 10, 11	Days 8–9	Day 13
Baseline	P (FS/Noise) = 2/6^A P (FS/$\overline{\text{Noise}}$) = 23/34^A	P (FS/Click) = 4/4^C	Click ^{A}
Summation	P (FS/Noise) = 2/6^A P (FS/$\overline{\text{Noise}}$) = 23/34^A	P (FS/Click) = 4/4^C	Click + Noise ^{A}

P = Probability
FS = Footshock

A = in Context A
B = in Context B
C = in Context C

FIG. 2.2. Paradigm for summation test to detect conditioned inhibitory properties of white noise using conditioned lick suppression as a behavioral index.

Overmier (1981), as well as by extensive prior work in our laboratory using the same apparatus and procedure.

The results were clear. As can be seen in Table 2.2, the two groups did not differ in lick latency before onset of the test stimuli. Following CS onset, the click produced suppression and the addition of the white noise significantly reduced this suppression.

To complement the summation test for CI, in a separate experiment we performed a retardation test. The experimental group this time was treated identically during Phases 1 and 2 to the animals in the preceding study. Additionally, two control groups were included. One received zero contingency training, which was achieved by increasing P(US/CS) from 0.33 to 0.67 so that it would match P(US/$\overline{\text{CS}}$). We chose to run this zero contingency group rather than one for which P(US/$\overline{\text{CS}}$) was lowered to the 0.33 value of P(US/CS), because the latter zero contingency group would have differed appreciably from the CI group in the total number of shocks received. This possibly would have produced facilitation of subsequent excitatory acquisition through some combination of: a) reduced blocking by the context, b) reduced nonassociative US-alone effects such as habituation, and, c) assuming the validity of the comparator hypothesis, a reduced value of the associative strength to which the CS's associative strength is presumably compared. The other control group never received the CS prior to the noise–shock pairings that constituted the retardation test; we controlled for nonassociative effects of the US by giving, in Context C, unsignaled shocks equal in number to the shocks received by the experimental group during Phase 1 (see Fig. 2.3). To determine if the experimental group would exhibit retarded acquisition, on Day 12 (Phase 3) all groups received two noise–shock pairings in Context A. By the time this study was performed, we already knew that the training context served as the associative comparator regardless of where testing occurs (these data are described later); hence, testing was performed in Context B to prevent any possibility of fear of the conditioning context (A) from summating with fear of the white noise. Testing in Context B, in which the animals had

TABLE 2.2
Outcome of Summation Test in Conditioned Inhibition Study

Group	*Mean Test Latency*[a] *to Emit Licks 0–25 (± Standard Error)*	*Mean Test Latency*[a,b] *to Emit Licks 25–50 (± Standard Error)*
Baseline	1.21 (± .13)	2.37 (± .11)
Summation	1.19 (± .12)	2.08[c] (± .10)

[a]Log sec.
[b]CS present during second 25 licks.
[c]Significantly less than baseline group, $p < .05$.

RETARDATION TEST FOR CONDITIONED INHIBITION

GROUP	PHASE 1	PHASE 2	PHASE 3	TEST
	Days 5,6,10,11	Days 8–9	Day 12	Day 13
CI	P(FS/Noise)=2/6^{A} P(FS/$\overline{\text{Noise}}$)=23/34^{A}	P(FS/Click)=4/4^{C}	P(FS/Noise)=2/2^{A}	Noise^{B}
ZC	P(FS/Noise)=4/6^{A} P(FS/$\overline{\text{Noise}}$)=23/34^{A}	P(FS/Click)=4/4^{C}	P(FS/Noise)=2/2^{A}	Noise^{B}
FSO	P(FS/Noise)=0/0^{C} P(FS/$\overline{\text{Noise}}$)=25/40^{C}	P(FS/Click)=4/4^{C}	P(FS/Noise)=2/2^{A}	Noise^{B}

CI=Conditioned Inhibition Group
ZC =Zero Contingency Control Group
FSO=Footshock Only Control Group
P=Probability

FS=Footshock
A=Context A
B=Context B
C=Context C

FIG. 2.3. Paradigm for retardation test to detect conditioned inhibitory properties of white noise using conditioned lick suppression as a behavioral index. See text for explanation of why testing occurred in Context B.

never been shocked, also eliminated the need for the reshaping day immediately prior to testing.

Not surprisingly, the CI group displayed significantly less excitatory acquisition on the retardation test than either of the control groups (see Table 2.3). Of course, passing the retardation test for CI still leaves open issues concerning the underlying mechanisms responsible for this phenomenon. As we presently argue, there is no need to postulate the existence of negative associations. The immediate point however is that our CI preparation, i.e., P(US/CS) = 0.33 and P(US/$\overline{\text{CS}}$) = 0.67, yields a conditioned inhibitor as measured by both summation and retardation tests.

TABLE 2.3
Outcome of Retardation Test in Conditioned Inhibition Study

Group	*Mean Test Latency*[a] *to Emit Licks 0–25 (± Standard Error)*	*Mean Test Latency*[a,b] *to Emit Licks 25–50 (± Standard Error)*
CI	0.96 (± .19)	1.32[c] (± .19)
ZC	1.27 (± .25)	1.78 (± .17)
FSO	0.80 (± .16)	1.85 (± .19)

[a]Log sec.
[b]CS present during second 25 licks.
[c]Significantly less than both control groups, $p < .05$.

Thus, as determined on the basis of excitation, summation, and retardation testing, the same CS, reinforced 33% of the time, appears able to serve as both a conditioned exciter and a conditioned inhibitor purely as a function of the density of unsignaled shocks being 0 or 67%, respectively. These observations are fully consistent with the contingency analysis of Rescorla (1968) and serve as the basis for the remainder of our analysis of CI.

Although P(US/CS) = 0.33 yielded excitation when P(US/$\overline{CS}$) = 0 and inhibition when P(US/$\overline{CS}$) = 0.67, the critical variable in this demonstration, P(US/CS), was confounded by the nature of the test performed. This problem is frequently overlooked in contrasts between CI and CE. The confound becomes apparent in a comparison of the paradigm in Fig. 2.1 with those of Fig. 2.2 and 2.3. The test for a conditioned exciter consists of simply looking for a CR following the CS presented alone, whereas the tests for a conditioned inhibitor consist of first pairing it with the US before presenting it alone (retardation) and of presenting it in the presence of a known exciter (summation). Thus, the traditional tests for a conditioned inhibitor are fundamentally different from those for a conditioned exciter; in principle a stimulus can simultaneously serve *both* as a conditioned inhibitor and as a conditioned exciter. Indeed, as is seen later, this situation sometimes arises.

To determine the degree to which a nominal conditioned exciter would also act as a conditioned inhibitor in our situation, we ran a study in which an exciter, i.e., P(US/CS) = 0.33 and P(US/$\overline{CS}$) = 0, was subjected to summation and retardation tests for CI. Not surprisingly, the target CS failed both of these tests. The excitatory group was accelerated rather than retarded compared to its appropriate control, and, on the summation test, the simultaneous compound of clicks plus white noise yielded only as much suppression as either the clicks or white noise alone. This absence of excitatory–excitatory summation despite the lack of a ceiling effect is suggestive of associative *averaging* rather than the often discussed but not always observed algebraic summation of associative strength influencing responding. Associative averaging is considered in more detail in the next section. The present point of consequence is that our putative conditioned inhibitor but not our conditioned exciter passed the summation and retardation tests for CI.

We also performed a test for excitation on a conditioned inhibitor, i.e., P(US/CS) = 0.33 and P(US/$\overline{CS}$) = 0.67. This group yielded a 25-lick mean suppression score in the presence of the white noise of 1.16 log sec in contrast to 1.17 log sec for a control group that experienced the same schedule of unsignaled shock but not white noise trials. As expected, the CI group exhibited no more suppression than the control group. (Admittedly a superior control group for this study would have been one that received, on each of the 4 conditioning days, shock following 2 out of 6 click trains in addition to the unsignaled shock. However, the 2 extra shocks and 6 click trains seem insignificant in view of the 23 unsignaled shocks. Consequently the study was not repeated with this small change.) The critical observation here is that our conditioned exciter but not our conditioned inhibitor passed a test for conditioned excitation. In toto, these data support the conclusion that P(US/CS) = 0.33 can yield both excitation and inhibition as a function of the associative strength of the context. Consistent with this position, the inhibitory power of a CS negatively correlated with shock decreases, despite no changes in the correlation of shock with the CS, if the shocks are signaled by a second CS that presumably overshadows the context (Baker, 1977).

Which Context Serves as the Comparator Stimulus?

The central questions raised by the comparator hypothesis are to what is the associative strength of the target CS compared, and when does this comparison take place? Turning initially to the former question, given our procedure, there are two likely possibilities. One is that the CS is compared to the conditioning context and the other is that the CS is compared to the test context.

Examining first the possible comparator role of the conditioning context, we performed a study in which four groups of animals received the same basic treatment as before, i.e., 2 out of 6 (33%) daily CS (white noise) presentations were paired with footshock on the conditioning days in Phase 1. For 2 of these 4 groups, 67% of the non-CS intervals in the conditioning context contained footshock, whereas for the other 2 groups there were no unsignaled shocks. Finally, half the groups were tested in the conditioning context (designated "A" in Fig. 2.4) and the other half were tested in the nonconditioning context (designated "B" in Fig. 2.4).

The results of this experiment are illustrated in Fig. 2.5. As can be seen, the critical factor was whether or not unsignaled shocks were given in the conditioning context. Provided unsignaled shocks were administered, CI was manifest in animals tested outside the conditioning context (Group HFA–TB) as well as those tested inside the conditioning context (Group HFA–TA). Clearly, the lack of unsignaled shocks administered in the place of testing did not influence the manifestation of CI. However, the question remained open as to whether or not the presence of unsignaled shocks in the test context would be of any consequence.

The next study was designed to examine the effects of unsignaled shocks in the context in which white noise conditioning did not occur, i.e., Context B. Again P(US/CS) = 0.33 with CS conditioning occurring in Context A, but now, rather than administer unsignaled shocks in Context A, for half the animals the probability of unsignaled shock in Context B = 0.67. As before, two groups were tested in Context A and two groups were tested in Context B. The design of this study is outlined in Fig. 2.6. As can be seen in Fig. 2.7, the CS, paired with shock on two out of six presentations in Context A, was excitatory (compared to the 1.20 log sec latency typical of naive subjects) both independent of the

CONDITIONED SUPPRESSION AS A FUNCTION OF TEST LOCATION AND ASSOCIATIVE VALUE OF THE CONDITIONING CONTEXT

GROUP	PHASE 1	PHASE 2	TEST
	Days 5, 6, 10, 11	Days 8–9	Day 13
HFA–TA	P (FS/Noise) = $2/6^{A}$ P (FS/$\overline{\text{Noise}}$) = $23/34^{A}$	P (FS/Click) = $4/4^{C}$	Noise ^{A}
LFA–TA	P (FS/Noise) = $2/6^{A}$ P (FS/$\overline{\text{Noise}}$) = $0/34^{A}$	P (FS/Click) = $4/4^{C}$	Noise ^{A}
HFA–TB	P (FS/Noise) = $2/6^{A}$ P (FS/$\overline{\text{Noise}}$) = $23/34^{A}$	P (FS/Click) = $4/4^{C}$	Noise ^{B}
LFA–TB	P (FS/Noise) = $2/6^{A}$ P (FS/$\overline{\text{Noise}}$) = $0/34^{A}$	P (FS/Click) = $4/4^{C}$	Noise ^{B}

P = Probability
HF = High Contextual Fear
LF = Low Contextual Fear
FS = Footshock

T = Test Location
A = in Context A
B = in Context B
C = in Context C

FIG. 2.4. Paradigm for determining the impact of associative value of the conditioning context and location of testing upon the potential of the white noise to induce conditioned lick suppression.

occurrence of unsignaled shocks in Context B and, as in the last experiment, independent of which context was used for testing. Despite the null results of this particular study, the data are worthy of attention owing to their contrast with the significant effect of unsignaled shock in the conditioning context, as seen in the highly similar study that preceded it.

The combined results of the last two experiments suggest that the training location as opposed to the test location plays the role of the associative com-

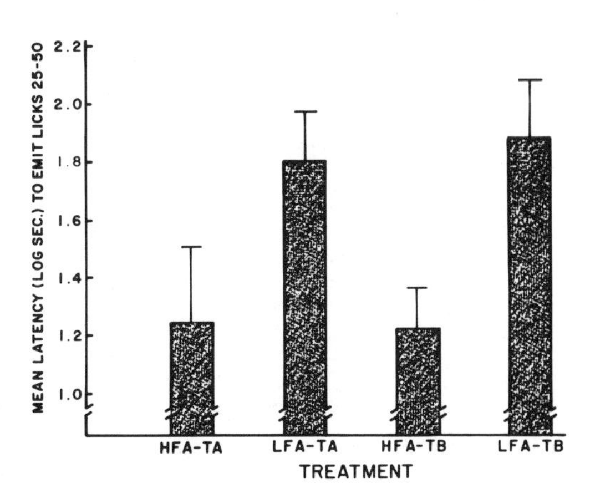

FIG. 2.5. Outcome of experiment outlined in Fig. 2.4. HFA = high fear of Context A, LFA = low fear of Context A, TA = test in Context A, TB = test in Context B.

CONDITIONED SUPPRESSION AS A FUNCTION OF TEST LOCATION AND ASSOCIATIVE VALUE OF THE NONCONDITIONING CONTEXT

GROUP	PHASE 1	PHASE 2	PHASE 3	TEST
	Days 5, 7, 12, 14	Days 6, 8, 13, 15	Days 10–11	Day 17
HFB–TA	P (FS/Noise) = $2/6^{A}$ P (FS/$\overline{\text{Noise}}$) = $0/34^{A}$	P (FS/$\overline{\text{Noise}}$) = $27/40^{B}$	P (FS/Click) = $4/4^{C}$	Noise ^{A}
LFB–TA	P (FS/Noise) = $2/6^{A}$ P (FS/$\overline{\text{Noise}}$) = $0/34^{A}$	P (FS/$\overline{\text{Noise}}$) = $0/40^{B}$	P (FS/Click) = $4/4^{C}$	Noise ^{A}
HFB–TB	P (FS/Noise) = $2/6^{A}$ P (FS/$\overline{\text{Noise}}$) = $0/34^{A}$	P (FS/$\overline{\text{Noise}}$) = $27/40^{B}$	P (FS/Click) = $4/4^{C}$	Noise ^{B}
LFB–TB	P (FS/Noise) = $2/6^{A}$ P (FS/$\overline{\text{Noise}}$) = $0/34^{A}$	P (FS/$\overline{\text{Noise}}$) = $0/40^{B}$	P (FS/Click) = $4/4^{C}$	Noise ^{B}

P = Probability
HF = High Contextual Fear
LF = Low Contextual Fear
FS = Footshock

T = Test Location
A = in Context A
B = in Context B
C = in Context C

FIG. 2.6. Paradigm for determining the impact of associative value of the non-conditioning context and location of testing upon the potential of the white noise to induce conditioned lick suppression.

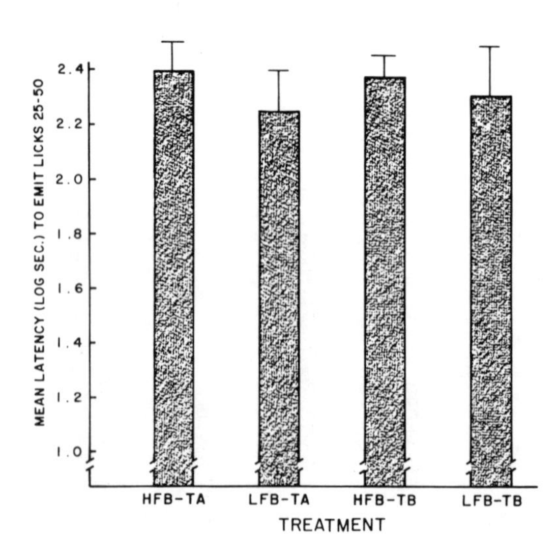

FIG. 2.7. Outcome of experiment outlined in Fig. 2.6. HFB = high fear of Context B, LFB = low fear of Context B, TA = test in Context A, TB = test in Context B. Contrast these null results with Fig. 2.5.

parator in determining whether or not a CS will be an effective conditioned inhibitor or conditioned exciter. Durlach (1983) has recently reported results similar to ours based on autoshaping data from pigeons conditioned and tested in the same or different contexts, in which contextual associative strength was varied by either signaling or not signaling the "unsignaled" US presentations with a nontarget stimulus.

Although the focal distinction between the last two experiments was whether the unsignaled shock was given in the context used to condition the white noise, i.e., Context A, or the other lick context, i.e., Context B, the two studies also differed in the times at which the unsignaled shocks were administered relative to the white noise presentations. In the first study, unsignaled shock in Context A was given on the same days as the noise–shock pairings, whereas in the second study, unsignaled shock in Context B was administered on an alternating day schedule with respect to the noise–shock pairings (compare Fig. 2.4 and 2.6). Kleiman and Fowler (1984) have recently presented data suggesting that unsignaled shocks delivered in moderately close temporal proximity to a nonreinforced stimulus (but not so close as to produce excitatory conditioning) will increase the effective inhibitory strength of this stimulus relative to that produced by unsignaled shocks delivered in the middle of the intervals between presentations of the target stimulus. This observation gives rise to the possibility that the lack of impact of the unsignaled shocks in Context B seen in the last study arose from their temporal separation rather than, or in addition to, their spatial separation from the noise presentations. To determine if time rather than place of the unsignaled shocks actually produced the differential results of the last two experiments, we performed a study in which all the conditions from the two immediately preceding studies were included except that, for the groups scheduled to receive unsignaled shock, the unsignaled shock was always given on the noise–shock conditioning days regardless of whether the unsignaled shocks were given in Context A or B. The procedure was the same as that illustrated in Fig. 2.4 except that on Days 5, 6, 10, and 11 subjects were placed in each of the 3 contexts for 20 min with the order of context exposure alternated daily between ACB and BCA. The data from this experiment were consistent with our prior finding that, independent of test location, unsignaled shocks in the context used to condition the white noise attenuated lick suppression to the white noise, whereas unsignaled shocks in the other lick context did not. However, even in this study there was an inequity in the timing of unsignaled shock. Subjects receiving unsignaled shock in Context A had their white noise conditioning trials within the same 20-min session as that in which the unsignaled shocks occurred, whereas subjects receiving unsignaled shock in Context B had their white noise conditioning trials in a 20-min session separated by 20 min from the beginning or end of the session containing the unsignaled shocks.

Completely eliminating this temporal inequity while keeping the CSs interspersed among the unsignaled shock and avoiding excessive handling has proven to be a difficult design problem, but we currently have in progress a study that we believe meets these requirements; the initial data are supportive of our earlier conclusions. Alternatively, the interspersion could be omitted and the unsignaled USs given prior to the CS presentations. Using pigeons in an auto-shaping preparation, Rescorla, Durlach, and Grau (in press) performed such a study and found that responding to the excitatory CS varied inversely with the excitatory value of the conditioning context but was little influenced by the excitatory strength of the test context.

Based on all currently available evidence, it appears that the context used for conditioning the target stimulus serves as the comparator baseline. However, we later qualify this when we discuss the effect of the intervals between the target stimulus and unsignaled reinforcers.

Temporal Location of the Comparison

Having tentatively established the role of the conditioning context as a comparator baseline for CSs, we next asked whether the comparator baseline was the excitatory value of the conditioning context at the time of conditioning or at the time of testing. To answer this question, we performed two experiments in which, for some animals, the associative value of the context was inflated or deflated during the retention interval. Considering the deflation manipulation first, we made the white noise a conditioned inhibitor for all subjects, i.e., $P(US/CS) = 0.33$, $P(US/\overline{CS}) = 0.67$. Then, to extinguish the excitatory context-US associations, half the animals received eight 60-min exposures to the conditioning context alone, i.e., $P(US/\overline{CS}) = 0$. The experimental design is outlined in Fig. 2.8. Note that summation testing of the white noise in compound

CONDITIONED SUPPRESSION AS A FUNCTION OF POSTCONDITIONING DEFLATION OF CONTEXTUAL FEAR

GROUP	PHASE 1	PHASE 2	PHASE 3	TEST
	Days 5, 6, 10, 11	Days 8–9	Days 12–19	Day 21
DA	P (FS/Noise) = 2/6^A P (FS/$\overline{\text{Noise}}$) = 23/34^A	P (FS/Click) = 4/4^C	P (FS/$\overline{\text{Noise}}$) = 0/120^A	Click + Noise^B
NDA	P (FS/Noise) = 2/6^A P (FS/$\overline{\text{Noise}}$) = 23/34^A	P (FS/Click) = 4/4^C		Click + Noise^B

P = Probability
D = Deflation
ND = No Deflation
FS = Footshock

A = in Context A
B = in Context B
C = in Context C

FIG. 2.8. Paradigm for determining the effects of postconditioning associative deflation of the conditioning context upon the potential of the white noise to induce conditioned lick suppression.

with the clicks occurred in the nonconditioning context, Context B, to prevent possible associative summation of the CSs and test context, a possibility that would obviously have contributed more suppression to the animals lacking contextual extinction than those receiving it. The test trial consisted of measuring latencies to complete 25 licks in the conjoint presence of the white noise and the click.

The results were clear: Deflation of the conditioning context significantly reduced the effective inhibitory value of the target CS. Subjects receiving contextual deflation exhibited a mean latency for 25 licks of 2.05 log sec compared to 1.73 log sec for subjects lacking the deflation treatment. This result is somewhat counterintuitive in that we made responding to the CS change in the excitatory direction without presenting either the CS or the US. Moreover, *the finding that the comparator baseline is the current associative value of the conditioning context rather than the associative value of the conditioning context at the time of conditioning implies that the critical comparison does not occur until the time of testing. Thus the information retained over the retention interval is apparently the independent associative strengths of the CS, i.e., P(US/CS), and the conditioning context, i.e., (P(US/*$\overline{CS}$*), rather than solely the outcome of the comparison, i.e., P(CR/CS).*

Our deflation results are consistent with those of Bottjer (1982) and Kaplan and Hearst (in press, a), who used an appetitive autoshaping preparation with pigeons. Moreover, Bottjer found that a novel stimulus presented just before the CS could restore inhibitory power to the CS as indexed by withdrawal from the CS (akin to Pavlovian disinhibition), suggesting that the animal retains the past associative history of the context but ordinarily responds with regard to its present associative value following contextual deflation. Further evidence consistent with our deflation data is Fowler and Lysle's (1982) observation that, following A+/AX− conditioning, manifest CI by X is reduced by extinguishing A and the context but not reduced by extinguishing X along with A and the context. Our interpretation of Fowler and Lysle's results assumes that X had acquired a small degree of excitatory strength (perhaps through within-compound X-A associations, see Rescorla, 1982), which was extinguishable simultaneously with deflation of the comparator term. In an article seemingly inconsistent with the preceding results, Marlin (1982) reported a decrease in responding to an excitatory CS when the conditioning context was deflated, despite the test occurring outside of the conditioning context. However, she administered very brief pretraining exposure to the conditioning context that favored both CS-context associations and context-US associations. Thus, the response-enhancing effects of deflating the comparator stimulus in her study probably were outweighed by the decrement in second-order associative strength mediated by the context.

Our next study was designed to inflate the associative value of the conditioning context during the retention interval that followed excitatory CS conditioning. To enhance sensitivity to possible decreases in manifest excitation, we set P(US/CS) = 0.50 rather than 0.33; thus, on the four target conditioning days, 3 out of 6 white noise presentations were reinforced with footshock. During conditioning of the white noise, no unsignaled shocks were given, i.e., P(US/$\overline{\text{CS}}$) = 0. But prior to testing, half the subjects had the associative value of the conditioning context inflated on each of 6 retention interval days through the administration of 27 unsignaled shocks during forty 30-sec intervals, i.e., P(US/$\overline{\text{CS}}$) = 0.67. The design of this experiment is outlined in Fig. 2.9.

When we tested these animals by measuring latencies to emit 25 licks in the presence of the white noise, the CS proved to be equally excitatory in both groups, $\bar{X}$s = 1.87 and 1.88 log sec. We then conducted a second, similar study, this time employing 8 inflation days and higher intensity footshock on the inflation days than on the conditioning days; the results again indicated no effect of postconditioning inflation of the training context, $\bar{X}$s = 1.93 and 1.91 log sec. These outcomes were not due to a failure of the inflation subjects to learn the conditioning context-US association, as strong suppression to the conditioning context alone was observed before the first shock was administered on the second and subsequent inflation sessions.

The present seeming lack of effect of US-alone presentations, i.e., contextual inflation, following CS conditioning is consistent with prior research that employed conditioned taste aversion (Brookshire & Brackbill, 1976; Holman, 1976; Riley, Jacobs, & LoLordo, 1976), conditioned lick suppression (e.g., Ayres & Benedict, 1973), and autoshaped key pecking (e.g., Jenkins & Lambos, 1983; Kaplan & Hearst, in press, a). The recurrence of this observation begins to create an inductive argument for the null result and suggests that no amount of contextual *inflation* will affect the comparator role of the conditioning context. However, this tentative conclusion is inconsistent with scalar expectancy theo-

CONDITIONED SUPPRESSION AS A FUNCTION OF POSTCONDITIONING INFLATION OF CONTEXTUAL FEAR

GROUP	PHASE 1	PHASE 2	PHASE 3	TEST
	Days 5, 6, 10, 11	Days 8–9	Days 12–17	Day 18
IA	P (FS/Noise) = 3/6^A P (FS/$\overline{\text{Noise}}$) = 0/34^A	P (FS/Click) = 4/4^C	P (FS/$\overline{\text{Noise}}$) = 27/40^A P (FS/$\overline{\text{Noise}}$) = 0/40^B	Noise ^{B}
NIA	P (FS/Noise) = 3/6^A P (FS/$\overline{\text{Noise}}$) = 0/34^A	P (FS/Click) = 4/4^C	P (FS/$\overline{\text{Noise}}$) = 0/40^B	Noise ^{B}

P = Probability
I = Inflation
NI = No Inflation
FS = Footshock

A = in Context A
B = in Context B
C = in Context C

FIG. 2.9. Paradigm for determining the effects of postconditioning associative inflation of the conditioning context upon the potential of the white noise to induce conditioned lick suppression.

ry's (Gibbon & Balsam, 1981) prediction that contextual inflation should attenuate responding to the discrete CS when the US-alone presentations occur after as well as before and during conditioning of the CS. Numerous experiments have shown that US-alone presentations either *before* (e.g., Holman, 1976) or *during* CS conditioning (e.g., the present research) do degrade responding to the CS. Consistent with this overall pattern of results is the finding of Jenkins and Lambos (1983) that, despite the lack of effect of postconditioning US-alone presentations on an initial CS test trial, responding on subsequent CS test trials was attenuated, presumably because the test trials themselves (which consisted of noncontingent presentations of the CS and US) acted as further CS conditioning trials (now zero contingency) that transformed the US-alone presentations into events embedded in the overall conditioning of the CS.

We currently do not have any reasonable hypothesis as to why contextual inflation appears to be ineffective, whereas contextual deflation does influence responding. Randich and Haggard (1983) reported reduced excitatory responding following postconditioning contextual inflation, but the effect was weak and short-lived. This last factor prompted Randich to view his inflation effect as being largely nonassociative. Moreover, Fowler and Lysle (1982) using an A+/AX− paradigm observed enhanced inhibition to X after inflation of A and the context, but this followed postconditioning deflation of A and the context, the two of which together likely served as the associative comparator. Whether the discrepancy between their results and ours is due to their A+/AX− procedure or the preinflation extinction of the comparator stimulus is not immediately clear, but we are inclined to view the preinflation extinction manipulation as the more important difference.

In addition to the possible asymmetry between the effects of comparator inflation and comparator deflation, the comparator hypothesis is in need of further research concerning how contextual events are weighted with respect to their temporal proximity of CS training. Minimally, the comparator hypothesis predicts that as the interval between a CS and US increases, the US will contribute progressively less to P(US/CS) and more to P(US/$\overline{\text{CS}}$), provided the unsignaled USs occur within the same session as the CS; it remains unclear why unsignaled USs following CS conditioning apparently fail to augment CI, whereas USs during CS conditioning do produce CI.

In contrast with our failure to obtain contextual inflation following CS conditioning, our deflation data demonstrate that nonreinforcement of the context, even long after CS conditioning, can be of considerable consequence. Consistent with this finding that the comparator stimulus can sometimes be effectively revalued by events far removed in time from CS presentations, a CS can acquire inhibitory properties even when the CS and US are presented on alternate days, i.e., with the unsignaled USs are occurring during CS conditioning but 24 hr removed from a CS presentation (Baker, 1977). Additionally Rescorla et al. (this volume) observed CRs inversely related to the associative strength of the conditioning context even when all the unsignaled USs preceded any of the CS presentations. However, Kleiman and Fowler (1984) have reported that long interstimulus intervals in the forward direction (operational trace conditioning) and backward direction (operational backward conditioning) yield CI, whereas USs midway between CS presentations do not (also see Wagner & Larew, this volume). Presumably the CI seen with the trace conditioning procedure was due to inhibition of delay (Pavlov, 1927; Rescorla, 1967). These observations as well as our inflation data and Kaplan and Hearst's (in press, b) recent analysis of trace conditioning speak for the importance of temporally local context as an associative comparator. There are obviously a number of unresolved questions here, and perhaps they will have to be answered separately for preconditioning contextual inflation, preconditioning contextual deflation, postconditioning contextual inflation, postconditioning contextual deflation, and contextual conditioning during conditioning of the discrete CS.

The Comparator Hypothesis and "Inhibitory" Test Behavior

The standard tests for CI, i.e., summation and retardation tests, are predicated on either the assumption of negative associative strengths (Wagner & Rescorla, 1972) or the assumption that a conditioned inhibitor is a stimulus that elevates the threshold for reactivating the US representation (Konorski, 1948; Rescorla, 1979). In rejecting both these conceptualizations of CI, the comparator hypothesis is obligated to provide alternative accounts of how a putative conditioned inhibitor comes to "pass" summation and retardation tests.

Summation Tests. Traditional explanations of summation tests either posit that the assumed negative associative strength of the conditioned inhibitor, when summated with the positive associative strength of a known exciter, yields an effective excitatory strength that is less than the exciter alone, or they argue that the conditioned inhibitor raises the associative threshold for reactivation of the US trace so that the exciter exceeds it by less than it would in the absence of the inhibitory stimulus. In rejecting both these explanations, the comparator hypothesis might explain summation test behavior as follows: The subject responds not to the sum of the associative strengths of the CSs present at the time of testing, but to a US expectation based on the *average* associative value of all cues present, presumably weighted by their attention-controlling saliencies. Thus, when a strong exciter is presented in compound with a putative conditioned inhibitor (which the comparator hypothesis views as a weak exciter relative to the conditioning context), these two stimuli collectively act as a moderate exciter of lower strength than the strong exciter alone.

The prediction of associative averaging as opposed to associative summation requires some sort of underlying theoretical mechanism. One likely possibility is that the putative conditioned inhibitor distracts some of the subject's attention from the strong exciter. This distraction from the exciter could either take the form of divided attention or switching of focused attention between the exciter and putative inhibitor at a rate so high that the switching is obscured by averaging over the duration of the test trial. To explain why a conditioned inhibitor attenuates excitatory behavior on a summation test far more than does a novel stimulus (Pavlov, 1927), the comparator hypothesis must assume that a functional conditioned inhibitor is of higher saliency than an associatively equal, novel stimulus, due to the inhibitor's temporal–spatial proximity to primary or secondary reinforcement during training, a proximity presumably great enough to increase the attention-attracting saliency of the stimulus but inadequate to imbue it with appreciable associative strength relative to the conditioning context. It may appear unparsimonious to postulate changes in saliency without changes in associative strength, but such processes appear to be demanded by phenomena other than CI, e.g., latent inhibition (Lubow, Weiner, & Schnur, 1981).

This averaging interpretation of summation tests is consistent with Cotton, Goodall, and Mackintosh's (1982) observation that an A-large shock/AX-small shock procedure renders X effectively inhibitory as measured in a summation test with a previously conditioned exciter (B) when B had previously been paired with the large shock, but *not* when B had been paired with the small shock. Thus, it appears that a CS can simultaneously be inhibitory when measured in summation with a strong exciter, and noninhibitory when measured in summation with a weak exciter. This observation is concordant with associative averaging but not associative summation. Furthermore, the comparator hypothesis would predict that the same stimulus would prove mildly excitatory when tested in summation with a stimulus completely lacking in associative strength, provided of course that P(US/CS) was greater than zero. Indirect evidence consistent with this prediction comes from Mackintosh and Cotton's (this volume) observation that a weakly reinforced conditioned inhibitor summated with an extinguished exciter elicited more responding than the extinguished exciter alone. Moreover, owing to associations between a conditioned inhibitor and the excitatory conditioning context (or A in an A+/AX− preparation), even a conditioned inhibitor that is never paired with the US may well exhibit these tendencies.

When A and B are independently reinforced and then presented simultaneously, initial responding is often greater than to A or B alone (e.g., Rescorla & Wagner, 1972). At first glance this may appear to demonstrate true algebraic summation of associative strength rather than averaging. However, on an A-alone test trial, the subject for example may be responding 70% to the highly excitatory CS (A) and 30% to the low excitatory context. Then when B is compounded with A, B possibly diverts some attention from both A and the context such that 45% of responding on the AB trial is attributed to A, 45% to B, and 10% to context. With 90% of the controlling stimuli now being highly excitatory, more responding is anticipated on the AB test trial than on an A-alone test trial. Thus the heightened responding sometimes (but surprisingly rarely) seen when two strong exciters are presented simultaneously is not necessarily *inconsistent* with the averaging interpretation of a summation test.

If associative averaging actually occurs and divided attention is a contributing factor, averaging as opposed to algebraic summation of the associative strengths of simultaneous stimuli ought to occur most readily when attentional capacity is heavily taxed; large demands on attention should favor dividing available attention between the known exciter and added putative conditioned inhibitor on a summation test, rather than increasing the total amount of attention to accommodate the putative conditioned inhibitor without decreasing attention to the known exciter. Thus, one might expect that, as the complexity of the stimuli increases, limited attentional capacity will force a progressive shift from algebraic summation to averaging. Admittedly this attentional interpretation of summation tests for CI is in need of further development.

A second explanation of behavior on a summation test is based on the configuring of simultaneously presented stimuli and the processes responsible for stimulus generalization decrement, a position developed by Pearce (1984). Modifying Pearce's model to make it compatible with the comparator hypothesis, we suggest that if A is a known exciter and X a putative conditioned inhibitor, a weaker response will be elicited by the configured AX stimulus than by A alone owing to generalization decrement going from A to AX. To assure greater generalization decrement when X is an effective conditioned inhibitor than when X is a neutral stimulus, the comparator hypothesis must assume that the degree of generalization decrement depends directly upon the saliency of X and that inhibition training with X increases the saliency of X. Clearly this explanation has much in common with the associative averaging proposal; moreover, assuming that configuring of A and X is not complete, the two explanations are not mutually exclusive.

Retardation Tests. Traditional explanations of a presumed conditioned inhibitor's resistance to becoming an evident conditioned exciter when it is paired with a reinforcer posit that negative associations must be overcome before positive associations can be formed, or, alternatively, that more excitatory strength is necessary to yield responding to a conditioned inhibitor than to a neutral stimulus because of an elevated threshold for responding to the inhibitor. However, without recourse to either negative associations or elevated response thresholds, there are at least four additional distinct, but not mutually exclusive, explanations of CI's effect on retardation tests. The comparator hypothesis would favor some combination of these latter explanations.

First among these possibilities, the high associative value of the context following CI treatment might act as an elevated *comparator* term, over which the associative strength of the putative conditioned inhibitor must rise in order to be manifest as an exciter. If valid, retardation from this source should also apply to novel stimuli. (As stated before, this mechanism is in some respects similar to the notion of elevated reactivation threshold but differs in that the present threshold is the associative value of the comparator stimuli.) Second, the high associative value of the context might *block* subsequent excitatory acquisition by the putative conditioned inhibitor. Again, retardation based on this mechanism should hold for novel stimuli as well as conditioned inhibitors. A third possible factor that might contribute to the phenomenon of retardation emphasizes the CS rather than the context. Specifically, between-group differences in *latent inhibition* are probable in cases in which the CI group experiences P(US/CS) = 0 and the control group receives no CS. Latent inhibition may appear contrary to distraction contributing to summation tests; however, we are not convinced that the two effects are mutually exclusive. For example, using our own variation of the Pearce and Hall (1980) model, CI treatments could simultaneously decrease associability and increase salience-with-respect-to-performance, thereby giving rise simultaneously to retardation and summation, respectively. Moreover, the Pearce and Hall model predicts latent inhibition to all experienced stimuli including those for which P(US/CS) > 0. A fourth possible contribution to retardation consists of nonassociative US effects, particularly habituation. This last factor, i.e., *nonassociative US effects,* should also retard acquisition to novel stimuli as well as to the putative conditioned inhibitor (except if the novel stimuli produced dishabituation). The relative contribution of each of these factors will of course depend on the specific CI training procedure, the choice of control groups, and the experimental parameters used.

Three of the four factors that we suspect of contributing to the retardation of a putative conditioned inhibitor predict equal retardation of a novel stimulus and the conditioned inhibitor. Therefore, we performed an experiment to look for transfer of retardation to a novel stimulus. The particular paradigm was selected to minimize latent inhibition to the putative conditioned inhibitor, latent inhibition being the one factor that should yield more retardation of the conditioned inhibitor than of a novel stimulus. Latent inhibition was minimized by setting P(US/CS) = 0.33, rather than 0, as is so often the case in studies of CI. Pearce, Kaye, and Hall (in press) have recently demonstrated that partial reinforcement schedules tend to minimize latent inhibition to the CS. Figure 2.10 outlines our paradigm. For Group Novel Stimulus, the white noise was novel on the Phase 3 acquisition trials, whereas Group Conditioned Inhibitor received our standard CI treatment with respect to the white noise prior to the Phase 3 acquisition trials. Note that testing occurred in Context B rather than in Context A, thereby eliminating the need for a lick recovery day prior to testing. In addition, testing

TRANSFER OF RETARDATION TO A NOVEL STIMULUS

GROUP	PHASE 1	PHASE 2	PHASE 3	TEST
	Days 5, 6, 10, 11	Days 8–9	Day 12	Day 13
Novel Stimulus	$P\ (FS/Click) = 2/6^A$ $P\ (FS/\overline{Click}) = 23/34^A$	$P\ (FS/Click) = 4/4^C$	$P\ (FS/Noise) = 2/2^A$ $P(FS/\overline{Noise}) = 0/38^A$	Noise ^{B}
Conditioned Inhibitor	$P\ (FS/Noise) = 2/6^A$ $P\ (FS/\overline{Noise}) = 23/34^A$	$P\ (FS/Click) = 4/4^C$	$P\ (FS/Noise) = 2/2^A$ $P\ (FS/\overline{Noise}) = 0/38^A$	Noise ^{B}

P = Probability
FS = Footshock

A = in Context A
B = in Context B
C = in Context C

FIG. 2.10. Paradigm to investigate transfer to a novel stimulus of retardation resulting from conditioned inhibition training.

outside the conditioning context removed the possibility of the associative strength of the conditioning context summating with that of the CS.

Both groups in this experiment yielded far less suppression than we ordinarily see following two such noise–shock pairings in otherwise naive animals, that typically yield mean log latencies for 25 licks of 2.30 ± .15 log sec in the presence of the white noise. However, significantly *more* suppression (excitation) was seen in the CI animals (1.73 ± .17 log sec) than in the novel stimulus animals (1.25 ± .16 log sec). Although we have proposed three sources of retardation of acquisition in the novel stimulus group, only one source, the comparator hypothesis, is able to explain the greater retardation in the novel stimulus group than in the conditioned inhibitor group. Specifically, the comparator hypothesis assumes that all associations are positive. Consequently the 33% reinforcement of the white noise during "inhibitory" training of the CI group made the CS modestly excitatory and gave it a head start relative to the novel stimulus group in accruing associative strength during Phase 3 of the study. This observation supports the notion that CI is not incompatible with low levels of excitatory strength. Rescorla (1969a) ran similar groups, but with P(US/CS) = 0. In his study the putative conditioned inhibitor yielded more retardation than the novel stimulus. Tentatively, we assume that this arose from latent inhibition to his conditioned inhibitor, which likely occurred in his study with P(US/CS) = 0, but not in our study in which P(US/CS) = 0.33.

Relationship Between Tests for CI. To certify the occurrence of CI, both summation and retardation tests have traditionally been required because passage of either test alone was recognized as being subject to alternative explanations. In contrast to traditional views of CI, the comparator hypothesis of CI views passage of summation and retardation tests as the result of two (or more) different mechanisms. However, except for the elevated comparator's contribution to retardation tests, none of these mechanisms are new; there are phenomena independent of CI that appear to require the existence of each of the other mechanisms. Moreover, although the comparator process itself can be regarded as novel, there are sufficient data presented in both the remainder of this chapter and elsewhere (e.g., Fowler, Kleiman, & Lysle, this volume; Kaplan & Hearst, in press, a) to support the existence of some comparator-like process even if one or more of the older explanations of retardation are retained.

With respect to the approach–withdrawal measure of CI, Hearst and Franklin (1977) and Wasserman, Franklin, and Hearst (1974) argue that approach to a CS that is negatively correlated with an aversive reinforcer (shock) and withdrawal from a CS that is negatively correlated with an appetitive reinforcer (food) are indicative of CI. Such indices are possibly a perfectly good direct measure of negative associative strength, but it is also possible that this locomotor behavior merely reflects relative frustration in the appetitive case and relative relief in the aversive case. Alternatively stated in the terminology of contemporary conditioning theory, the context may well be more excitatory than the saliency-weighted average associative value of the context plus the putative conditioned inhibitor. Thus in the appetitive case, the animal may opt for the context alone over the context plus the conditioned inhibitor to maximize second-order reinforcement. Just the opposite would hold for the aversive case.

Extinction

One of the more baffling aspects of CI is that, unlike the extinction that ordinarily results from CS-only presentations of an excitatory cue, postconditioning CS-only presentations of an effectively inhibitory cue have been reported to have little consequence upon behavioral control by the stimulus (e.g., Owren & Kaplan, 1981, cited in Kaplan & Hearst, in press, a; Zimmer–Hart & Rescorla, 1974). As is presently evident, the comparator hypothesis fares well in dealing with this asymmetry between extinction of exciters and apparent inhibitors.

Consider first the excitatory case in which the discrete CS is more excitatory than the conditioning context. Typically, the extinction manipulation reduces the effective associative value of the CS, thereby making it closer to the low or zero excitatory value of the conditioning context. According to the comparator hypothesis, this in turn decreases conditioned responding elicited by the CS, thereby giving rise to the phenomenon of behavioral extinction of excitatory conditioning. Ordinarily, the associative strength of the conditioning context does not change during extinction of an exciter, both because of a floor effect and because deflation of the conditioning context is usually a slow process that requires prolonged exposure to the context.

Consider now the inhibitory case. In, for example, Zimmer–Hart and Rescorla's (1974) failure to obtain extinction of a nominal conditioned inhibitor, the

CS was never reinforced, i.e., $P(US/CS) = 0$ during initial conditioning. Thus, the comparator hypothesis would argue that floor effects prevented CS-alone presentations from altering the associative strength of the CS. Exposure to the conditioning context during extinction might have partially deflated the comparator (Zimmer–Hart and Rescorla used an A+/AX− paradigm), but the non-extinguished control animals received the same exposure to the conditioning context, thereby preventing group differences from originating with contextual deflation.

The comparator hypothesis posits that all CS–US and context-US associations are excitatory and that CS-alone or context-alone presentations can only reduce their value. Thus in contrast to the usual expectation that CS-alone presentations will cause the manifest associative value of a putative conditioned inhibitor to regress toward zero, the comparator hypothesis predicts that, if floor effects are avoided and the associative value of the conditioning context is not also reduced during the extinction treatment, CS-alone presentations should *increase* the effective inhibitory value of the CS.

To test this prediction, we performed a study in which all subjects experienced $P(US/CS) = 0.33$ and $P(US/\overline{CS}) = 0.67$. Then half the animals received extensive CS-alone presentations. The procedure used in this study is depicted in Fig. 2.11. Note that the extinction treatment occurred outside the conditioning context to minimize potential changes in the associative value of the conditioning context. Additionally, testing occurred in the extinction context to avoid the attenuation of extinction effects often seen when context is changed between extinction and testing (Bouton & Bolles, 1979). The outcome of this experiment was quite consistent with the comparator hypothesis. Operational extinction of the conditioned inhibitor made it more inhibitory on a summation test with the

EXTINCTION OF A CONDITIONED INHIBITOR

GROUP	PHASE 1	PHASE 2	PHASE 3	TEST
	Days 5, 6, 10, 11	Days 8–9	Days 12–19	Day 20
E	P (FS/Noise) = $2/6^{A}$ P (FS/$\overline{\text{Noise}}$) = $23/34^{A}$	P (FS/Click) = $4/4^{C}$	P(FS/Noise) = $0/12^{B}$ P (FS/$\overline{\text{Noise}}$) = $0/68^{B}$	Click + Noise ^{B}
NE	P (FS/Noise) = $2/6^{A}$ P (FS/$\overline{\text{Noise}}$) = $23/34^{A}$	P (FS/Click) = $4/4^{C}$	P (FS/$\overline{\text{Noise}}$) = $0/80^{B}$	Click + Noise ^{B}

P = Probability
E = Extinction
NE = No Extinction
FS = Footshock

A = in Context A
B = in Context B
C = in Context C

FIG. 2.11. Paradigm for determining the effects of CS-alone presentations outside the conditioning context on a conditioned inhibitor that had a negative, but not perfect, correlation with the US.

excitatory clicks (1.16 ± 0.16 mean log sec to emit 25 licks) relative to animals lacking the extinction treatment (1.85 ± 0.12 mean log sec to emit 25 licks).

Our analysis of extinction of a CS has assumed that, even after nonreinforced CS exposure outside the conditioning context, i.e., in Context B, the conditioning context, i.e., Context A, continues as the sole comparator baseline. An alternative view still within the comparator framework posits that the extinction treatment constitutes a second conditioning treatment, this time with the extinction context serving as the comparator baseline. If so, during subsequent testing, the effective comparator might be expected to be some sort of average associative strength of Context A and Context B, which would necessarily be less than that of Context A alone because footshock was never given in Context B. Although this averaging of comparators predicts a decrease in CI, this reduction in the effective comparator may or may not be outweighed by the decreased value of P(US/CS) in the animals given CS-alone exposure. Thus these two variants of the comparator hypothesis cannot be differentiated on the basis of the present data.

A third interpretation of our results, also consistent with the comparator hypothesis, is that with CS presentations occurring in two locations, i.e., Context A during conditioning and Context B during extinction, the subject has the opportunity to form a discrimination between contexts that would result in $P(US/CS)_A$ being compared to $P(US/\overline{CS})_A$ when testing occurs in Context A, and $P(US/CS)_B$ being compared to $P(US/\overline{CS})_B$ in Context B. However, such a discrimination predicts that our extinguished CS would have been neutral rather than inhibitory when tested in Context B, which is contrary to what was observed. Hence, based on our present data, it appears that Context A, the context in which CS–US pairings and unsignaled USs occurred, remains a major component of the comparator for the CS despite extinction treatment and testing in Context B. This tentative conclusion leaves open the question of the nature of the comparator term if either CS–US pairings or unsignaled USs had occurred in Context B following initial conditioning in Context A.

Finally, the observed increase in manifest CI following CS-alone exposures could conceivably be a consequence of second-order conditioning in addition to or instead of factors dependent on the comparator process. Specifically, Cunningham (1981) and Rescorla (1982) have pointed out that, in an A+/AX− paradigm, X will not only become a conditioned inhibitor, but it will also enter into association with A (the context in our case) making it a second-order exciter. Hence, the full impact of X as an effective conditioned inhibitor may be masked by second-order excitation. Presentation of X alone may then extinguish the X–A association, reducing the second-order positive reinforcing value of X and consequently revealing the true degree of CI previously accrued by X. However, this view is no more congenial to CI reflecting negative associations than is the comparator hypothesis. Conditioned inhibition, when viewed as a negative association, fails to explain why X's conditioned inhibitory qualities do not extinguish along with its second-order excitatory qualities, whereas the comparator hypothesis predicts the frequently observed lack of extinction of CI and also provides two plausible mechanisms that might contribute to the present increase in CI produced by CS-alone presentations.

Despite our reservations concerning the importance of X–A associations, we performed a study to examine the contribution of second-order associations (CS-context, context-US) to the increase in CI observed after CS-alone presentations. The experiment consisted of setting P(US/$\overline{CS}$) = 0.67 and examining the effects of postconditioning CS-alone treatment (in the nonconditioning context) upon subjects for whom P(US/CS) = 0 as well as upon those for whom P(US/CS) = 0.33. Provided that behavioral floor effects are avoided, the second-order excitation position predicts similar effects in the two cases, whereas effects due exclusively to decreases in P(US/CS) should not increase manifest CI in the P(US/CS) = 0 group because the CS will have little or no excitatory strength to lose. The various treatments used in this study are illustrated in Fig. 2.12. In both extinction groups, manifest CI as seen on a summation test was actually *increased* relative to appropriate nonextinguished control subjects (as in Table 2.4). These results are consistent with the view that at least part of the increase in manifest CI following CS-alone exposures is due to extinction of CS-context associations. However, there was also a nonsignificant tendency toward a greater increase in CI due to CS-alone presentations in the P(US/CS) = 0.33 group than in the P(US/CS) = 0 group. Both this tendency toward greater extinction-induced CI in P(US/CS) = 0.33 animals and the failure of CS-alone presenta-

EXTINCTION OF CONDITIONED INHIBITION AS A FUNCTION OF P(US/CS)

GROUP	PHASE 1	PHASE 2	PHASE 3	TEST
	Days 5,6,10,11	Days 8–9	Day 12–19	Day 20
2/6–E	P(FS/Noise)=2/6^{A} P(FS/$\overline{Noise}$)=23/34^{A}	P(FS/Click)=4/4^{C}	P(FS/Noise)=0/12^{B} P(FS/$\overline{Noise}$)=0/68^{B}	Click+^{B} Noise
2/6–NE	P(FS/Noise=2/6^{A} P(FS/$\overline{Noise}$)=23/34^{A}	P(FS/Click)=4/4^{C}	P(FS/$\overline{Noise}$)=0/80^{B}	Click+^{B} Noise
0/6–E	P(FS/Noise)=0/6^{A} P(FS/$\overline{Noise}$)=23/34^{A}	P(FS/Click)=4/4^{C}	P(FS/Noise)=0/12^{B} P(FS/$\overline{Noise}$)=0/68^{B}	Click+^{B} Noise
0/6–NE	P(FS/Noise)=0/6^{A} P(FS/$\overline{Noise}$)=23/34^{A}	P(FS/Click)=4/4^{C}	P(FS/$\overline{Noise}$)=0/80^{B}	Click+^{B} Noise

E=Extinction
NE=No Extinction
P=Probability
FS=Footshock

A=Context A
B=Context B
C=Context C

FIG. 2.12. Paradigm for determining the effects of CS-alone presentations upon conditioned inhibition as a function of the frequency with which the CS was initially reinforced.

TABLE 2.4
Outcome of Summation Test as a Function P(US/CS) and Extinction of CS

Group	*Mean Test Latency*[a] *to Emit Licks 0–25* *(± Standard Error)*	*Mean Test Latency*[a,b] *to Emit Licks 25–50* *(± Standard Error)*
2/6-E	.67 (± .05)	1.25[c] (± .16)
2/6-NE	.65 (± .07)	1.90 (± .20)
0/6-E	.69 (± .07)	1.06[d] (± .19)
0/6-NE	.69 (± .08)	1.59 (± .15)

[a]Log sec.
[b]White noise plus click present during second 25 licks.
[c]Significantly less than Group 2/6-NE, $p < .05$.
[d]Significantly less than Group 0/6-NE, $p < .05$.

tions to attenuate CI suggest that CS-context associations are not sufficient to completely explain the anomalous effects of operational extinction upon a conditioned inhibitor. However, the comparator hypothesis and the occcurrence of CS-context associations together appear able to fully explicate the effects of CS-alone treatments on a conditioned inhibitor. Further research on the extinction of conditioned inhibition is in progress in our laboratory.

Other Phenomena

There are several phenomena in the CI literature beyond those already discussed that defy explanation by the currently prevailing theories. We now consider the ability of the comparator hypothesis to address these effects when it is appended to existing theories as a rule for response generation.

Baker (1974) found, contrary to the Rescorla–Wagner (1972) model, that a nonreinforced pairing of a neutral stimulus (X) with an established conditioned inhibitor (B) does not make the neutral stimulus excitatory. The comparator explanation of this observation is that the associative strength of the conditioned inhibitor was near zero, and its inhibitory powers arose from the relative excitatory value of its nominal "conditioning" context. Floor effects would prevent the associative value of the conditioned inhibitor from appreciably decreasing on the BX− trials, and without reinforcement there is no basis for the neutral stimulus to gain associative strength on such trials. Hence, in Baker's study neither the nominal conditioned inhibitor nor the neutral stimulus would be expected to change in value as a result of their nonreinforced pairings.

In a more recent study, Baker and Baker (this volume) report making a target stimulus (X) behave like a conditioned inhibitor by using an unsignaled shock procedure similar to ours; moreover, they found that the CI was attenuated when another cue (nontarget stimulus) signaled the explicitly unpaired shocks, a finding also reported by Fowler et al. (this volume). Baker and Baker found that this attenuation was particularly pronounced when long intervals between shocks were employed. The attenuation in CI produced by signaling the shocks with a nontarget stimulus is consistent with the added signal's overshadowing the conditioning context, thereby degrading the excitatory strength of the context and ultimately reducing the context's superiority in associative strength over the CI. Moreover, the long intershock intervals presumably facilitated between-trial extinction to the context. The overshadowing of the context by stimulus A corresponds to an A+/X− procedure that typically yields weaker CI than A+/AX− or explicitly unpaired paradigms (in which the context effectively plays the role of A). In the latter two cases, A and the context, respectively, likely serve as strong associative comparators, whereas in the case of A+/X− the context is a weaker comparator by virtue of A's interfering with the context accruing manifest excitatory value.

Kohler and Ayres (1982) have addressed a paradox arising from the role of context-US associations in trace conditioning. Context-US associations appear to increase as a direct function of the interval between the nominal CS and US in contrast to the decrease in CS–US associations seen with increases in the interstimulus interval (Marlin, 1981). The paradox is that, during testing, the associative strength of the context should summate with that of the CS to yield full-strength responding. This paradox also arises with respect to the comparator hypothesis. If, through the comparator principle, an excitatory context degrades responding to a discrete CS, why doesn't the very same excitatory strength of the context simultaneously enhance responding by directly summating (or at least "averaging") with the response eliciting quality of the discrete CS? Surely in some situations, strong context-US associations do enhance responding in the presence of associatively weak nomimal CSs. We, among others, have published several papers demonstrating such summation effects, although we did not employ trace conditioning to enhance the context-US associations (Balaz et al., 1981; Balaz et al., 1982).

Seemingly, both the summation principle and the comparator hypothesis are valid; however, these two factors just happen to cause context-US associations to both enhance and attenuate responding by summation and comparison, respectively. The question is, when will summation predominate and when will comparison predominate? Based upon between-experiment comparisons, we expected that the comparator hypothesis would predominate if CS onset occurred long (but not too long) after the animal is introduced into the context on the test trial. In this case, direct context-US associations should not have appreciably extinguished, but the animal should have acclimated to the level of US expectation predicted by the context, thereby preventing associative summation of the context with the CS from being evident in performance. Such associative adaptation presumably does not impair the comparator role played by the context provided the exposure to the context is not so prolonged as to produce extinction. On the other hand, if the CS is presented soon after the animal is introduced into the context during testing, acclimation to the US expectation announced by the context should not have had time to occur; thus, associative summation should result and if sufficiently large it will mask the comparator effect. To examine these possibilities, we recently performed a preliminary study in which, using our standard preparation for CI (see Fig. 2.2), we presented the tone *and lick tube* on the test trial either immediately upon placement in the training/test apparatus or 15 min after placement. Consistent with summation predominating in the first instance, CE was observed in the former case (2.78 log sec) but CI was seen in the latter case (0.95 log sec). Appropriate control groups indicated that 15 min of exposure to the context resulted in insufficient extinction of the context to account for this difference. Thus, it appears that simultaneous presentation of the discrete CS and training/test context maximizes associative summation and thereby may mask manifestation of the comparator effect.

Several researchers (e.g., Baker & Mackintosh, 1977; Halgren, 1974; Rescorla, 1971) have reported that preconditioning CS-alone presentations produce retarded acquisition of both CE and CI. (In fact this was the key observation, along with the failure of latent inhibitors to pass a summation test for CI, that led to the conclusion that "latent inhibition" was not truly inhibitory.) This finding suggests a symmetry between CE and CI that the comparator hypothesis denies. How then does the comparator hypothesis explain the dual retardation of CE and CI produced by latent inhibition? If the prevailing explanation of latent inhibition is correct, i.e., pretraining CS-alone exposures decrease attention to the CS (Lubow et al., 1981; Mackintosh, 1975), the CS will be retarded in becoming an exciter because it is not being attended to during acquisition, and, according to the comparator hypothesis, the CS will be similarly retarded in coming to pass a summation test for CI because lack of attention to it (relative to a novel stimulus) works against it's receiving much weight in the associative averaging that the comparator hypothesis posits to underlie summation tests.

Preconditioning exposure to the US alone is known to retard subsequent excitatory conditioning in the same context. This effect has been attributed by Baker, Mercier, Gabel, and Baker (1981) and Randich and Haggard (1983) to a combination of nonassociative habituation to the US and associative blocking of the CS by the context. Without questioning the relative contributions of these two factors, the comparator hypothesis suggests a third possible mechanism. Specifically, the preconditioning US exposures might inflate the comparator term making the CS effectively less excitatory. Notably both this potential comparator mechanism and the blocking by context, whereas detrimental to excitatory conditioning of the CS, should have decidedly different effects on inhibitory conditioning. Whereas preconditioning habituation to the US, by rendering the US less effective, should yield retardation of CI as well as CE, blocking by the context should occur only in the case of CE, and the very same comparator mechanism that retards CE should facilitate CI. Such a facilitatory effect upon CI of preconditioning US-alone exposures has in fact been reported by Hinson (1982).

The Relationship Between CI and Contextual Potentiation

Beyond the context's role as a comparator and direct responding to the context, there is a third, independent associative role of the context that we have explored. Specifically, testing in the conditioning context appears to potentiate retrieval of, or at least behavioral control by, associations between the discrete CS and US. This function (at various times called contextual potentiation, contextual facilitation, occasion setting, and conditional discrimination), based on studies in which the associative strength of the context was attenuated by latent inhibition or extinction, does not seem to depend on context-US associations per se; rather the context appears to be modulating the effectiveness of the CS–US relationship (Balaz et al., 1981; Miller & Schachtman, in press; Rescorla, this volume). Because contextual potentiation of retrieval appears to depend on higher order contextual associations to the CS–US link rather than on direct context-US associations, the comparator hypothesis views contextual potentiation, however important in its own right, as probably having little impact upon the associations per se (as opposed to their retrieval) that underlie CI. Better contextual potentiation of retrieval of the CS–US (excitatory) relationship should elevate the effective CS–US associative strength, but so will it also enhance the effective context-US association. In the case of CI, the comparator hypothesis holds that the latter association is the larger; hence making it fully effective should not only negate potentiation of the effective CS–US association but might well result in increased CI. Thus, contextual potentiation is viewed with respect to the comparator hypothesis as an enhancer of whatever behavior is determined by the target and comparator associations.

In contrast to the view that contextual potentiation is a phenomenon independent of the comparator process that we believe underlies CI, Holland (this volume), Jenkins (this volume), and Rescorla (this volume) have all noted that a type of conditional discriminative stimulus (called a "facilitator" by Rescorla and an "occasion setting" event by Holland), be it a context or a discrete stimulus, can attenuate responding to the target CS as well as enhance it as a function of whether the compound of the discriminative stimulus and target stimulus is selectively nonreinforced or selectively reinforced. A conditional discriminative stimulus that attenuates conditioned responding obviously passes the summation test for CI. Holland presents data suggesting that behavior-attenuating discriminative stimuli presented simultaneously with an excitatory CS during conditioning will also pass a retardation test for CI, whereas behavior-attenuating discriminative stimuli presented prior to the target CS during conditioning (serial presentation) will not pass a retardation test. However, Jenkins reports studies indicating that the use of simultaneous rather than serial cues is alone not sufficient to ensure passage of a retardation test for CI. Holland's review of the many differences between traditional conditioned inhibitors and conditional discriminative stimuli nevertheless strongly suggests the existence of two distinctly different phenomena. Defining CI broadly, Holland, Rescorla, and Jenkins have all suggested that both conditional discriminative stimuli and traditional conditioned inhibitors be regarded as conditioned inhibitors because they attenuate responding to conditioned exciters. However, because of the numerous differences between conditional discriminative stimuli and traditional conditioned inhibitors that extend far beyond behavior on retardation tests (see Holland, this volume), it appears highly unlikely that common processes mediate these two types of discriminative responding. Thus, calling both types of discriminative stimuli conditioned inhibitors reduces the concept of CI to a synonym for attenuation of behavior on a summation test.

We prefer to retain the narrower and more traditional definition of CI that requires passage of both summation and retardation tests. With this narrower definition, only discriminative stimuli that pass both the retardation and summation tests qualify as conditioned inhibitors, whereas conditional discriminative stimuli appear to modulate behavior by much the same process as is seen in contextual potentiation. The observation that the modulation of a target association can produce decrements as well as increments in conditioned behavior indicates that neither "potentiators" nor "facilitators" are ideal names for this class of discriminative stimuli.

One of the most notable distinctions between what Rescorla (this volume) calls the opposite of a facilitator and a traditional conditioned inhibitor hinges on the importance of direct associations of the US to the stimulus in question. Balaz et al. (1981), Jenkins (this volume), and Rescorla (1981) have all demonstrated that the modulatory role of a stimulus is independent of the associative value of that stimulus. Facilitation and its negative counterpart clearly depend on the modulatory role, which we view as highly similar if not synonymous with conditional discriminative stimuli, whereas the phenomena traditionally called CI and the comparator hypothesis depend on the associative value of stimuli, specifically the conditioning context in the current data.

Some Unresolved Issues and Limitations

The comparator hypothesis is basically a rule for response generation; it is silent concerning how either the nominal CS or context accrue excitatory associative strength; hence, it is not offered in direct opposition to any of the currently popular models of learning. With various degrees of modification, most contemporary models of associative learning could be made compatible with the comparator hypothesis. This identification of contingency with response generation rather than acquisition is consistent with the conclusions of Damianopoulos

(1982) and Kaufman and Bolles (1981), who, working from data sets highly dissimilar to ours, conclude that learning is based on contiguity and that responding depends on contingency. Despite the comparator hypothesis being a response rule, application of it requires knowledge of the associative strength of both the CS and the comparator term. For purposes of consistency with Rescorla (1968, 1969a), we have thus far implied that the associative comparator depends on $P(US/\overline{CS})$, but as a CS presentation is necessarily embedded in a context, the more plausible associative comparator is the association based on P(US/context). This possibility gives central importance to the question of whether or not a US administered during (or immediately following) a CS contributes to the contextual comparator term as well as to the predictive power of the CS; that is, is the associative strength of the conditioning context a function of only the unsignaled USs delivered in it, literally $P(US/\overline{CS})$, or is the associative strength a function of all the USs delivered there? Alternatively stated, is the context subject to blocking and overshadowing by the discrete CS as predicted by the Rescorla–Wagner (1972) model, or is the context free of such effects as is predicted by Scalar Expectancy Theory (Gibbon & Balsam, 1981) and the Relative Waiting Time Hypothesis (Jenkins et al., 1981)? Disturbingly enough, there are considerable data supporting each view. In our own laboratory, we have found overshadowing and blocking of the conditioning context to be reliable, but far from universal, phenomena. With appropriate parameters, we can consistently get blocking and overshadowing, but with other parameters for which the Rescorla–Wagner model would predict these phenomena, we equally consistently fail to obtain blocking and overshadowing. Currently we have only rough intuitions concerning what the critical variables are.

Another problem for the comparator hypothesis, as well as most other contemporary theories, is the need to differentiate between a well-learned, low correlation of CS and US, and a poorly learned, high correlation of CS and US. Somehow P(US/CS) must be regarded as an asymptotic limit that, to predict behavior, has to be multiplied by the degree of learning (or "confidence" in associative strength). Malleability of associative status over repeated trials is presumably a function of degree of learning but not P(US/CS). The notion of associative confidence obviously has much in common with Pearce and Hall's (1980) concept of associability, except that associative confidence is CS–US specific rather than a parameter exclusively of the CS as is the case with associability. (Although both Pearce and Hall's concept of associability and our concept of associative confidence can explain both retarded manifestation of CS-strong shock associations following CS-weak shock pairings [Hall & Pearce, 1979] and retarded manifestation of A-strong shock/AX-no shock associations following A-weak shock/AX-no shock treatment [Hall, Kaye, & Pearce, this volume], only CS–US-specific associative confidence can explain why manifestation of CS-ice water associations are not as strongly retarded following CS-weak shock pairings as are CS-strong shock associations [Kasprow, Schachtman, & Miller, in press).] One of the immediate benefits of the concept of associative confidence is its ability to explain learned irrelevance without recourse to CS-only and US-only effects exclusively (Baker & Mackintosh, 1979; Mackintosh, 1973). Specifically, the apparent retardation in subsequent excitatory and inhibitory acquisition following truly random CS–US presentations likely reflects higher associative confidence in the truly random relationship than occurs in the appropriate control group, rather than the two groups starting the retardation phase of the experiment with appreciably different values of P(US/CS) or P(US/$\overline{\text{CS}}$).

A further issue that deserves attention is the generality of the comparator hypothesis to CI produced by procedures other than negatively correlated CS/US presentations embodied in the present procedure. For example, we know that A+/X− trials will produce CI although the effect is weak presumably because of A overshadowing the conditioning context. (In fact the CI produced by A+/X− is often so slight that this treatment is sometimes used as a nonassociative control, e.g., Holland, this volume.) A recent chapter by Kaplan and Hearst (in press, a) on the role of context reports that an excitatory context is as essential to the expression of CI created with the A+/X− procedure as it is to CI produced with the +/X− explicitly unpaired procedure. Thus, in the A+/X− paradigm there is reason to doubt that associations to A serve as part of the comparator to which X is compared; however, this conclusion has yet to be directly tested. (The lack of data likely stems from the relative difficulty in obtaining CI from the A+/X− procedure.) However, Baker and Baker (this volume) attenuated CI by signaling the explicity unpaired shock with a nontarget cue; if this cue had served as part of the comparator term, CI should not have been reduced. In contrast, A+/AX− procedure yields strong CI, often stronger than the explicitly unpaired procedure (+/X−), and consequently is the most commonly used means of obtaining CI. In this case, as with the A+/X− procedure, A may be expected to partially overshadow the context. But presenting A in compound with X likely makes A, as well as the context, part of the comparator term, thereby accounting for the robust CI that is ordinarily observed with this procedure. Evidence supporting a comparator role for both A and the conditioning context given to the A+/AX− procedure comes from a number of studies, perhaps the most compelling being the reduction in CI seen following postconditioning deflation of either A or the context (Fowler et al., this volume; Fowler & Lysle, 1982). However it should be noted that Rescorla and Holland (1977) failed to attenuate CI by extinguishing A following A+/AX− training. The difference between this study and that of Fowler and his colleagues likely originates with the relative contributions to the comparator term of A and the context, in addition to the generally greater resistance to extinction of contexts than discrete stimuli.

Just as A is likely part of the comparator term following Pavlovian CI treatment, based on the similarity of the procedures for blocking (A+/AX+) and Pavlovian CI (A+/AX−), it is also probable that A is part of the comparator term following Phase 2 of blocking (and also following AX+ overshadowing of X). If so, this contribution to the comparator term could account for part or even all of the lack of responding to X following blocking (or overshadowing) of X by A. Provided A plays such a role, pretest extinction of A should restore responding to X. Kaufman and Bolles (1981) provide data consistent with this prediction for overshadowing, although they offer a very different, highly cognitive interpretation of the phenomenon. Mackintosh (personal communication) reports preliminary data from a blocking paradigm that is consistent with this prediction. In our own laboratory, the effects of pretest extinction of blocking and overshadowing stimuli to date have been unreliable. Further research to investigate this novel interpretation of blocking and overshadowing is necessary.

Although the present discussion has repeatedly referred to P(US/CS) and P(US/$\overline{\text{CS}}$), the comparator hypothesis states that acquired behavior depends on the comparison of two associations and not the comparison of two probabilities. We have used probability notation because probabilities of events can be operationally defined and are readily quantifiable whereas associative values are not. We choose to presume no more than that associative strength increases monotonically with the probability of reinforcement. Equating P(US/CS) with P(US/$\overline{\text{CS}}$) at some value larger than zero asymptotically will leave the CS associatively excitatory and behaviorally neutral. However, due to CS and $\overline{\text{CS}}$ having differing associabilities (αs in terms of the Rescorla–Wagner (1972) model), during the preasymptotic period the CS could be either behaviorally excitatory or behaviorally inhibitory as a function of whether the associability of the CS were greater or less, respectively, than that of the comparator.

In conclusion, we believe that CI may usefully be regarded as a performance effect based on a comparison between two excitatory associations; we question the need for either the concept of negative associations or the concept of elevated reactivation thresholds to explain CI. Additionally, the critical comparison, at least in our preparation, appears to be between the associative value of the *conditioning* context regardless of where testing occurs. The power of the comparator hypothesis has been overlooked in recent years owing to the overuse of explicitly unpaired conditioned inhibitors, i.e., a P(US/CS) = 0, which invites floor effects with respect to many otherwise illuminating phenomena. A more general understanding of CI and the comparator process will come from the study of putative conditioned inhibitors with varied predictive values of reinforcement rather than just one in which the target CS is consistently unpaired with reinforcement. Moreover, the comparator hypothesis, in postulating excitatory associations exclusively, cannot help but add to our understanding of manifest conditioned excitation as well as manifest conditioned inhibition.

ACKNOWLEDGMENTS

The research reported in this chapter was supported by National Institute of Mental Health Grant MH 33881. Thanks are due Andrea Brown, Doreen Catterson, Linda Feldman,

Laura Gladstein, Jay Greenwald, Wesley J. Kasprow, and Joseph Serwatka for their assistance in collecting the data. We are also grateful to Peter Balsam, Eliot Hearst, John Pearce, and Stanley Scobie for their constructively commenting on an early version of this chapter.

REFERENCES

Ayres, J. J. B., & Benedict, J. O. (1973). US-alone presentations as an extinction procedure. *Animal Learning and Behavior,* 1, 5–8.

Baker, A. G. (1974). Conditioned inhibition is not the symmetrical opposite of conditioned excitation: A test of the Rescorla–Wagner model. *Learning and Motivation, 5,* 369–379.

Baker, A. G. (1977). Conditioned inhibition arising from a between-sessions negative correlation. *Journal of Experimental Psychology: Animal Behavior Processes, 3,* 144–155.

Baker, A. G., & Mackintosh, N. J. (1977). Excitatory and inhibitory conditioning following uncorrelated presentations of CS and UCS. *Animal Learning and Behavior, 5,* 315–319.

Baker, A. G., & Mackintosh, N. J. (1979). Preexposure to a CS alone, US alone, or CS and US uncorrelated: Latent inhibition, blocking by context, or learned irrelevance? *Learning and Motivation, 10,* 278–294.

Baker, A. G., Mercier, P., Gabel, J., & Baker, P. A. (1981). Contextual conditioning and the US preexposure effect in conditioned fear. *Journal of Experimental Psychology: Animal Behavior Processes,* 7, 109–128.

Balaz, M. A., Capra, S., Hartl, P., & Miller, R. R. (1981). Contextual potentiation of acquired behavior after devaluing direct context-US associations. *Learning and Motivation, 12,* 383–397.

Balaz, M. A., Capra, S., Kasprow, W. J., & Miller, R. R. (1982). Latent inhibition of the conditioning context: Further evidence of contextual potentiation of retrieval in the absence of context-US associations. *Animal Learning and Behavior, 10,* 242–248.

Balsam, P. (in press). Relative time in trace conditioning. In J. Gibbon & L. Allan (Eds.), *Timing and time perception.* New York: New York Academy of Sciences.

Bottjer, S. W. (1982). Conditioned approach and withdrawal behavior in pigeons: Effects of a novel extraneous stimulus during acquisition and extinction. *Learning and Motivation, 13,* 44–67.

Bouton, M. E., & Bolles, R. C. (1979). Contextual control of the extinction of conditioned fear. *Learning and Motivation, 10,* 445–466.

Brookshire, K. H., & Brackbill, R. M. (1976). Formation and retention of conditioned taste aversions and UCS habituation. *Bulletin of the Psychonomic Society,* Z, 125–128.

Cotton, M. M., Goodall, G., & Mackintosh, N. J. (1982). Inhibitory conditioning resulting from a reduction in the magnitude of reinforcement. *Quarterly Journal of Experimental Psychology, 34B,* 163–180.

Cunningham, C. (1981). Association between the elements of a bivalent compound stimulus. *Journal of Experimental Psychology: Animal Behavior Processes, 7,* 425–436.

Damianopoulos, E. N. (1982). Necessary and sufficient factors in classical conditioning. *Pavlovian Journal of Biological Science, 17,* 215–229.

Durlach, P. J. (1983). Effect of signaling intertrial unconditioned stimuli in autoshaping. *Journal of Experimental Psychology: Animal Behavior Processes, 9,* 374–389.

Fantino, E. (1969). Choice and rate of reinforcement. *Journal of the Experimental Analysis of Behavior, 12,* 723–730.

Fowler, H., & Lysle, D. T. (1982). *Internal inhibition as a "slave" process: Deactivation of conditioned inhibition through extinction of conditioned excitation.* Paper presented at Eastern Psychological Association Meetings, Baltimore.

Gibbon, J., & Balsam, P. (1981). Spreading association in time. In C. M. Locurto, H. S. Terrace & J. Gibbon (Eds.), *Autoshaping and conditioning theory.* New York: Academic Press.

Gibbon, J., Berryman, R., & Thompson, R. (1974). Contingency spaces and measures in classical and instrumental conditioning. *Journal of the Experimental Analysis of Behavior, 21,* 585, 605.

Halgren, C. R. (1974). Latent inhibition in rats: Associative or nonassociative? *Journal of Comparative and Physiological Psychology, 86,* 74–78.

Hall, G., & Pearce, J. M. (1979). Latent inhibition of a CS during CS–US pairings. *Journal of Experimental Psychology: Animal Behavior Processes, 5,* 31–42.

Hearst, E. (1972). Some persistent problems in the analysis of conditioned inhibition. In R. A. Boakes & M. S. Halliday (Eds.), *Inhibition and learning.* London: Academic Press.

Hearst, E., & Franklin, S. R. (1977). Positive and negative relations between a signal and food: Approach–withdrawal behavior to the signal. *Journal of Experimental Psychology: Animal Behavior Processes, 3,* 37–52.

Hinson, R. E. (1982). Effects of UCS preexposure on excitatory and inhibitory rabbit eyelid conditioning: An associative effect of conditioned contextual stimuli. *Journal of Experimental Psychology: Animal Behavior Processes, 8,* 49–61.

Holman, E. W. (1976). The effect of drug habituation before and after taste aversion learning in rats. *Animal Learning and Behavior, 4,* 329–332.

Hull, C. L. (1943). *Principles of behavior.* New York: Appleton–Century–Crofts.

Jenkins, H. M., Barnes, R. A., & Barrera, F. J. (1981). Why autoshaping depends on trial spacing. In C. M. Locurto, H. S. Terrace, & J. Gibbon (Eds.), *Autoshaping and conditioning theory.* New York: Academic Press.

Jenkins, H. M., & Lambos, W. A. (1983). Tests of two explanations of response elimination by noncontingent reinforcement. *Animal Learning and Behavior, 11,* 302–308.

Kaplan, P. S., & Hearst, E. (in press, a). Contextual control and excitatory vs. inhibitory learning: Studies of extinction, reinstatement, and interference. In P. D. Balsam & A. Tomie (Eds.), *Context and learning.* Hillside, NJ: Lawrence Erlbaum Associates.

Kaplan, P. S., & Hearst, E. (in press, b). Trace conditioning, contiguity, and context. In M. Commons, R. Herrnstein, & A. R. Wagner (Eds.), *Harvard symposium on quantitative analysis of behavior:* (Vol. 3) *Acquisition.* Cambridge, MA: Ballinger.

Kasprow, W. J. Schachtman, T., & Miller, R. R. (in press). Associability of a previously conditioned stimulus as a function of qualitative changes in the US. *Quarterly Journal of Experimental Psychology.*

Kaufman, M. A., & Bolles, R. C. (1981). A nonassociative aspect of overshadowing. *Bulletin of the Psychonomic Society, 18,* 318–320.

Kleiman, M. C., & Fowler, H. (1984). *Variations in explicitly unpaired training are differentially effective in producing conditioned inhibition. Learning and Motivation, 15,* 127–155.

Kohler, E. A., & Ayres, J. J. B. (1982). Blocking with serial and simultaneous compounds in a trace conditioning procedure. *Animal Learning and Behavior, 10,* 277–287.

Konorski, J. (1948). *Conditioned reflexes and neural organization.* Cambridge, UK: Cambridge University Press.

Konorski, J. (1967). *Integrative activity of the brain: An interdisciplinary approach.* Chicago: University of Chicago Press.

Lubow, R. E., Weiner, I., & Schnur, P. (1981). Conditioned attention theory. In G. H. Bower (Ed.), *The psychology of learning and motivation* (Vol 15). New York: Academic Press.

Mackintosh, N. J. (1973). Stimulus selection: Learning to ignore stimuli that predict no change in reinforcement. In R. A. Hinde & J. Stevenson–Hinde (Eds.), *Constraints on learning.* New York: Academic Press.

Mackintosh, N. J. (1975). A theory of attention: Variations in the associability of stimuli with reinforcement. *Psychological Review, 82,* 276–298.

Marlin, N. A. (1981). Contextual associations in trace conditioning. *Animal Learning and Behavior, 9,* 519–523.

Marlin, N. A. (1982). Within-compound associations between the context and the conditioned stimulus. *Learning and Motivation, 13,* 526–541.

Miller, R. R., & Schachtman, T. R. (in press). The several roles of context at the time of retrieval. In P. D. Balsam & A. Tomie (Eds.), *Context and learning.* Hillsdale, NJ: Lawrence Erlbaum Associates.

Owren, M. J., & Kaplan, P. S. (1981). *On the failure to extinguish Pavlovian conditioned inhibition: A test of a reinstatement hypothesis.* Paper presented at Midwestern Psychological Association Meeting, Detroit, MI.

Patterson, J., & Overmier, J. B. (1981). A transfer of control test for contextual associations. *Animal Learning and Behavior, 9,* 316–321.

Pavlov, I. P. (1927). *Conditioned reflexes.* London: Oxford University Press.

Pearce, J. M. (1984). *A model for stimulus generalisation in Pavlovian conditioning.* Manuscript submitted for publication.

Pearce, J. M., & Hall, G. (1980). A model for Pavlovian learning: Variations in the effectiveness of conditioned but not of unconditioned stimuli. *Psychological Review, 87,* 532–552.

Pearce, J. M., Kaye H., & Hall, G. (in press). Predictive accuracy and stimulus associability: Development of a model of Pavlovian conditioning. In M. Commons, R. Herrnstein, & A. R. Wagner (Eds.), *Quantitative analysis of behavior:* (Vol. 3) *Acquisition.* Cambridge, MA: Ballinger.

Randich, A., & Haggard D. (1983). Exposure to the unconditioned stimulus alone: Effects on retention and acquisition of conditioned suppression. *Journal of Experimental Psychology: Animal Behavior Processes, 9,* 147–159.

Rescorla, R. A. (1967). Inhibition of delay in Pavlovian fear conditioning. *Journal of Comparative and Physiological Psychology, 64,* 114–120.

Rescorla, R. A. (1968). Probability of shock in the presence and absence of CS in fear conditioning. *Journal of Comparative and Physiological Psychology, 66,* 1–5.

Rescorla, R. A. (1969a). Conditioned inhibition of fear resulting from negative CS–US contingencies. *Journal of Comparative and Physiological Psychology, 67,* 504–509.

Rescorla, R. A. .(1969b). Pavlovian conditioned inhibition. *Psychological Bulletin, 72,* 77–94.

Rescorla, R. A. (1971). Summation and retardation tests of latent inhibition. *Journal of Comparative and Physiological Psychology, 75,* 77–81.

Rescorla, R. A. (1979). Conditioned inhibition and extinction. In A. Dickinson & R. A. Boakes (Eds.), *Mechanisms of learning and motivation.* Hillsdale NJ: Lawrence Erlbaum Associates.

Rescorla, R. A. (1981). Within-signal learning in autoshaping. *Animal Learning and Behavior, 9,* 245–252.

Rescorla, R. A. (1982). Some consequences of associations between the excitor and the inhibitor in a conditioned inhibition paradigm. *Journal of Experimental Psychology: Animal Behavior Processes, 8,* 288–298.

Rescorla, R. A., Durlach, P. J., & Grau, J. W. (in press). Contextual learning in Pavlovian conditioning. In P. D. Balsam & A. Tomie (Eds.), *Context and learning.* Hillsdale, NJ: Lawrence Erlbaum Associates.

Rescorla, R. A., & Holland, P. C. (1977). Associations in Pavlovian conditioned inhibition. *Learning and Motivation, 8,* 429–447.

Rescorla, R. A., & Wagner, A. R. (1972). A theory of Pavlovian conditioning: Variations in the effectiveness of reinforcement and nonreinforcement. In A. H. Black & W. F. Prokasy (Eds.), *Classical conditioning II: Current research and theory.* New York: Appleton–Century–Crofts.

Riley, A. L., Jacobs, W. J., & LoLordo, V. M. (1976). Drug exposure and the acquisition and retention of a conditioned taste aversion. *Journal of Comparative and Physiological Psychology, 90,* 799–807.

Wagner, A. R., & Rescorla, R. A. (1972). Inhibition in Pavlovian conditioning: Application of a theory. In R. A. Boakes & M. S. Halliday (Eds.), *Inhibition and learning*. London: Academic Press.

Wasserman, E. A., Franklin, S. R., & Hearst, E. (1974). Pavlovian appetitive contingencies and approach versus withdrawal to conditioned stimuli in pigeons. *Journal of Comparative and Physiological Psychology, 86* 616–627.

Zimmer–Hart, C. L., & Rescorla, R. A. (1974). Extinction of Pavlovian conditioned inhibition. *Journal of Comparative and Physiological Psychology, 86*, 837–845.

3 Conditioned Inhibition from Reinforcement Reduction

N. J. Mackintosh
University of Cambridge

M. M. Cotton
University of Newcastle

THEORIES OF INHIBITORY CONDITIONING

There was a time when Western students of conditioning, by which we mean, as they themselves did in those far-off days, excitatory conditioning, spent much of their time and energy disputing whether conditioning consisted solely in the acquisition of responses to stimuli, or whether it depended on the formation of associations between stimuli. As with all disputes that drag on with no apparent sign of resolution, some people soon began to argue that the distinction had never been a real one in the first place. Such a solution is in danger of losing as much as it gains. A more sensible suggestion is that the distinction, although not always easy to draw, is real enough—but that it is not of fundamental ideological significance. A satisfactory account of conditioning, in other words, can find room for both classes of association.

One version of such an account might be that provided by Konorski (1967), who distinguished between preparatory and consummatory conditioning. Although the fit is not perfect, one could argue that preparatory conditioning corresponds to the establishment of associations between a representation of the conditioned stimulus (CS) and some set of the emotional or affective reactions elicited by the unconditioned stimulus (US) with which it has been paired. Consummatory conditioning, on the other hand, can be regarded as dependent on an association between the CS and a presentation of the sensory attributes of the US. Preparatory conditioning, then, is a form of stimulus–reponse learning—in particular of the kind appealed to by two-factor theories, whereas consummatory conditioning is a form of stimulus–stimulus learning.

According to Konorski, most conditioning paradigms will engage both proceses, although there will be variables such as the affective value of the US or the temporal interval separating CS and US that will favor one process rather than the other. One could argue, for example, that a preparation such as that routinely used to study conditioned suppression or the conditioned emotional response is well named, because it emphasizes the association between the CS and the affective reaction of fear or anxiety elicited by the US. Witness the wide range of stimuli (shocks, air blasts, loud noises) that can serve as the US to support conditioned suppression, and the good transfer between them (e.g., Bakal, Johnson, & Rescorla, 1974). Other preparations, such as the conditioned nictitating membrane response in rabbits, almost certainly produce an association between CS and a more precise sensory representation of the US (cf. Mackintosh, 1983, pp. 59ff.). Certainly, there are numerous other experiments that require one to credit animals with a more exact knowledge of the US signaled by a CS than the affective reaction it elicits. Food and sucrose pellets are not just two appetitive reinforcers capable of eliciting elation or hope; they have discriminably different sensory attributes that are clearly associated with the CSs that signal their occurrence (Trapold, 1970). And five pellets of food do not simply condition a stronger emotional reaction than that conditioned by a single pellet: It is evident that rats associate CSs paired with these USs with specifically different sensory consequences (e.g., Capaldi, Hovancik, & Friedman, 1976; Carlson & Wielkiewicz, 1976; see also the earlier work of Spear, 1967).

These are hardly novel observations and are worth making here only to stress the contrast with the case of inhibitory conditioning. Whereas many theorists would probably allow that excitatory conditioning may involve associations between the CS and both the sensory attributes of, and the reactions elicited by, the US with which it is paired, no such liberal attitude has been displayed toward inhibitory conditioning. One popular account of inhibitory conditioning, indeed, allows no associations between an inhibitory CS (hereafter CS−) and its consequences at all. According to Konorski (1948) a CS− acts by directly inhibiting (or possibly raising the threshold for activation of) the center for the US whose absence it signals. Both Rescorla (1979) and Wagner (Wagner, Mazur, Donegan, & Pfautz, 1980) have interpreted the Rescorla–Wagner (1972) model of inhibitory conditioning in these sorts of terms.

There have, of course, been associative theories of inhibitory conditioning that assume that the inhibitory properties of a CS− depend on its association (that is to say, excitatory association) with certain consequences. The most venerable Western account of inhibitory conditioning, competing-response theory, is one such. According to competing-response theory (e.g., Wendt, 1936), a CS− suppresses the responses normally elicited by a CS+ because it elicits its own responses that compete with those to the CS+. More modern descendants of competing-response theory, such as Amsel's frustration theory or Denny's relaxation theory (Amsel, 1958; Denny, 1971), have moved some way from the original contention, inspired by Guthrie (1935), that competing responses are simply overt skeletal responses physically incompatible with the CR elicited by a CS+. And in the hands of Mowrer (1960) and Dickinson and Dearing (1979), the emphasis in such theories is entirely on the central, emotional, or affective reactions elicited by a CS−. But their ancestry can still be detected. These latter accounts assume that a CS− acts by becoming associated with, and hence eliciting, the affective reaction initially elicited by the omission of an expected US. This affective reaction directly suppresses activity in the center for that US (or the emotional reaction elicited by it), thus enabling the CS− to counteract the CRs normally elicited by a CS signaling that US.

In terms of the distinction we drew earlier for the case of excitatory conditioning, these are all stimulus–response accounts of inhibitory conditioning. They assume that a CS− is associated only with the responses or reactions (overt and skeletal or covert and affective) elicited by the omission of a US. The only theorist to have questioned this assumption is Konorski himself, whose later account of inhibitory conditioning introduces what appears to be a parallel to the distinction he had drawn between preparatory and consummatory conditioning in the case of excitatory conditioning. According to Konorski (1967), a CS− is associated not only with an "antidrive" center, but also with a "no-US" center. The former may be taken as a form of preparatory conditioning and, if this were all that were being suggested, would yield a theory rather similar to Mowrer's or to Dickinson and Dearing's. But an association with a no-US center seems to constitute a form of consummatory conditioning in which a CS− evokes rather more precise knowledge of its consequences.

The question in which we are interested and to which this lengthy discussion has been leading, then, is this: What are the implications of this distinction that Konorski, and he alone, has drawn for the case of inhibitory conditioning? Is there any reason to believe that a CS− is associated with a no-US center or to question the assumption of other associative theories that a CS− evokes only emotional or overt competing responses? The nature of a no-US center is not immediately apparent and Konorski himself is not particularly helpful. It may be easier therefore if we start by considering what appear to be the restrictions on an animal's knowledge about a CS− implied by all other theories of inhibitory conditioning. Our understanding is that such knowledge must be confined to information about the broad class (appetitive or aversive) to which the omitted reinforcer belongs, and the magnitude of the reduction in reinforcement signaled by the CS−.

Theories such as Mowrer's or Dickinson and Dearing's, and by implication Amsel's and Denny's, all distinguish between two classes of emotional reaction potentially elicited by a CS−, between frustration or disappointment on the one hand and relaxation or relief on the other. These are presumed to be generated by the omission of appetitive and aversive USs, respectively, and this distinction in affective reaction therefore provides information whether the omitted US was appetitive or aversive. But there is no provision for any more precise knowledge of the nature of the omitted US. If the US was appetitive, the organism remains in ignorance whether it was food, water, sucrose solution, or a mate; or if the US was aversive, whether it was shock (and if so where it was delivered), a puff of air to the cornea, immersion in cold water, sickness, or nausea.

The only other source of information allowable by such theories must be provided by the intensity of the conditioned emotional reaction. At first sight, this would seem to be an appropriate correlate of the magnitude or intensity of the omitted US. But this is not quite accurate. The intensity of conditioned frustration will in fact be correlated not with the magnitude of an omitted reward, but with the magnitude of the reduction in reward signaled by a CS−. Thus a stimulus signaling a reduction from two pellets of food to none would presumably be indistinguishable, in terms of its conditioned properties, from one signaling a reduction from 10 pellets to some smaller number. Of course, we do not know what precise number would generate an equivalent reduction—that remains a matter for empirical enquiry. The point is that there must be *some* reduction that would produce a level of conditioned frustration indistinguishable from that produced by the reduction from two pellets to zero.

There appear, therefore, to be several testable implications of the S–R assumption that a CS− is associated only with the reactions elicited by the omission of a US, whether those reactions be overt competing responses or covert emotional states. Some of these implications, indeed, seem rather striking. It does not seem particularly plausible to suppose that a CS− provides no more knowledge of the nature of the omitted US than whether it was appetitive or aversive, and Konorski's assumption that the CS− is associated with a no-US center very clearly implies the opposite. But there is little or no experimental evidence that would allow one to decide the question. Some of these, and related, issues are considered by other contributors to this volume (e.g., by LoLordo and Fairless and by Miller and Schachtman).

The specific question that we have attempted to answer is whether a CS− provides information only about the extent of the reduction in reinforcement that it signals, or whether animals also have some knowledge of the absolute level to which that reduction is taken. If, for example, rats were trained on a discrimination between a tone signaling the availability of five pellets of food and a tone–light compound signaling a single pellet, does the light simply cancel the expectation of five pellets signaled by the tone or elicit a conditioned frustration reaction? Or is it, on the other hand, associated with a representation of the one pellet actually delivered in its presence?

EVIDENCE OF INHIBITORY CONDITIONING FROM REINFORCEMENT REDUCTION

Indirect Evidence

A variety of experiments has suggested that animals will learn discriminations between stimuli signaling different magnitudes or probabilities of reinforcement even when neither stimulus signals the complete omission of reinforcement. And some of these studies have pointed to the possible involvement of inhibitory conditioning to the stimulus signaling the lower, but nonzero value. Thus pigeons will learn a discrimination between key lights correlated with different variable-interval schedules of food reinforcement, and subsequent generalization tests reveal a U-shaped gradient of responding around the stimulus signaling the lower value (e.g., Weisman, 1969). Rats will discriminate between different runways in one of which running is rewarded with eight pellets and in the other with only one pellet, and their speed of running for this single pellet is less than that displayed by rats receiving a single pellet in both runways (e.g., Bower, 1961). And Daly (1974) has summarized a number of studies that suggest that a reduction from a large to a small number of food pellets will generate a variety of effects in rats normally attributed to conditioned frustration.

Although the results of these studies are consistent with the suggestion that a reduction from a high to a lower, but nonzero, value of reinforcement is sufficient to produce inhibitory conditioning to the stimulus signaling that reduction, the criteria they employed for assessing such effects are rather different from the summation and retardation tests widely accepted as defining the concept of inhibitory conditioning in studies of classical conditioning (Hearst, 1972; Rescorla, 1969). Our first task, therefore, was to see whether a stimulus signaling a reduction in magnitude of reinforcement would indeed act as a conditioned inhibitor at least on a standard summation test. Only if we were successful in this initial endeavour could we proceed to the further task of elucidating what animals learned about this stimulus.

Evidence from Summation Tests

We used two quite different conditioning preparations: classical aversive conditioning, the conditioned suppression of lever pressing; and a discriminated operant, lever pressing for food rewards on variable-interval schedules. The basic design of the experiment was the same in both cases. A 2-KHz tone signaled the occurrence of a strong shock in the first study or the availability of five pellets of food in the second. Such trials alternated, for the discrimination group (called Group TL+), with trials to a tone–light compound that signaled a weaker shock or only one pellet of food. Once the discrimination between tone and tone–light compound had been learned in the sense that animals' rates of responding on the two kinds of trial differed reliably, they were then trained with a third stimulus, a 10-Hz train of clicks that signaled the same reinforcer as that signaled by the tone alone. They finally received a series of nonreinforced trials to the clicker alone and to the clicker–light compound, this constituting a summation test of conditioned inhibition to the light.

The procedures employed in the study of conditioned suppression were standard ones and the study has been published elsewhere (Cotton, Goodall, & Mackintosh, 1982). It is unnecessary to dwell on the details, other than to mention that the strong shock signaled by tone and clicker was 1.0 ma for 0.5 sec (from a Grason–Stadler generator), and the weak shock signaled by the tone–light compound was 0.4 ma for 0.5 sec. There were also two further groups in the study treated identically to the discrimination group TL+ described previously, except on trials when the latter were exposed to the tone–light compound signaling the weak shock. Both of these additional groups received the same weak shock on these trials, but for one (Group T+) they were signaled by the same tone that occurred on strong-shock trials, whereas for the other (L+) they were signaled by the light alone. For Group T+, therefore, the light was novel at the time of the summation test with the clicker (except for two nonreinforced preexposure trials); whereas for Group L+, the light, although paired with the weak shock, had not signaled the absence of the stronger shock otherwise signaled by the tone. In effect, Group T+ provides a control for the unconditioned effects of the light, whereas Group L+ is a conservative control that may be expected to show less inhibitory conditioning to the light than Group TL+.

Suppression ratios to the clicker and to the clicker–light compound are shown in Fig. 3.1. There was no alleviation of suppression to the clicker by the presentation of the light in Groups T+ and L+, but a clear and statistically significant effect in Group TL+. One of the standard procedures for producing inhibitory conditioning to a stimulus that signals the absence of reinforcement, and one of the standard procedures for assessing that inhibitory conditioning, produce quite standard results even when the stimulus signals no more than a reduction in the magnitude of a reinforcer.

The results of the appetitive counterpart of this study were very similar. The procedure employed was a compromise between a free-operant multiple schedule and a discrete-trial discrimination. Trials were separated by intertrial intervals varying between 60 and 90 sec. The start of a trial, marked for example by the

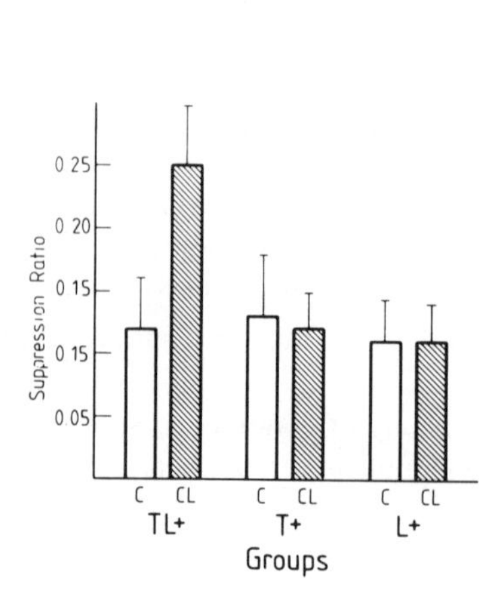

FIG. 3.1. Results of summation tests for conditioned inhibition to a clicker (C) previously paired with a strong shock and a clicker–light compound (CL). Group TL+ had been trained on a discrimination between tone signaling strong shock and tone–light compound signaling weak shock; for Group L+ the weak shock had been signaled by the light alone; and for Group T+ the light was relatively novel. The scores shown are suppression ratios of the form (Rate of responding to CS) ÷ (Rate of responding to CS + Rate of responding in 30 sec immediately preceding CS). A score of .00 therefore indicates complete suppression and one of .50 no suppression.

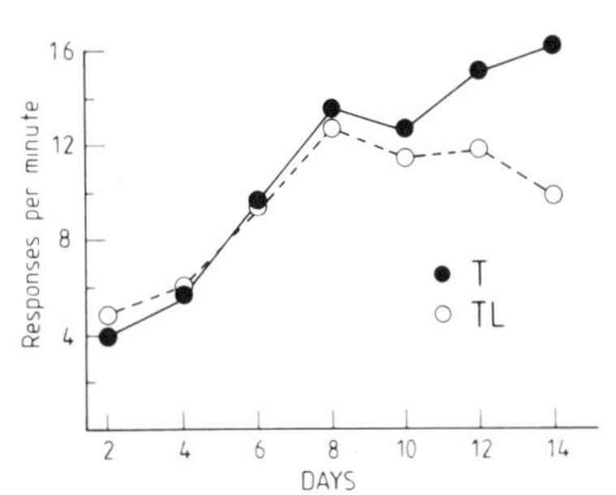

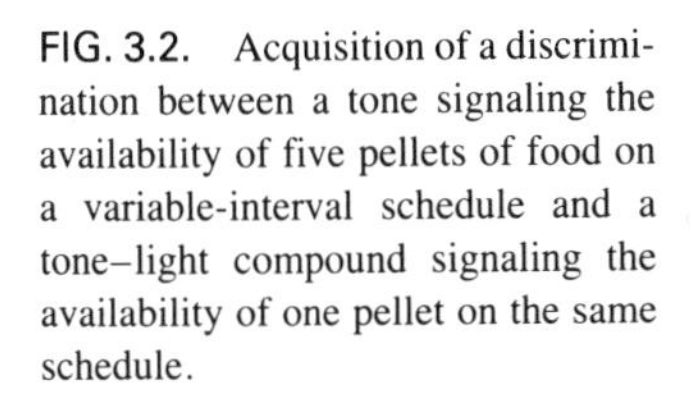

FIG. 3.2. Acquisition of a discrimination between a tone signaling the availability of five pellets of food on a variable-interval schedule and a tone–light compound signaling the availability of one pellet on the same schedule.

onset of the tone, set into effect a variable-interval 40-sec schedule of reinforcement, but the lever press that earned the first reinforcer set up by the schedule also terminated the trial. Trials were thus of variable length and ended after a single reinforcer (in the absence of a response the trial terminated and the reinforcer was cancelled 30 sec after it was first set up). The intended virtues of this procedure were that it should allow a measure of rate of responding over relatively long periods, uncontaminated by the time taken to consume a reinforcer and by any transient satiating effects its consumption might have, and that it should reduce the extent to which traces of one reinforcer on a trial could serve as a discriminative stimulus signaling the reinforcement contingencies in effect for the remainder of the trial.

After preliminary training[1] 24 rats were divided into 3 groups of 8 for 14 days of discrimination training. Each day's session consisted of 8 trials, signaled by the tone, on which a large reward (5 pellets) was available on a variable-interval 40-sec schedule, and 24 trials on which a single-pellet reward was available according to the same schedule. These small reward trials were signaled by the tone–light compound (Group TL+), by the tone alone (Group T+), or by the light alone (Group L+). At the completion of this phase of the experiment, all animals received three sessions of training in which the clicker signaled the availability of the large reward. There were 12 trials in each of these sessions. Finally, animals received a series of nonreinforced test trials to the clicker and clicker–light compound: there were 8 trials of each type presented in a double alternating sequence and each trial was 1 min long.

The course of discrimination learning by Group TL+ is shown in Fig. 3.2, and the results, for all three groups, of test trials to the clicker and clicker–light

[1]Preliminary training consisted of the following phases: 2 sessions of magazine training with both 5- and 1-pellet rewards; 1 session of consistently reinforced lever-press training; and 3 sessions in which lever pressing was reinforced on an ordinary variable-interval 30-sec schedule (1-pellet reward). In the first and second of these sessions, there was no intertrial interval and the tone sounded continuously in one session and the clicker in the second; in the final session, there were sixteen 60-sec trials separated by intervals of 20 sec, with the trials and thus the availability of food being signaled by the tone, the clicker, and the tone–light and clicker–light compounds (4 of each kind of trial).

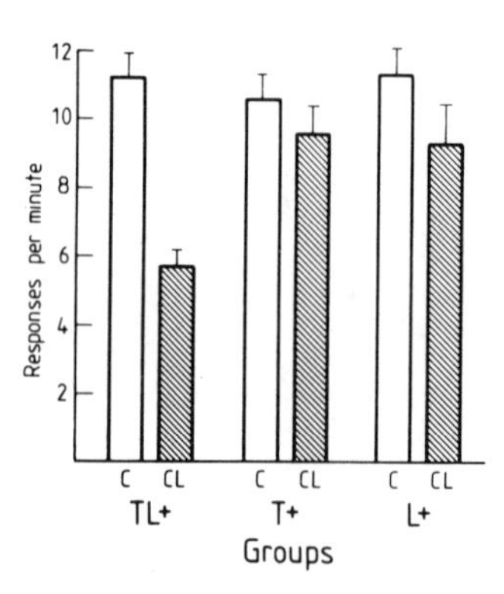

FIG. 3.3. Results of summation tests for conditioned inhibition to a clicker (C) previously signaling five pellets of food and a clicker–light compound (CL). Group TL+ had been trained on the discrimination shown in Fig. 3.2; for Group T+ the light was novel; and Group L+ had learned a discrimination between a tone signaling five pellets and a light signaling one pellet.

compound, in Fig. 3.3. Not surprisingly, the single-pellet reward maintained a reasonable rate of responding to the tone–light compound throughout discrimination training, and the discrimination between tone and tone–light may not seem very impressive. All animals, however, responded more rapidly to the tone than to the tone–light on the last day of training, and all but one animal had done so for the preceding 2 days. In Group T+, by contrast, five of eight animals were responding more rapidly on small-reward than on large-reward trials on the last day of training. Even Group L+ did not show very much better evidence of discrimination.

Further evidence that Group TL+ had learned that the light signaled a reduction in reward is provided by the test results shown in Fig. 3.3. In this group, and in this group only, the addition of the light very substantially suppressed the level of responding otherwise maintained by the clicker. There is some suggestion of suppression in Group L+, but little or none in Group T+. An analysis of the difference scores (between clicker and clicker–light trials) revealed a significant group effect ($F^{2,21} = 7.68$, $p < 01$), and Newman–Keuls tests revealed that Group TL+ differed from the other two groups that did not differ from one another.

The results of this experiment, therefore, are in close agreement with those of the earlier study on conditioned suppression. A visual stimulus signaling a reduction in the magnitude of reinforcement, but not the complete omission of reinforcement, can act as a conditioned inhibitor on a summation test when it is presented in conjunction with another stimulus previously established as a signal for a large reinforcer.

EVIDENCE OF LEARNING ABOUT THE REINFORCER PAIRED WITH A CS– SIGNALING A REDUCTION IN REINFORCEMENT

We are now in a position to ask whether the light is acting as a signal only for this reduction, or whether it might also be associated with some representation of the actual events that occur in its presence. Is a reduction from five pellets to one, for example, simply a less frustrating experience than a reduction from five to zero, and does the light that signals one simply elicit a weaker frustrative reaction than the light that signals the other? Or is there any suggestion that animals learn not only that a stimulus can signal a given degree of reduction in reinforcement but also to what level that reduction is taken?

Preliminary Experiments

Our initial intuition was that, if animals trained on the discrimination between tone and tone–light compound had learned that the light signaled both a reduction in reinforcement and a reduction to a specific small reinforcer, the light, although capable of suppressing the conditioned responding maintained by the clicker when it was established as a CS for the large reinforcer, would be less likely to act as an inhibitor if the clicker had itself been paired with the same small reinforcer as that previously signaled by the tone–light compound. This expectation was tested in a pair of companion experiments, identical in all respects to the two already reported (indeed run, in two replications, at the same time as those two), save that after the completion of discrimination training the clicker was paired with the small reinforcer (0.4 ma or 1 pellet) instead of the large. In both experiments an additional group, TL−, was trained on a discrimination between the tone signaling the large reinforcer and the tone–light compound signaling the complete omission of reinforcement. The purpose of this group was to determine whether evidence of inhibition could be obtained on a summation test when the clicker itself might be expected to maintain only modest levels of conditioning. The only other change was that in the discriminated-operant experiment, owing to a shortage of apparatus, Group L+ was dropped.

The results of major interest in both experiments, those of test trials to the clicker and clicker–light compound, are shown in Figs. 3.4 and 3.5. Although the details of the results are slightly different, they suggest very much the same conclusion: There was no evidence that the light, in Group TL+, acted as a

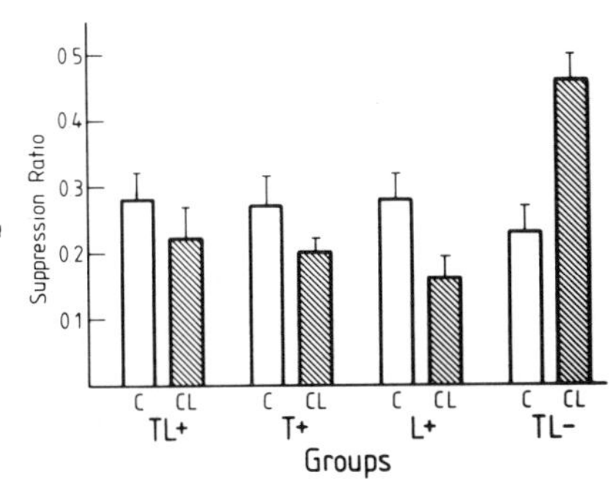

Fig. 3.4. Results of summation tests for conditioned inhibition to clicker and clicker–light compound, when the clicker had previously been paired with weak shock. Groups as in Fig. 3.1; Group TL− had been trained on a discrimination between tone paired with strong shock and tone–light compound paired with no shock.

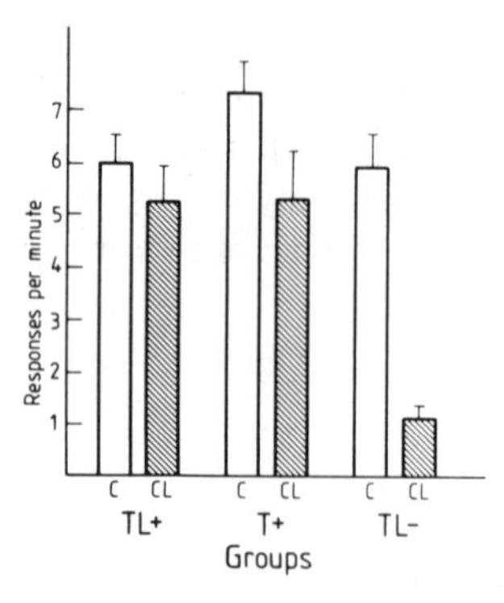

FIG. 3.5. Results of summation tests for conditioned inhibition to clicker and clicker–light compound when the clicker had previously signaled the availability of one pellet of food. Group designations are the same, *mutatis mutandis,* as in Fig. 3.4.

conditioned inhibitor. In the experiment on conditioned suppression, the clicker maintained a moderate level of suppression on its own, and this suppression was attenuated by the light only in Group TL−. In the discriminated-operant experiment, the clicker maintained moderate rates of lever pressing, which were suppressed to various degrees by the light in all three groups. An analysis of variance showed that the three groups differed in the extent to which the light suppressed responding to the clicker ($F^{2,21} = 7.05$, $p < .01$), and the Newman–Keuls tests showed that this suppression was most marked in Group TL−, and that it did not differ in Groups TL+ and T+. There is no reason to believe, therefore, that the difference in rate of responding to clicker and clicker–light compound observed in Group TL+ represents anything more than the unconditioned disruptive effects of the light on the relatively fragile level of responding maintained in extinction by the clicker.

Alternative Analysis

The results shown in Figs. 3.4 and 3.5 are essentially negative. They provide no evidence of inhibition in that rates of responding to clicker and clicker–light compound are the same in animals for whom the light signaled a reduction in magnitude of reinforcement and those for whom the light was relatively novel. But although the performance of Group TL− shows that it is possible to obtain unequivocal inhibitory results from our testing procedures, someone might argue that these procedures remain too insensitive to pick up the inevitably smaller effect that any theory would have expected to see in Group TL+. One problem for such an account is to explain just why it should be more difficult to detect an inhibitory effect of the light in Group TL+ when the clicker is associated with a small reinforcer than when it is associated with a larger one (as in Figs. 3.1 and 3.3). All that we know about ceiling effects suggests, on the contrary, that it should have been easier rather than harder to detect a disruption of the intermediate levels of conditioning maintained by the clicker in Figs. 3.4 and 3.5. Nevertheless, these doubts should not be dismissed out of hand, and they are strengthened by the results of a study of conditioned suppression in rats by Wagner et al. (1980) that, employing much the same general design as we had, found significant evidence of inhibition. In Wagner et al.'s experiment, rats were initially conditioned to two stimuli, A and C, associated respectively with a strong (3.0 ma) and a weak (0.6 ma) shock. They then received discriminative conditioning between A still paired with the 3.0-ma shock and A + B paired with the 0.6-ma shock. Test trials to C and C + B, interspersed with this discriminative conditioning, revealed that B gradually acquired the capacity to attenuate the suppression maintained by C alone.

There are several major differences between Wagner et al.'s procedures and ours. One that seemed as if it might be important was that in their experiment conditioning of C, the CS+ paired with the weak shock, preceded discriminative conditioning between A and A + B. In our experiment, however, conditioning to the clicker (the equivalent of C) followed discriminative conditioning to tone and tone–light (A and A + B). The results of a further experiment (Cotton et al., 1982, Experiment 5) suggested that this may well have been the most important single difference. This new experiment was modeled relatively closely on our earlier study; it contained two groups, differing only in the order in which they received conditioning to the clicker (paired with weak shock) and discriminative conditioning to tone (paired with strong shock) and tone–light compound (paired with weak shock). On this occasion, when, as before, conditioning to the clicker followed discriminative conditioning, the light significantly *increased* the rather modest level of suppression maintained by the clicker; when conditioning to the clicker preceded discriminative conditioning, the light had no effect. Although we were unable to replicate Wagner et al.'s observation that the light could significantly attenuate the suppression maintained by the clicker, this marked difference in outcome was consistent with the suggestion that it might be easier to find such an effect with their order of conditioning trials rather than ours.

Why might this be so? One possibility is that we should recognize that the subject's estimate of the magnitude of the reinforcer may not always perfectly match the reinforcer actually delivered by the experimenter. The subjective intensity of shock, it has often been supposed, may decline with repeated exposure as the subject "habituates" to the shock (Randich & LoLordo, 1979). If this happened, the order in which subjects experienced the clicker and the tone–light compound might affect the subjective intensity of the shock associated with each. If conditioning to the clicker came first, animals would associate it with a subjectively stronger shock than that associated with the light, and it would not be surprising, therefore, if the light was able to alleviate the suppression maintained by the clicker.[2]

---

[2]An alternative possibility, suggested by Miller and Schachtman's comparator hypothesis (this volume), is that the order of trials might have affected the associative value of contextual cues. When

Subjective estimates of magnitude of reinforcement may also be affected by contrast effects. Relatively rapid alternation between large and small rewards may generate contrast effects that exaggerate the apparent difference between them (Flaherty, 1982). There is reason to believe that such an effect may have occurred in our discriminated operant experiment where, unlike the conditioned suppression experiments, trials to tone signaling a large reward and tone–light compound signaling a small reward were interspersed in the same session. A contrast effect would have decreased the subjective magnitude of the reward associated with the light, thus possibly increasing the chances of observing an inhibitory effect when the light was compounded with the clicker. In a further, small-scale study, following up our earlier discriminated operant experiments, we found that, if trials to the tone and tone–light compound were given in separate sessions, discriminative conditioning proceeded much more slowly, and that the light then tended to increase rather than decrease the rate of responding maintained by a clicker also associated with the small reward. A contrast effect would also explain why, in Fig. 3.3, Group L+ shows some inhibitory effect of the light when it is compounded with the clicker paired with the large reward.

Further Experiments

Although our initial experiments provided evidence entirely consistent with the idea that animals learn that a CS− may signal a specific small reinforcer rather than, or in addition to, learning that it signals a reduction in reinforcement, that evidence is essentially negative. It rests on a failure to find inhibitory effects in Group TL+ when the light is presented in conjunction with a clicker that had itself been paired with the smaller reinforcer. The argument that the light is associated with some representation of the small reinforcer actually occurring in its presence would be greatly strengthened by additional positive evidence.

Direct Measurement of Responding to CS− On the face of it, the simplest possibility would seem to be, having given discriminative conditioning between the tone signaling a large reinforcer and the tone–light compound signaling a small reinforcer, to measure responding to the light alone. We have indeed done this, but the results, although generally consistent with our expectations, are not

---

conditioning to the clicker paired with the weak shock intervened between discriminative conditioning and the final test, a long interval separated test trials from any conditioning trial on which a strong shock had been delivered. This may have been sufficient to reduce the associative value of contextual cues and thus to increase that of the light *relative to* those contextual cues. Continued reinforcement of the tone with a strong shock up to the moment of testing, on the other hand, may have maintained a higher associative value for the contextual cues and thus decreased that of the light *relative to* those cues. The hypothesis assumes that the inhibitory effect of the light is determined by its associative value relative to that of other cues.

very convincing. There appear to be at least two problems: One is that the level of responding may be too low to be readily detected; the other is to find an appropriate control group against which to assess any behavioral effects of excitatory conditioning to the light in Group TL+. We chose to compare Group TL+ with Group T+, i.e., animals for whom the light was relatively novel, having been seen only on preexposure trials before the start of conditioning. In a conditioned suppression experiment, we found only very slight evidence of greater suppression on the first trial to the light alone in Group TL+ than in Group T+: The suppression ratios were .37 and .43, respectively, and the difference was not significant. But here the novelty of the light for Group T+ may have produced some unconditioned suppression. In a companion experiment employing the discriminated operant procedure, we found a somewhat larger, and this time statistically significant, difference in rate of responding to the light alone: Over four 1-min nonreinforced test trials, Group TL+ responded at a rate of 3.67 responses per minute, whereas Group T+ responded at a rate of 0.92 responses per minute ($p < .05$). But here the novelty of the light for Group T+ may have worked in our favor by producing a spurious suppression of responding.

Summation with an Extinguished CS+. In an attempt to surmount at least one of these problems, specifically that associated with the low levels of conditioned responding observed in both these experiments, we returned to our earlier reliance on summation tests in which the light is presented in conjunction with an auditory stimulus, but in this case an auditory stimulus that although initially associated with the large reinforcer undergoes extinction before the final test.

We used the discriminated operant procedure. After the usual preliminary training, animals were divided into 2 groups of 12. As before, Group TL+ was trained on the discrimination between tone signaling 5 pellets and tone–light compound signaling 1 pellet, receiving 4 trials to the tone and 16 to the tone–light compound each day for 16 days. Each session also contained 4 trials on which the clicker signaled the availability of the large reward. Group T+ was treated identically except that the 16 small rewards were signaled by the tone alone rather than by the addition of the light to the tone. At the end of this phase, both groups received 4 days of extinction to 1 of the 2 auditory stimuli. For half the animals in each group the extinguished stimulus was the tone; for the other half it was the clicker. For both groups, each extinction session consisted of 12 nonreinforced trials to the stimulus being extinguished interspersed with 4 reinforced trials to the other auditory stimulus. By the end of this phase, all animals were responding more rapidly to the stimulus that was still being reinforced than to that which was being extinguished.

The final phase of the experiment consisted of a single test session containing 16 nonreinforced test trials, 4 to the tone, 4 to the clicker, and 4 each to the tone–

light and clicker–light compounds. The expectation was that in Group TL+ the light might act as an inhibitor to suppress the responding maintained by the reinforced auditory stimulus, but as an excitor to elevate rate of responding to the extinguished auditory stimulus. The results, shown in Fig. 3.6 averaged across tone and clicker subgroups, confirm this expectation. Both Groups TL+ and T+ responded more to the reinforced than to the extinguished auditory stimulus; but although the addition of the light had little or no effect on the rate of responding to these stimuli in Group T+, it had significant effects in Group TL+, suppressing responding to the reinforced stimulus, but elevating responding to the extinguished stimulus. The critical statistical comparisons were between the difference scores (between auditory stimulus alone and auditory–visual compound) of Groups TL+ and T+. When the light was compounded with the unextinguished auditory stimulus, it suppressed responding significantly more in Group TL+ than in Group T+ ($t^{22} = 3.29, p < .01$); when compounded with the extinguished auditory stimulus, the light led to a greater increase in rate of responding in Group TL+ than in Group T+ ($t^{22} = 2.96, p < .01$).

Summation with a CS+ Associated with an Even Smaller Reinforcer. In a final experiment, we used a different procedure in an attempt to provide the light with greater positive value than the clicker with which it was eventually compounded on test. Instead of extinguishing the clicker, we paired it from the outset with a reinforcer of even less value than that signaled by the light. Rats were trained on a discrimination between tone signaling a high value of reinforcement and tone–light compound signaling a smaller, intermediate value. The clicker was then independently associated with a yet lower value, and the effect of the light on responding to the clicker was assessed in a final series of test trials. The

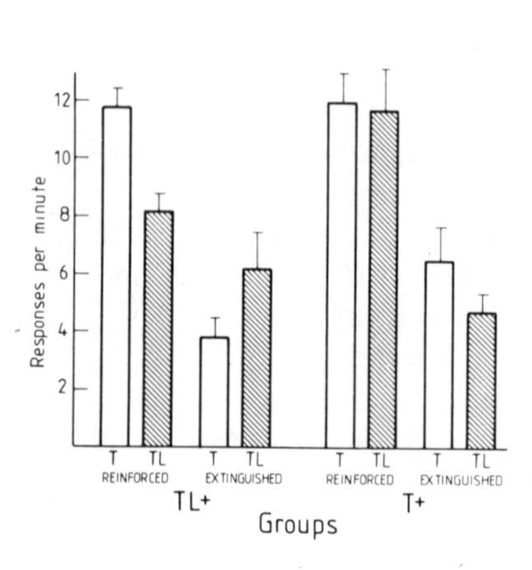

Fig. 3.6. Results of summation tests to tone, clicker, and tone–light and clicker–light compounds. Before testing, all animals received extinction trials to either the tone or the clicker: Responding in test to the extinguished stimulus, either alone or in compound with the light, is shown separately from test responses to the unextinguished or reinforced stimulus. To save space the auditory stimulus alone is referred to as T (whether it was tone or clicker) and the compound as TL (whether it was tone plus light or clicker plus light). In Phase 1, Group TL+ had been trained with tone signaling five pellets and tone–light one, whereas for Group T+ both five-pellet and one-pellet trials had been signaled by the tone.

expectation was that the light should enhance the level of conditioned responding maintained by the clicker even though it had suppressed the responding maintained by the tone.

The experiment required us to find a source of reinforcement with three different quantities or values between which the rats could readily discriminate, that is which would support three clearly differentiable levels of responding. We doubted whether using our procedures, different levels of shock, or different numbers of pellets of food reward would avoid the problems of ceiling or floor effects, and therefore we used different concentrations of sucrose solution as the reward for lever pressing by hungry rats. The general procedure was much the same as that of the discriminated operant experiments varying size of reward. Rats pressed a lever for access to a 0.1-ml dipper of sucrose solution; discrete trials, separated by 60- to 90-sec intertrial intervals, were marked by the onset of a tone, tone–light compound, or clicker, and signaled the availability of sucrose on a variable-interval 40-sec schedule; the trial terminated as soon as the first reinforcer set up by the schedule had been collected or, in the absence of a response, 30 sec after it had been set up. Because contrast effects appear to be rather marked when rats are exposed to rapid alternation between different concentrations of sucrose (Flaherty, 1982), the main difference from the earlier experiments on reward size was that a given session contained trials with only one type of reinforcer. After preliminary training, rats received two 15-trial sessions each day, separated by about 2 hours.

In a preliminary study, 8 rats received magazine and lever-press training followed by 7 sessions in which the tone signaled the availability of 20% sucrose solution, followed in turn by a further 40 sessions on half of which the tone continued to signal the 20% solution and on the other half the tone–light compound signaled the availability of an 8% solution. There was 1 session of each type each day. Once this discrimination had been learned, they were given 4 sessions of conditioning to the clicker signaling the 20% solution followed by a single test session consisting of 16 nonreinforced trials each 60-sec long, half to the clicker and half to the clicker–light compound. The aim of this study was to establish first that animals would respond at different rates to the tone and tone–light compound when these signaled 20% and 8% solutions, and secondly that the light would then suppress the responding maintained by a clicker that had also been associated with the 20% solution. The results are shown in Fig. 3.7. Both aims were satisfied. Averaged over the last 4 days of training, all eight animals responded faster to the tone alone than to the tone–light compound. On test trials, the addition of the light reduced the level of responding maintained by the clicker ($t_7 = 2.31$, $p < .05$, 1-tail). Although there is no control group comparable to Group T+ in earlier experiments, the results from the next experiment, shown in Fig. 3.8, suggest that the light will not necessarily have this effect in all cases.

The main experiment followed exactly the same general procedure: All animals learned a discrimination between tone and tone–light compound, were then

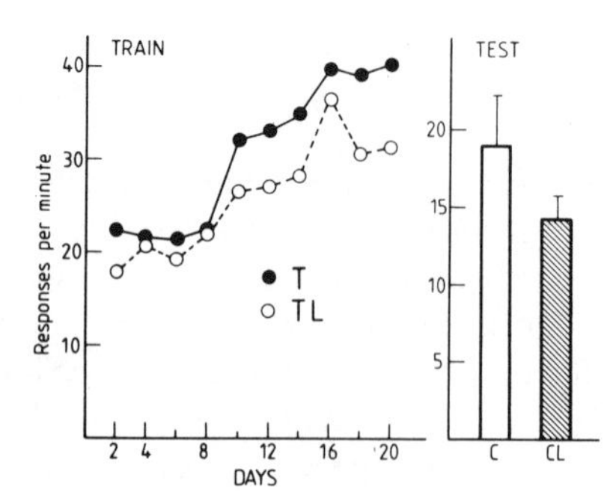

FIG. 3.7. The left-hand panel shows acquisition of a discrimination between a tone signaling the availability of a 20% sucrose solution and a tone–light compound signaling an 8% solution. The right-hand panel shows the results of a summation test to clicker and clicker–light compound when the clicker had also signaled the 20% solution.

conditioned to the clicker, and finally received test trials to clicker and clicker–light compound. There were three groups of eight rats each, differing only in the consequence signaled by the tone–light compound during discriminative conditioning. For Group TL 8, this was the intermediate 8% solution; for Group TL 2, it was a weak 2% solution; and for Group TL 0, the tone–light compound signaled no reinforcement at all. For all three groups, in the following phase of the experiment the clicker signaled the availability of the weak 2% solution.

All three groups learned to discriminate between tone and tone–light compound, although, as one would expect, Group TL 8 maintained a higher rate of responding on tone–light trials than Group TL 2, which in turn continued to respond more rapidly than Group TL 0. Over the final 4 days of training the average difference in speed of responding to tone and tone–light compound was 6.4, 21.1, and 26.0 responses per minute in Groups TL 8, TL 2, and TL 0, respectively. No animal in any group responded faster to the tone–light than to the tone alone over these 4 days. The critical results are those shown in Fig. 3.8: It can be seen that the light suppressed responding to the clicker in Group TL 0, had relatively little effect in Group TL 2, and accelerated responding in Group TL 8. An analysis of variance on the difference scores (between clicker and clicker–light trials) revealed a significant effect of groups ($F^{2,21} = 9.69$, $p < .01$), and Newman–Keuls tests established that each group differed from the other two.

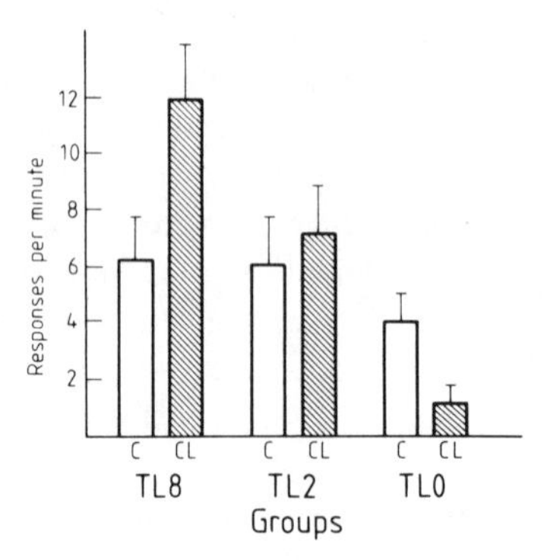

FIG. 3.8. Results of summation tests to clicker and clicker–light compound. For Groups TL 8, TL 2, and TL 0, the tone–light compound had signaled, respectively, the availability of 8% or 2% sucrose solution or no sucrose at all. For all groups the clicker signaled the availability of a 2% sucrose solution.

The results for Groups TL 2 and TL 0 essentially confirm, for the case of sucrose reinforcers, the conclusions drawn from our earlier experiments on intensity of shock and size of reward: The responding maintained by a clicker paired with a weak reinforcer is disrupted by a light that has previously signaled the complete omission of reinforcement, but a light that has signaled a reduction from a stronger reinforcer to the same weak one subsequently associated with the clicker has little or no effect on responding to the clicker. The new finding is that such a light can significantly increase rate of responding to a clicker associated with a yet weaker reinforcer. If we compare the results of Group TL 8 with those obtained in the preliminary study (and although described as preliminary, animals were actually run at the same time), we can say that the same light signaling the same consequence can act either as an inhibitor or as an excitor of responding to the clicker, depending on whether the clicker is associated with a stronger or with a weaker reinforcer than that signaled by the light.

THEORETICAL IMPLICATIONS

The evidence presented here confirms that rats will learn a discrimination between different intensities or magnitudes of reinforcement, and that a stimulus signaling a reduction from a large to a small reinforcer will act as a conditioned inhibitor on a summation test by suppressing the responding maintained by another stimulus associated with the large reinforcer. There is nothing very surprising about this finding: It confirms a central implication of most modern theories of inhibitory conditioning, one made quite explicit in Rescorla and Wagner's (1972) formal model of conditioning, that inhibitory conditioning is generated by any negative discrepancy between expected and obtained reinforcement.

Our concern has been to consider a rather different question that may be asked about inhibitory conditioning: What does an animal learn about a CS− that signals this reduction from large to small reinforcer? Our understanding of most accounts of inhibitory conditioning is that they regard such a CS− simply as a weaker inhibitor than one that had signaled the complete omission of the large reinforcer. Thus a CS− signaling a reduction from, say five pellets to one, although a weaker inhibitor than one signaling the complete omission of the five pellets, might actually be a stronger inhibitor than one signaling the omission of one pellet. This seems to be the implication of the large majority of modern theories of inhibitory conditioning, both of those like Konorski's (1948) account that a CS− acts by cancelling an expectation of reinforcement, and of those such as Mowrer's (1960) and Dickinson and Dearing's (1979) theories, which assume that a CS− elicits an affective reaction opposite in sign to that elicited by a CS+ signaling the occurrence of reinforcement.

In both cases, inhibitory conditioning is assumed simply to endow a CS− with a greater or lesser capacity to counteract a CS+, and the extent of this capacity is a function only of the magnitude of the negative discrepancy it signals. Insofar as it addresses this kind of question, the Rescorla–Wagner model makes this quite clear. In the equation,

$$\Delta V_A A = \alpha,\beta\ (\lambda - \bar{V}),$$

inhibitory conditioning to CS_A occurs whenever λ is less than $\bar{V}$, and the asymptotic inhibitory value of CS_A will therefore depend solely on the quantities λ and $\bar{V}$. At the start of conditioning to CS_A, the value of $\bar{V}$ will be determined by the associative value of other stimuli present, in general being greater the larger the reinforcer associated with them. Thus the strength of inhibitory conditioning depends directly on the magnitude of the omitted reinforcer. The value of λ is determined by the nature of the reinforcing event occurring on a conditioning trial, being conventionally set at zero when no reinforcer occurs and at increasing positive values as the magnitude of the reinforcer increases. Thus the strength of inhibitory conditioning to CS_A will be an inverse function of the magnitude of the reinforcer occurring in its presence. But that reinforcer acts simply by decreasing the magnitude of the discrepancy between λ and $\bar{V}$.

Our results do not accord with this view, for they suggest that when a light signals a reduction from a large to a small reinforcer, it is associated with some representation of the small reinforcer occurring in its presence. Our initial observation that the light failed to inhibit responding to another stimulus previously paired with the same small reinforcer was certainly consistent with this conclusion, although not perhaps forcing its acceptance. But our subsequent experiments make it difficult to reject it, because it is apparent that the light can simultaneously act to suppress responding when presented with a stimulus previously associated with a larger reinforcer, but to increase conditioned responding when presented in conjunction either with an extinguished stimulus or with one previously associated with a yet smaller reinforcer. Such positive effects imply a positive, excitatory value for the light.

We should stress, however, that the inhibitory conditioning resulting from a reduction in magnitude of reinforcement is not easily understood solely in terms of an excitatory association between the CS− and the reinforcer occurring in its presence. A glance at the data shown in Figs. 3.1 and 3.3 should make it clear why this suggestion raises as many problems as it solves. In these experiments, the light was tested in conjunction with a clicker that had signaled the large reinforcer. In Group TL+, trained on a discrimination between the tone signaling the large reinforcer and the tone–slight compound signaling the small reinforcer, the light acted as an effective conditioned inhibitor. But in Group L+, trained on the discrimination between tone signaling large and light signaling small reinforcer, the light had little or no inhibitory effect. There can be no reason to doubt that the light was associated with the small reinforcer in Group

L+, but it is apparent that such an association is not sufficient to reproduce the inhibitory effect seen in Group TL+. As others have found before us (e.g., Rescorla & Holland, 1977), more profound inhibitory conditioning is produced by a stimulus that signals the absence of reinforcement when presented in conjunction with a CS+ otherwise predicting reinforcement than by a stimulus simply presented on its own without reinforcement.

We have evidence, then, that a light signaling a reduction in the magnitude of reinforcement is associated with the smaller reinforcer occurring in its presence, but that this association is not sufficient to explain its inhibitory properties. How is this apparent contradiction to be resolved? There are at least two possible answers, one rather more radical than the other. The more conservative supposes that in the conditions experienced by Group TL+, the light becomes both an inhibitor, signaling a negative discrepancy between expected and obtained reinforcement, and an excitor, signaling the occurrence of the small reinforcer. This is, in effect, the interpretation we should like to put on Konorski's (1967) account of inhibitory conditioning, with its distinction between the associations formed by a CS− with an antidrive center and those with a no-US center. The former, preparatory association with the affective reaction generated by the reduction in reinforcement provides the inhibitory effect, and the latter, consummatory association (here with a small US) provides the excitatory effect. Thus, in the discrimination between a tone signaling a 20% sucrose solution and a tone–light compound signaling an 8% solution, the light is associated both with the frustration elicited by the reduction in concentration of sucrose and with some representation of the actual sucrose earned in its presence. It is the former association that enables the light to suppress responding maintained by another stimulus signaling a high concentration, and the latter that enables it to increase the rate of responding maintained by another stimulus signaling a yet lower concentration. Presumably the frustration reaction tends to oppose the expression of excitation in the latter case, just as the association with the reinforcer occurring in its presence may oppose some of the inhibitory effect that would otherwise be produced by the conditioned frustration (a point discussed in more detail by Cotton et al., 1982). If this is correct, we should have seen a greater enhancement of responding to the clicker paired with the weak sucrose solution in a group trained on the simple tone–light discrimination (a new Group L+) than in Group TL+.

There is, however, a more radical possibility that should be considered. This would allow that the conditions of training experienced by Groups TL+ and L+ in our experiments ensured that both groups associated the light with the small reinforcer occurring in its presence, the sole difference between the two being that Group TL+ would learn that the light signaled a small reinforcer *at a time when another stimulus signaled a larger reinforcer*. The distinction between excitatory and inhibitory conditioning would thus be confined to the specification of the other conditions holding at the time when a particular CS signaled a particular outcome. The idea is clearly related to that advanced by Miller and Schachtman (this volume) as the "comparator hypothesis." For such an account to explain the difference between Groups TL+ and L+, it is necessary that the comparator term be applied at the time of initial learning. We should have to assume that Groups TL+ and L+ learn different things about the light: Group TL+ that the light signals a small reinforcer at a time when another stimulus present signals a larger reinforcer; Group L+ that the light signals a small reinforcer at a time when the only other stimuli present (contextual cues) are only very weakly predictive of a larger reinforcer.

This is not the place to pursue this line of reasoning further, if only because we are not clear in our own minds what further predictions it may generate and are certainly not aware of any data that would distinguish between our conservative and radical proposals. We conclude, instead, by taking up two subsidiary points. Our data imply that an excitatory association is formed between the light and some representation of the small reinforcer occurring in its presence. But there is nothing in those data that requires us to suppose that this association is with a *sensory* representation of the reinforcer, rather than with the affective reaction such a reinforcer normally elicits. Nevertheless, the latter possibility seems an uncomfortable one, for it would require us to assume that the light simultaneously activated affective reactions of opposite sign—in Mowrer's (1960) terminology both hope and disappointment. If one accepts the analysis of counterconditioning advanced by Dickinson and Dearing (1979), such opposed affective reactions should cancel each other out. But our data seem to imply that the two coexist. It therefore seems more plausible to suppose that the association is with some sensory representation of the small reinforcer that could readily be assumed to coexist with a frustration reaction. More direct evidence bearing on this possibility would be provided by a revaluation experiment in which the value of the small reinforcer was somehow changed after discriminative conditioning to the tone and tone–light compound was complete. If such revaluation worked, it would imply that the light had been associated with a sensory representation of the small reinforcer.

Finally, although our main concern has been to understand the content of inhibitory learning rather than the conditions under which learning, whether excitatory or inhibitory, occurs, it is nevertheless evident that if a light signaling a reduction from a large to a small reinforcer is assumed to be associated with a representation of the small reinforcer, this has implications for accounts of the conditions under which associations are formed. On the face of it, the occurrence of excitatory conditioning to the light in these circumstances runs counter to the predictions of a model such as Rescorla and Wagner's (1972). According to that model, excitatory conditioning to the light would require the presence of a positive discrepancy between λ and $\bar{V}$. But the only discrepancy arising on tone–light trials is negative—between the small value of λ (determined by the small reinforcer presented) and the large value of $\bar{V}$ (determined by the strong associative value of the tone associated with the large reinforcer).[3]

There are several problems here, not all equally serious. The main difficulty, however, arises from the model's central assumption that conditioning to the light, whether excitatory or inhibitory, requires a discrepancy between λ and the total associative value of all stimuli present. That discrepancy can hardly be both positive and negative at the same time. Or rather, the price paid for allowing this possibility is a high one and may cause one to look with more favor at other theories of conditioning (e.g., Mackintosh, 1975; Pearce & Hall, 1980) that do not make the same central assumption as the Rescorla–Wagner model. According to these other accounts, conditioning to the light would depend on the discrepancy between λ and its own associative value and would be only indirectly affected (through α, its learning rate parameter) by the associative value of the other stimuli present. They do not face quite the same difficulty, therefore, in allowing that a stimulus could gain both excitatory and inhibitory value on the same conditioning trial.

For Rescorla and Wagner to allow this possibility would seem to require a rather radical departure from the assumptions common to most formal models of conditioning. They would have to reject the assumption that reinforcers differing in magnitude or intensity can be ranked along a single dimension, represented simply by variations in the parameter λ. In fact, as we have already noted, there is good evidence that the difference between five pellets of food and one pellet is not solely a matter of quantitative difference in their reinforcing or response strengthening properties (Capaldi et al., 1976; Carlson & Wielkiewicz, 1976). Although most theories of conditioning have chosen to make the simplifying assumption that the difference in their reinforcing properties at least is a purely quantitative one, perhaps this also is false. It may be necessary to take Logan's (1960) micromolar approach more seriously, and to assume that reinforcers differing in quantity also have qualitatively different reinforcing properties. Rescorla and Wagner could then say that the presentation of one pellet of food when five were expected generated both a negative discrepancy (the absence of five pellets) and a positive discrepancy (the occurrence of one). But the implications of this assertion would need more consideration than we have the space or inclination to devote to them here.

---

[3]This will be particularly true when animals have received extensive prior conditioning with the tone alone signaling the large reinforcer before the introduction of the tone–light compound signaling the small reinforcer. This was the procedure in our sucrose experiments. Where discriminative conditioning begins after only brief preliminary training, as in the experiments varying size of reward, there will be scope for some initial excitatory conditioning to the light, although the Rescorla–Wagner model's assumption of independence of path implies that this will be irrelevant to the final inhibitory status of the light.

ACKNOWLEDGMENTS

This research was supported by grants from the U.K. Science and Engineering Research Council. We are grateful to Glyn Goodall and Lawrence Gardner for assistance with some of the experiments reported here, to Paul Garrud for helping to develop the procedures employed, and to Anthony Dickinson and the editors of this volume for their numerous helpful comments on an earlier draft.

REFERENCES

Amsel, A. (1958). The role of frustrative nonreward in noncontinuous reward situations. *Psychological Bulletin, 55,* 102–119.

Bakal, C. W., Johnson, R. D., & Rescorla, R. A. (1974). The effect of change in US quality in the blocking effect. *Pavlovian Journal of Biological Sciences, 9,* 97–103.

Bower, G. H. (1961). A contrast effect in differential conditioning. *Journal of Experimental Psychology, 62,* 196–199.

Capaldi, E. D., Hovancik, J. R., & Friedman, F. (1976). Effects of expectancies of different reward magnitudes in transfer from noncontingent pairings to instrumental performance. *Learning and Motivation, 7,* 197–210.

Carlson, J. G., & Wielkiewicz, R. M. (1976). Mediators of the effects of magnitude of reinforcement. *Learning and Motivation,* 1976, *7,* 184–196.

Cotton, M. M., Goodall, G., & Mackintosh, N. J. (1982). Inhibitory conditioning resulting from a reduction in the magnitude of reinforcement. *Quarterly Journal of Experimental Psychology, 34B,* 163–180.

Daly, H. B. (1974). Reinforcing properties of escape from frustration aroused in various learning situations. In G. H. Bower (Ed.), *The psychology of learning and motivation* (Vol. 8). New York: Academic Press (pp. 187–231).

Denny, M. R. (1971). Relaxation theory and experiments. In F. R. Brush (Ed.), *Aversive conditioning and learning.* New York: Academic Press (pp. 235–295).

Dickinson, A., & Dearing, M. F. (1979). Appetitive-aversive interactions and inhibitory processes. In A. Dickinson & R. A. Boakes (Eds.), *Mechanisms of learning and motivation.* Hillsdale, N. J.: Lawrence Erlbaum Associates (pp. 203–231).

Flaherty, C. F. (1982). Incentive contrast: A review of behavioral changes following shifts in reward. *Animal Learning and Behavior, 10,* 409–440.

Guthrie, E. R. (1935). *The psychology of learning.* New York: Harper.

Hearst, E. (1972). Some persistent problems in the analysis of conditioned inhibition. In R. A. Boakes & M. S. Halliday (Eds.), *Inhibition and learning.* London: Academic Press (pp. 5–39).

Konorski, J. (1948). *Conditioned reflexes and neuron organization.* Cambridge: Cambridge University Press.

Konorski, J. (1967). *Integrative activity of the brain.* Chicago: University of Chicago Press.

Logan, F. A. (1960). *Incentive.* New Haven: Yale University Press.

Mackintosh, N. J. (1975). A theory of attention: Variations in the associability of stimuli with reinforcement. *Psychological Review, 82,* 276–298.

Mackintosh, N. J. (1983). *Conditioning and associative learning,* Oxford: Oxford University Press.

Mowrer, O. H. (1960). *Learning theory and behavior.* New York: Wiley.

Pearce, J. M., & Hall, G. (1980). A model for Pavlovian learning: Variations in the effectiveness of conditioned but not of unconditioned stimuli. *Psychological Review, 87,* 532–552.

Randich, A., & LoLordo, V. M. (1979). Associative and nonassociative theories of the UCS preexposure phenomenon: Implications for Pavlovian conditioning. *Psychological Bulletin, 86,* 523–548.

Rescorla, R. A. (1969). Pavlovian conditioned inhibition. *Psychological Bulletin, 72,* 77–94.

Rescorla, R. A. (1979). Conditioned inhibition and extinction. In A. Dickinson & R. A. Boakes (Eds.), *Mechanisms of learning and motivation.* Hillsdale, NJ: Lawrence Erlbaum Associates (pp. 83–110).

Rescorla, R. A., & Holland, P. C. (1977). Associations in Pavlovian conditioned inhibition. *Learning and Motivation, 8,* 429–447.

Rescorla, R. A., & Wagner, A. R. (1972). A theory of Pavlovian conditioning: Variations in the effectiveness of reinforcement and nonreinforcement. In A. H. Black & W. F. Prokasy (Eds.), *Classical conditioning II: Current research and theory.* New York: Appleton–Century–Crofts (pp. 64–99).

Spear, N. E. (1976). Retention of reinforcement magnitude. *Psychological Review, 74,* 216–234.

Trapold, M. A. (1970). Are expectancies based upon different positive reinforcing events discriminably different? *Learning and Motivation, 1,* 129–140.

Wagner, A. R., Mazur, J. E., Donegan, N. H., & Pfautz, P. L. (1980). Evaluation of blocking and conditioned inhibition to a CS signaling a decrease in US intensity. *Journal of Experimental Psychology: Animal Behavior Processes, 6,* 376–385.

Weisman, R. G. (1969). Some determinants of inhibitory stimulus control. *Journal of the Experimental Analysis of Behavior, 12,* 443–450.

Wendt, G. R. (1936). An interpretation of inhibition of conditioned reflexes as competition between reaction systems. *Psychological Review, 43,* 258–281.

4 Factors Affecting the Acquisition and Extinction of Conditioned Inhibition Suggest a "Slave" Process

Harry Fowler, Morris C. Kleiman, and Donald T. Lysle
University of Pittsburgh

Our research on conditioned inhibition (CI) came about somewhat incidentally, for it was the result of several unsuccessful attempts to produce a phenomenon that depended on the presence of CI. In the face of those failures, we concluded that we had not generated sufficiently strong CI even though we had followed prescribed, and presumably reliable, procedures for establishing a robust effect. The first part of this chapter describes our subsequent attempts at producing substantial CI effects. By illustrating our successes and failures, we intend to point up those factors that control the development of CI. Because our research in this vein is still ongoing, some of our conclusions must necessarily be tentative, but we believe we can strengthen our main conclusions by the findings that are reported in the second part of this chapter. That part highlights factors controlling the maintenance and extinction of CI. From our perspective, the second part illustrates an ironic aspect of CI because, in spite of whatever difficulties one has in developing CI, once produced, it cannot be eliminated by the conventional extinction procedure of presenting the conditioned stimulus (CS) by itself. We argue on the basis of the findings from both parts that CI is a process that does not operate autonomously.

PROCEDURAL DIFFERENCES AFFECTING THE ACQUISITION OF CI

Our approach to assessing factors that control the development of CI was shaped by a review of the numerous procedures that reportedly produce CI. Because

LoLordo and Fairless (this volume) have described these procedures and their respective effects in reasonable detail, our accounting of them can be reduced to an enumeration, for there are more than half a dozen procedures just for producing first-order CI. They include: (1) backward conditioning, in which the CS is contingent upon termination of the unconditioned stimulus or US (e.g., Moscovitch & LoLordo, 1968; Weisman & Litner, 1971); (2) cessation training, in which the CS is presented during the US but overlapping its termination (e.g., Moscovitch & LoLordo, 1967; see also Maier, Seligman, & Solomon, 1969); (3) delay or trace conditioning, in which the US is contingent upon the CS but only after an extended delay or trace interval (e.g., Hinson & Siegel, 1980; Rescorla, 1967, 1968); (4) "explicitly unpaired" training, in which the US is contingent upon the absence of the CS (e.g., Rescorla, 1969a; Weisman & Litner, 1969); (5) differential conditioning, in which the US is contingent upon one CS (A+) but not upon another (B−), the putative inhibitor (e.g., Weisman & Litner, 1969, 1971); (6) "Pavlovian" CI training, in which one CS (A+) is reinforced by the US but the compound of that CS and another (AB−) is not reinforced (e.g., Rescorla & Wagner, 1972; Wagner & Rescorla, 1972); and (7) excitatory extinction, in which the putative inhibitory CS is compounded with an excitatory CS during the latter's extinction (e.g., Rescorla, 1979). This listing could be extended by including a procedure that produces second-order CI (Rescorla, 1976).

What we found disturbing was that with so many CI procedures available, there was relatively little systematic research assessing their differential effectiveness. Some investigators seemed to take it for granted that the Pavlovian procedure (A+, AB−) worked best because of the conditioned excitation that was mediated by A when that stimulus was nonreinforced in compound with B (e.g., Wagner & Rescorla, 1972). However, apart from a single study by Baker (1977), there were no explicit evaluations of the role that conditioned, or even unconditioned, excitation played in producing CI. Furthermore, the few comparisons of the different procedures that were available were restricted at best. Some investigators had employed different CI procedures but in different experiments of a study, thereby precluding a direct comparison (e.g., Rescorla & LoLordo, 1965; Weisman & Litner, 1971). Others had used different CI procedures in the same experiment but did not offer a comparison (e.g., Weisman & Litner, 1969). Still others offered questionable comparisons, such as comparing the single-element inhibitory CS of a differential procedure with the inhibitory compound of a Pavlovian procedure (e.g., Mahoney, Kwaterski, & Moore, 1975; Marchant & Moore, 1974). As a result, we were not very far from the conclusion that Rescorla (1969b) had drawn in his earlier review of CI: "As yet, there is not sufficient experimental analysis to make final assertions about the critical conditions for producing conditioned inhibition" (p. 92).

The paucity of comparisons of different CI procedures is probably related to the fact that there also has been relatively little research assessing variations within a single procedure, and therefore we cannot know which variant of a particular procedure should be used in a comparison of different procedures. For that reason, our approach has been to assess the effectiveness of different procedures within the context of a seemingly more general procedure; that is, where different procedures are cast as but variants of one procedure. Consider, for example, the explicitly unpaired procedure. The only requirement of that procedure is that the putative inhibitory CS be unpaired with the US; that is, not placed in an immediate, forward relationship to the US. However, that leaves open whether the unpaired CS should occur in a *backward* relationship to the US, or in a *trace* relationship to the US, or somewhere in between; or even whether the CS should occupy a small or large part of the inter-US interval. These variations in the locus and duration of an explicitly unpaired CS are important not merely because they afford a comparison of different CI procedures, but because they all satisfy Rescorla's (1967, 1969b) contingency rule for developing CI, namely, that the CS and US be negatively correlated. Hence, to the extent that such variations were differentially effective in producing CI, they would require a modification of that rule, and therefore our description of CI. Our first experiment explored these variations.

Variations in Explicitly Unpaired Training: Backward, Trace, and Mid-Locus CSs (Experiment 1)

The general methodology for our first experiment, as well as for all succeeding experiments, involved a conditioned-suppression paradigm for rats in which the baseline response was licking for water. The subjects were experimentally naive males of the Sprague–Dawley strain that weighed approximately 350 g and were 80–90 days old at the start of an experiment. Following acclimation to a home-cage deprivation schedule, which allowed water for 8 min per day and food ad libitum, the subjects received baseline training in which the daily session length was always 8 min and provided the subject's total water intake for the day. Thereafter, the subjects were given on-baseline habituation training to attenuate any unconditioned suppression to the CSs to be used in the experiment. During habituation training, there were usually four to eight presentations of each CS, distributed over several days. Then, depending on the particular experiment, the subjects received on- or off-baseline inhibitory conditioning in the original (X) or a different (Y) chamber. (For off-baseline sessions, water was available for 8 min per day in the animal's home cage following the session.) The X and Y chambers were of different types, with different dimensions, grid floors, and internal features, such as vertically striped black and white panels and an air freshener (Airwick Forest Pine) in Y but not in X. Following inhibitory conditioning, the animals received either a summation or a retardation test, or both, on-baseline in the original X chamber.

In our first experiment (Kleiman & Fowler, 1984, Experiment 1), which concerned the noted variations in CS locus and duration, explicitly unpaired training was conducted off-baseline for 8 days in chamber Y. Each daily session consisted of a 14.7-min period in Y during which a .5-sec, 1.25-mA shock US was presented on a 110-sec, variable-time (VT) schedule (range = 50–170 sec). Presentations of the CS (an overhead flashing light, L) were systematically varied across 5 groups. For 3 groups, the duration of L was held constant at 10 sec but its locus within the inter-US interval varied as follows: Group BK (backward) always received L 10 sec after termination of the US. Group TR (trace) always received L 30 sec prior to the US, resulting in a 20-sec gap between L's termination and the onset of the US. Group M10 (mid, 10-sec CS) always received L in the exact middle of the interval defined by the BK and TR conditions, i.e., between 10 sec after a US and 20 sec prior to the next US. The two remaining groups also received L in the exact middle of this interval, but the duration of L was extended equally in both backward and forward directions so that L always encompassed slightly more than one-half the interval for group M45 (mid, average L duration = 45 sec) and the entire interval for group M80 (mid, average L duration = 80 sec).

Following explicitly unpaired training, the groups were returned to baseline sessions in X where they subsequently received conditioned-suppression training to a 10-sec, 80-dB clicker (C) stimulus that was consistently reinforced by the shock US. Then, all groups received a series of summation tests in X in which each daily session consisted of one reinforced presentation of C and three non-reinforced presentations of a simultaneous LC compound. During this phase, the shock US was increased to 1.5 mA in anticipation that nonreinforced LC presentations would reduce suppression to C. Our purpose in using this reinforced summation test (actually, Pavlovian CI training) was to allow any small differences resulting from variations in explicitly unpaired training to be manifest in a "savings" test of CI. However, over the course of 16 test sessions, there was no evidence of CI; virtually complete suppression occurred to both C and LC. Kamin suppression ratios (CS/[CS + pre-CS], where CS represents licks in the 10-sec CS, and pre-CS represents licks in the 10-sec period immediately prior to the CS) showed that mean suppression to C and to LC over the last 4 sessions of reinforced summation testing was .06 and .08, respectively.

The absence of CI effects in the reinforced summation test was disconcerting, but additional research in both our lab and those of others has suggested at least three reasons why such effects would not have been apparent. First, extended training with our test procedure is normally required before substantial CI is observed to a salient L stimulus of the type employed (see Lysle, 1983). L apparently exerts unconditioned suppression that is disruptive in tasks requiring an active inhibitory response, e.g., licking at a drinking tube; however, in tasks requiring a passive inhibitory response, e.g., not running to avoid shock, L's suppressive effect can actually facilitate performance (see Jacobs & LoLordo,

1980). Second, an inhibitory CS that is developed in one context, or with one excitatory CS, does not readily transfer to a different context (e.g., Fowler & Lysle, 1982), or to a different excitor (e.g., Rescorla, 1982). Third, the use of a stronger shock US to sustain conditioned suppression to an excitatory CS in a summation test can "overpower" the effect of an inhibitory CS that is established with a weaker US. Indeed, in the present study, the inhibitory effect of L was obscured as well when all groups were subsequently given nonreinforced summation testing on C and LC. Only after 12 such sessions did an inhibitory effect of L begin to emerge, and even then differences among the groups were not fully evident until the 10-sec CSs were extended to 20 sec for 8 more sessions to hasten the extinction of suppression. The data for these last 8 sessions are presented in Fig. 4.1.

As indicated, the reduction of suppression to C produced by L in compound with C was reliably greater for the BK and TR groups than for the three mid-locus groups. Although differences among the latter failed to reach significance, ancillary comparisons of the difference between C and LC showed that the BK and TR groups differed reliably from the M10 and M45 groups but not from the M80 group. Virtually the same effects were obtained when L was subsequently presented by itself in a retardation test involving 50% reinforcement for that stimulus. In that test, the development of suppression to L was comparably retarded for the BK, TR, and M80 groups by comparison with the M10 group, with the M45 group falling between. Thus, both tests indicated that a mid-locus CS in an explicitly unpaired procedure failed to generate strong CI unless the duration of that stimulus was extended to encompass the temporal loci of the BK and TR CSs, as in the M80 case.

The particular mechanism for the present effects must remain obscure for the moment because of our failure to obtain CI in the reinforced summation test. For example, it could be argued that only latent inhibition (e.g., Lubow, 1973) or learned irrelevance (e.g., Mackintosh, 1973) developed during explicitly unpaired training, and that because these effects were most pronounced for the M10 CS, which was temporally dissociated from the US, that CS was most retarded in

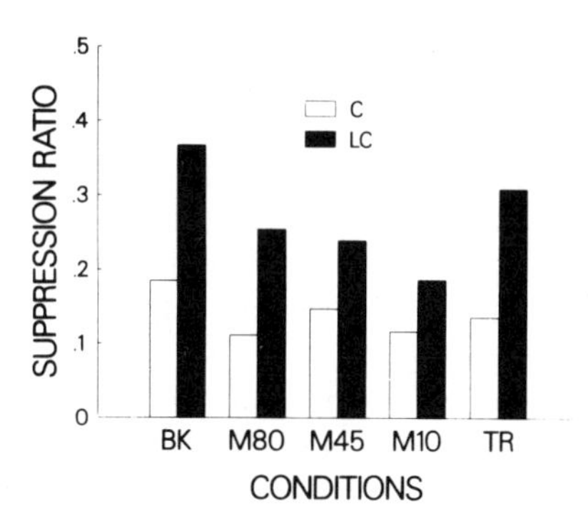

FIG. 4.1. Mean suppression to the 20-sec C and LC compound CSs for the BK, M80, M45, M10, and TR groups during the nonreinforced summation test. Pooled *SE* for the difference in suppression between C and LC is .030. (From Kleiman & Fowler, 1984.)

developing CI during the reinforced summation test. Although we would accept this possibility for the M10 CS, we would argue that the BK and TR conditions, as well as the M80 condition, actually promoted the development of CI, but that such effects were obscured for the reasons earlier indicated. We proffer this stronger conclusion on the basis of the findings to be reported for Experiment 2, and also those that have been reported for inhibitory conditioning with both a backward and a trace CS.

Both Plotkin and Oakley (1975) and Maier, Rapaport, and Wheatley (1976) found that a backward CS functioned as an inhibitor in a summation test if the US–CS interval during training had been short (.2-.5 and 3 sec, respectively), but not if that interval had been long (55 and 30 sec, respectively). The only apparent opposition to these findings are those by Moskovitch and LoLordo (1968). They found no significant difference in CI to a backward CS in a summation test if the US–CS interval had been either 1 or 15 sec; however, the direction of their difference clearly favored the 1-sec group. In view of the present outcome for the BK and M10 groups, for which the US–CS interval was, respectively, 10 sec and a mean of 45 sec (range = 15–75 sec), it would appear that CI develops with US–CS intervals up to 15 or so sec (where its development begins to strain), but not with intervals that are far more extended. It is noteworthy that Grelle and James (1981) reached the same conclusion regarding the temporal extent of an effective US–CS interval based on their backward pairing of a CS with either a shock US (22.5-sec interval) or a signal for that US (.5-sec interval). Their evidence for inhibitory conditioning with the long interval following the US was mixed, whereas that with the short interval following the signal for the US was clearly positive.

That a backward CS must be temporally close to the US (or a signal for the US) in order for CI to develop to that CS is also consonant with the results of studies that have tracked the course of conditioning to a backward CS. Although early investigations of this subject indicated no conditioning to a backward CS with extended training (see Mackintosh, 1974), more recent investigations have repeatedly shown that with a US–CS interval anywhere from 0–10 sec, the backward CS initially develops conditioned *excitation* (CE) in the first 10 or so trials (e.g., Burkhardt, 1980; Heth, 1976; Heth & Rescorla, 1973; Keith–Lucas & Guttman, 1975; Mahoney & Ayres, 1976; Shurtleff & Ayres, 1981; for the only exception, see Siegel & Domjan, 1971). However, with continued training, CE to the backward CS declines completely (e.g., Heth, 1976; Smith, Coleman, & Gormezano, 1969), and to the extent that appropriate assessments are then made, CI is observed in place of CE (e.g., Maier et al., 1976; Moscovitch & LoLordo, 1968; Plotkin & Oakley, 1975; Siegel & Domjan, 1971; Weisman & Litner, 1971).

Collectively, the preceding findings suggest a relatively simple interpretation. The development of CI to a backward CS is dependent on a perseverative effect of the US (e.g., rehearsal of the US in short-term memory) that persists in relatively strong form for 15 or so sec. Being contiguous with US rehearsal by virtue of a short US–CS interval, the backward CS is initially associated with excitation and can therefore develop CE. However, with continued training, the animal learns to discriminate the backward CS as a signal for nonreinforcement and thereby develops an inhibitory reaction to that CS. In the absence of any perseverative effect of the US, as with more extended US–CS intervals, such a process cannot commence and thus a short-duration mid-locus CS will not develop CI. If anything, that CS will be subject to latent inhibition or learned irrelevance, as suggested for the M10 CS of the present study.

Essentially the same interpretation can be applied to a trace CS. In this case, though, the basis for CI development would not be rehearsal of the US but rehearsal of the CS. With a relatively short gap between the CS and the US, a memory trace of the CS should be contiguous with the US. That should allow the development of CE both to the trace of the CS and, by mediated generalization, to the compound of nominal CS (*n*) and its memory trace (*t*). However, because *t* by itself is reinforced whereas the *nt* compound is not, the animal should learn to discriminate the nominal CS as a signal for nonreinforcement and should thereby acquire an inhibitory reaction to that CS. Although there is only one relevant investigation of trace conditioning bearing on this issue, it indicates the prescribed course of conditioning: With relatively little training, CE develops to the trace CS, but with extended training, it declines completely (Bolles, Collier, Bouton, & Marlin, 1978). Furthermore, to the extent that appropriate assessments are made following extended trace conditioning, CE is apparently replaced by CI (e.g., Hinson & Siegel, 1980; Rescorla, 1968).

Comparisons of Explicitly Unpaired (Trace and Mid) and Differential Procedures (Experiment 2)

Because of the far less substantial evidence on the course of CE and then CI development to a trace CS, we sought in Experiment 2 to monitor that development and to demonstrate that inhibitory conditioning to the CS did, in fact, depend on excitatory conditioning to its memory trace. We also sought to demonstrate that little, if any, CE or CI developed to a mid-locus CS in our explicitly unpaired procedure, or to a trace CS that was accompanied by an immediate signal for the US. With inclusion of the latter, i.e., a differential (B−, A+) procedure, we extended our comparisons of different CI procedures but for the purpose of having A overshadow excitatory conditioning to B's memory trace and thereby preclude inhibitory conditioning to B. In addition, because of our difficulty in developing CI to a flashing-light (L) CS in the prior study and in others (see Experiment 3), and also in demonstrating that effect in a reinforced summation test, we employed an 80-dB, white-noise (N) stimulus in place of L as the CS− and used only a nonreinforced summation test prior to a retardation test.

To monitor any CE to N during the initial portion of inhibitory conditioning, we gave the subjects of Experiment 2 (Kleiman & Fowler, 1984, Experiment 2) inhibitory training to N on-baseline in the X chamber where they had received baseline training and CS habituation. For reasons indicated later, ambient lighting in X during this phase was reduced so that the chamber was virtually dark. For 16 days, the daily session for all subjects consisted of an 8-min drinking period during which there were 4 presentations of a .5-sec, 1.25-mA shock US on a 2-min VT schedule (range = 60–180 sec). Presentations of N varied across 6 groups that were comprised of 2 experimental groups, 2 "associative" controls, and 2 "nonassociative" controls.

For the 2 experimental groups (T20 and T30), the 10-sec N stimulus was always presented either 30 or 40 sec prior to US onset so as to generate a constant trace interval of either 20 or 30 sec. For one of the associative controls (MID), N was dissociated from the US by having N always occur in the exact middle of the interval defined in Experiment 1, i.e., between 10 sec after a US and 20 sec prior to the next US. For the other associative control (DIF), the 10-sec N stimulus was always presented 40 sec prior to US onset so as to generate a constant trace interval of 30 sec, like that for group T30; but, in addition, a 10-sec L stimulus (the overhead flashing light) always occurred 10 sec prior to US onset. Because of the darkened X chamber during this training phase, L was a highly salient stimulus and could thus overshadow (or disrupt rehearsal of) N's memory trace as a signal for the US. Hence, little, if any, first-order CE or CI was expected to develop to N for the DIF group. It was conceivable, though, that some second-order CE, and ultimately CI, might develop to N for these subjects, despite the 20-sec gap between N and L. For that reason, in part, we had included the T20 group that received the same temporal gap but between N and the US.

In contrast to the preceding four groups, the two nonassociative controls did not receive any presentations of N during the inhibitory-conditioning phase. One of these controls (CS+) always received L 10 sec prior to US onset, and the other (USa) received presentations of the US alone. By employing these nonassociative controls, for which N would be a relatively novel CS in the subsequent summation and retardation tests, we could assess the absolute magnitude of CI development to N for the T20 and T30 groups, and also for their associative controls, DIF and MID. In that manner, we could reevaluate the outcomes observed for the TR and M10 groups of Experiment 1, because those two groups were virtually identical to the T20 and MID groups of the present experiment.

Throughout on-baseline inhibitory conditioning, Kamin suppression ratios of the standard form were calculated for each 10-sec N and each 10-sec L stimulus. In addition, for Groups T20, T30, and USa, suppression ratios were calculated for each 10-sec no-stimulus (or "dummy") interval immediately prior to US onset. For all groups except the MID group, the 10-sec pre-CS interval for N, L, or the dummy interval began 50 sec prior to US onset; for the MID group, the pre-CS interval was the 10-sec interval immediately prior to the onset of N. Because of the variable location of N and of the pre-CS interval for the MID group, recording of licks during the dummy interval for this group was not possible with the equipment available. In all subsequent phases of training and testing, suppression ratios were calculated according to the standard practice, i.e., with pre-CS referring to the 10-sec interval immediately prior to the CS.

Figure 4.2 presents the results of the on-baseline, inhibitory-conditioning phase. The bottom panel shows the development of conditioned suppression to the reinforced L stimulus for groups DIF and CS+, and to the comparable dummy or no-stimulus interval for groups T20, T30, and USa. As expected, the DIF and CS+ groups showed substantial conditioned suppression to L over the course of inhibitory training. However, Group T20 and, to a lesser extent, Group T30 also showed sustained and reliable conditioned suppression to the dummy interval by comparison with the USa control. These results suggest that the T20 and T30 subjects utilized a memory trace of N to time US delivery, and that their rehearsal of N was more pronounced or "vivid" (and hence subject to stronger CE), the shorter the trace interval.

The top panel of Fig. 4.2 depicts the course of conditioned suppression to the N stimulus itself for groups T20, T30, DIF, and MID. As indicated, over the first 24 trials of inhibitory conditioning, group T20 showed a fairly marked increase in suppression to N, followed by a decline, whereas group T30 showed a sustained amount of suppression to N, if not a modest increment. Each of these effects was reliable by comparison with those for the DIF and MID groups. They showed little, and certainly no sustained, suppression to N.

Because suppression to N for the T20 and T30 groups was proportionately less than that shown by these groups to the memory trace of N in the dummy interval (cf. the top and bottom panels of Fig. 4.2), the obtained suppression to N indicates a generalization effect that was mediated by the memorial processing of N.[1] However, that the T20 and T30 subjects were able to discriminate N from its

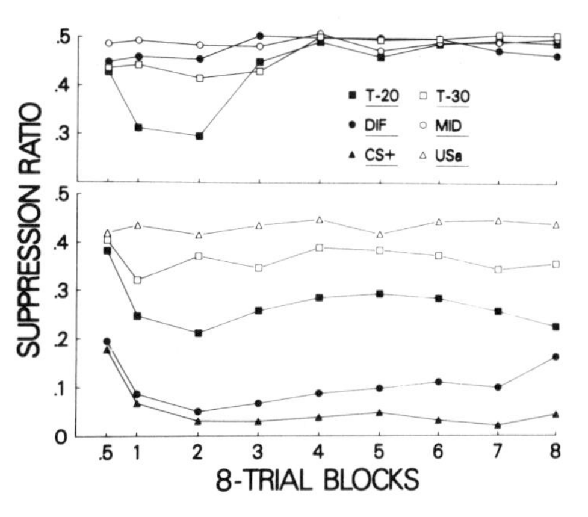

FIG. 4.2. Mean suppression to the N trace stimulus for groups T20, T30, DIF, and MID (top panel), and to either the L or "dummy" stimulus immediately prior to the US for groups T20, T30, DIF, CS+, and USa (bottom panel), during on-baseline inhibitory conditioning. Pooled *SE* for N over Trial-blocks 1–3 is .012; and for L and the dummy stimulus over Trial-blocks 1–8, .018. (From Kleiman & Fowler, 1984.)

[1]Note that if N were used solely to start an internal clock by which the animal timed US delivery, there would not necessarily be any excitatory generalization to N because that stimulus does not bear any similarity to the point in clock time where the US is judged to occur.

memorial representation is well attested by the fact that suppression to N completely extinguished over the course of additional inhibitory training, whereas that to the memory trace of N in the dummy interval was fully maintained. Such differential conditioning indicates that, in the presence of CE mediated by the memorial processing of N, N itself became a signal for nonreinforcement and presumably a conditioned inhibitor.

Following the inhibitory-conditioning phase, all subjects received baseline sessions in X with ambient lighting restored to its original brightness (to attentuate CE to contextual cues for the unsignaled shock groups), and then conditioned-suppression training to a 10-sec clicker (C) stimulus. However, in the present experiment, C was reinforced by the same US intensity and was subsequently subjected to a 50% schedule of reinforcement so that suppression to C would not be overpowering in the following summation test. The results of that summation test, which involved four nonreinforced presentations each of C and of a simultaneous NC compound over 2 successive days, are presented in Fig. 4.3.

As indicated, the reduction of suppression to C produced by N in compound with C was reliably greater for the T20 and T30 groups by comparison with either the associative controls (DIF and MID), or the nonassociative controls (CS+ and USa), or both. Furthermore, there were no reliable differences between the two sets of controls, whether they were treated as sets or were partitioned into subsets based on prior training with either a signaled US (DIF vs. CS+) or an unsignaled US (MID vs. USa). Hence, the reduction of suppression produced by N for the DIF and MID groups can be ascribed to an external-inhibition effect (e.g., Pavlov, 1927), analogous to that generated by the novel N stimulus for the CS+ and USa groups. Exactly the same effects were observed in the subsequent retardation test when N was presented by itself and was reinforced by the US on a 50% basis. By comparison with the controls, among which there were no reliable differences, the T20 and T30 groups showed significantly retarded suppression to N.

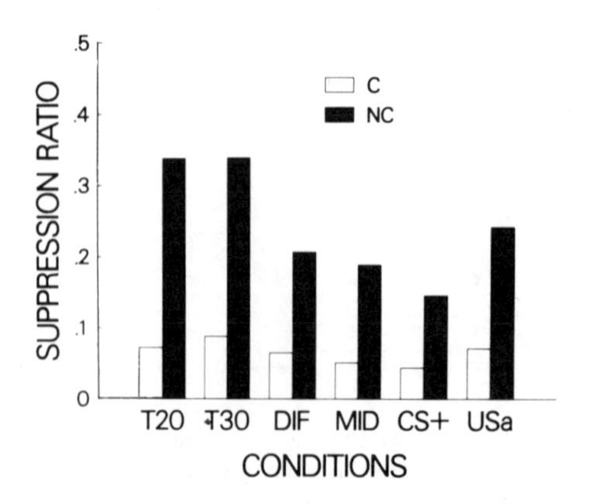

FIG. 4.3. Mean suppression to the C and NC compound CSs for the T20, T30, DIF, MID, CS+, and USa groups over the 2-day summation test. Pooled *SE* for the difference in suppression between C and NC is .018. (Adapted from Kleiman & Fowler, 1984.)

The present findings are fully consistent with our expectations that in explicitly unpaired training, CI would develop to a trace CS but not to a mid-locus CS. As such, the present findings support those of Experiment 1, indicating that relatively strong CI develops only when the explicitly unpaired CS is temporally close to the US (as in the case of a backward or a trace CS) and, of course, is followed by some period of nonreinforcement (see, e.g., Moscovitch & LoLordo, 1968; Weisman & Litner, 1971). That description of the basis for CI development must be qualified, though, because the present outcome for the DIF group showed that CI did not develop to a trace CS when there was another, qualitatively different CS that better signaled the US and thereby overshadowed excitatory conditioning to the memory of the trace CS. Taken together, the findings of the two experiments argue that the development of CI to an explicitly unpaired CS is dependent on that CS being contiguous with strong, *mediated* excitation, as is provided for a trace CS by the memorial processing of that CS, and for a backward CS by rehearsal of the US itself.

Although the principle of mediated excitation posited for CI development is certainly not foreign to theories of inhibitory conditioning (cf. Rescorla & Wagner, 1972; Schull, 1979), the suggested basis for that mediation is. For that reason, we believe that the present findings pose difficulty not only for contingency interpretations of CI (e.g., Rescorla, 1967, 1969b) but also for any contiguity interpretation that does not explicitly rely on the memorial processing of stimulus events. In particular, the fact that CI did not develop to a mid-locus CS, but did to both a backward and a trace CS, argues strongly that the mere specification of a negative correlation, or contingency, between the CS and US is not sufficient as an operational basis for the development of CI. The same fact argues equally strongly that the positing of excitatory mediation by contextual cues (e.g., Wagner & Rescorla, 1972) is not sufficient to account for the development of CI to an explicitly unpaired CS. Otherwise, it would have developed equally well for a mid-locus CS.

We are not denying that CI can develop in explicitly unpaired training on the basis of just excitatory contextual cues, for there are ample studies apparently demonstrating that fact (e.g., Rescorla, 1966, 1969a; Rescorla & LoLordo, 1965; Weisman & Litner, 1969). Rather, we would point up the necessity of using training conditions other than those employed in the present research. We would particularly stress the need for a fairly complete VT schedule of US presentations, and hence random presentations of the CS and the US, rather than the constrained VT schedule that was employed not only in the present two experiments, but in each of those earlier described studies that obtained CI to a backward CS (e.g., Grelle & James, 1981; Maier et al., 1976; Moscovitch & LoLordo, 1968; Plotkin & Oakley, 1975).

In order to assess the effect of a specific interval between the US and the CS, the noted backward-conditioning studies and the present two experiments all employed an alternating schedule of CS and US presentations by which the animal could well have learned not to fear receipt of an impending US until after the CS was presented. That would mean that a mid-locus CS (or a backward CS with an extended US–CS interval) would not only be temporally dissociated from the US but would also occur in the presence of relatively little fear. Such an effect was apparent in an analysis of the pre-CS data from the inhibitory-conditioning phase of the present experiment. Those data showed that licking in the presence of just background cues for the MID group at first decreased but then increased with continued on-baseline training, in evidence of reduced fear of the context. Given those circumstances, it is entirely gratuitous to assume that merely unpairing a CS and an unsignaled US will result in sufficiently strong CE to contextual cues to allow CI development to a CS that is distal to the US. That apparently was our mistake in the earlier research that led us to investigate factors controlling the development of CI.

Pretraining Variations in Differential Conditioning: Signaled Versus Unsignaled US Presentations (Experiment 3)

Along with our use of a differential procedure in the preceding experiment, we have begun to explore how CI might be generated with this procedure even though it does not typically produce CI when qualitatively different stimuli are employed as the CS+ and CS− (see Experiment 2; Baker, 1977; Desmond, Romano, & Moore, 1980; Holland, this volume; Wagner & Rescorla, 1972). Although our research in this vein is still in its incipient stage, we think it fitting to report the outcome of our first study because it illustrates how contextual cues can, indeed, mediate sufficiently strong excitation to a nonreinforced CS to promote its development as a conditioned inhibitor.

The design of our study was patterned after a study by Baker (1977) in which CI developed for explicitly unpaired subjects that received presentations of a shock US and of a nonreinforced CS in separate sessions. Based on the reasonable assumption that a discrete signal for the US would preclude CE to contextual cues, Baker gave other subjects the same "between-sessions" training but with the US consistently signaled by another CS (as in differential training) in order to assess the role of excitatory contextual cues for his explicitly unpaired subjects. For us, an important feature of Baker's study was his use of a pretraining phase that did not involve the CS− but entailed unsignaled shocks for his explicitly unpaired group and signaled shocks for his differential group. What we wanted to know was whether CI would develop for animals that received differential conditioning but after unsignaled shocks in pretraining. If unsignaled shocks in pretraining produced CI with this procedure, that would certainly be strong evidence for the mediation of CE by contextual cues—and it would allow us to suggest a specific mechanism for that mediation.

After baseline training and CS habituation in chamber X (which did not include presentations of a tone that was subsequently used as a CS+ to overshadow excitatory conditioning to the context), 6 groups of rats were assigned to one of the following pretraining treatments conducted off-baseline in chamber Y: 0, 4, or 12 unsignaled US presentations (designated 0, 4+, and 12+) or 0, 4, or 12 US presentations that were consistently signaled by a 10-sec, 80-dB, 1500-Hz tone (designated 0, 4T+, and 12T+). The two 0 groups were treated identically in pretraining but, as described later, differently in the differential-conditioning phase. Pretraining lasted 3 days and, for the appropriate groups, consisted of a daily 16-min session during which there were 4 presentations of a .5-sec, 1.4-mA shock US on a 4-min VT schedule (range = 2–6 min). The two 0 groups did not receive any exposure to Y, whereas the 4+ and 4T+ groups received only 1 pretraining day in Y (with that day being counterbalanced for subjects within these groups across the first and third day of pretraining for the 12+ and 12T+ groups). We purposely did not equate exposure to Y for the groups so as to avoid differential latent-inhibition effects to that context.

Differential conditioning lasted 8 days and, like pretraining, consisted of a daily 16-min, off-baseline session in Y. On each day, all groups, except 1 of the 0 groups, received 8 tone-signaled US presentations (T+) on a 2-min VT schedule (range = 60–180 sec) and 8 nonreinforced presentations of the overhead flashing light (L). For these groups, presentations of T+ and of L were random with the restriction that the mean gap between T+ and a succeeding L was 32.5 sec (range = 20–45 sec), and that between L and a succeeding T+, 42.5 sec (range = 30–55 sec). Thus, L occupied a variable mid-locus position and was temporally protected from both backward and trace (or higher order) conditioning effects. The other 0 group (hereafter designated 0/N) received the same schedule of T+ presentations but no presentations of L. In that this group received a novel L during the subsequent summation and retardation tests, it served as a control for evaluating the absolute magnitude of CI development to L for the other groups.

Because the present experiment was conducted concurrently with Experiment 1, it included both reinforced and nonreinforced summation tests that were identical to those described for Experiment 1, except for the following. After differential training in chamber Y, the subjects received 4 baseline days in chamber X, and then 3 days of conditioned-suppression training to a 10-sec clicker (C) stimulus that was reinforced by the US on only a 50% basis. Thereafter, all subjects received reinforced summation testing, i.e., Pavlovian CI training, in which there was one reinforced presentation of C and three nonreinforced presentations of an LC compound per session.

As in Experiment 1, there were no group differences or apparent inhibitory effects of L during the first 8 sessions of reinforced summation testing. However, group differences in CI did emerge during the next 8 sessions. The left panel of Fig. 4.4 shows that the reduction of suppression to C produced by L in com-

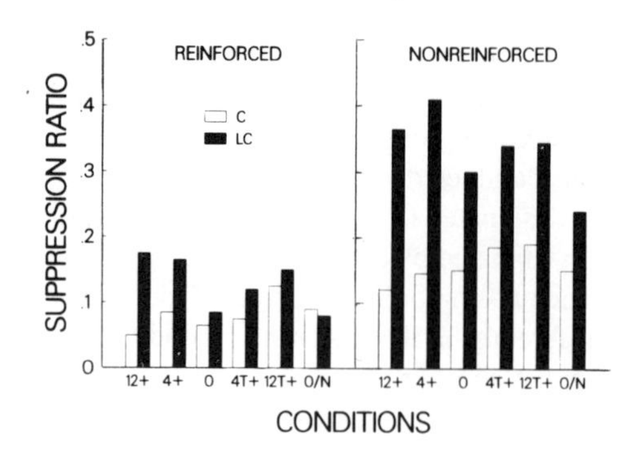

FIG. 4.4. Mean suppression to the C and LC compound CSs for the 12+, 4+, 0, 4T+, 12T+, and O/N groups during the last eight sessions of reinforced summation testing (left panel); and during the last eight sessions of nonreinforced summation testing (right panel). Pooled *SE* for the difference in suppression between C and LC is .030 during the reinforced test, and .031 during the nonreinforced test.

pound with C was greatest for the 12+ group, intermediate for the 4+ group, and minimal for the remaining groups. These differences, which were reliable, cannot be ascribed to the development of latent inhibition to L during differential conditioning. If that were the case, the development of CI to L during reinforced summation testing would have been retarded for the 0, 4T+, and 12T+ groups by comparison with the 0/N control that received a novel L during summation testing. Hence, the results of the reinforced summation test indicate that unsignaled US presentations during pretraining facilitated the development of CI to L during differential training, and, more so, the greater the number of those presentations.

Basically, the same effects were apparent during subsequent nonreinforced summation testing when the US for C was omitted to allow any overpowering suppression to C to extinguish. The right panel of Fig. 4.4 presents the data for the last 8 of 16 such nonreinforced summation-test sessions. As indicated, L in compound with C continued to produce a reliably greater reduction of suppression to C for the 12+ and 4+ groups by comparison with the other groups, among which there were no reliable differences (including that between the 0/N control and the 0, 4T+, and 12T+ groups). The retardation test, which followed, showed the same effects in that there was significantly slower acquisition of suppression to L for the 12+ and 4+ groups, and no apparent differences among the remaining groups.

We believe that the present results provide strong evidence for the role that excitatory contextual cues can play in producing CI. Indeed, a single pretraining session with just four unsignaled US presentations was sufficient to promote CI to a mid-locus CS in a differential procedure where CI is not typically observed (nor was observed for subjects that received as many if not more presentations of a signaled US during pretraining). That outcome, however, only prompts additional questions about the nature of contextual mediation, particularly in light of the outcomes of Experiments 1 and 2 that showed that CI did not develop to a mid-locus CS in explicitly unpaired training. We argued previously that the lack of CI to a mid-locus CS in those experiments, as well as in each of the noted backward-conditioning studies, was due to the use of an alternating schedule of CS and US presentations. But that cannot be the sole answer. In the study by Baker (1977), which prompted the present experiment, substantial CI was observed for explicitly unpaired subjects that received unsignaled US presentations and nonreinforced CS presentations on *alternate* days. Is there something special, then, about the separate sessions with an unsignaled US that Baker used in his study, and that we used in the pretraining phase of the present study? And, if so, by what means do excitatory contextual cues function as mediators to promote CI? The mere fact that they do is certainly not sufficient to account for CI.

A nonreinforced CS that occurs in the presence of excitatory contextual cues is also followed by the same excitatory cues. Extant accounts of CI that touch upon this subject (e.g., Wagner & Rescorla, 1972) do not offer any explicit basis by which chronic fear of the context preceding a CS will be reduced following the CS, to allow that CS to develop as an inhibitor. Furthermore, we do not think it reasonable that chronic contextual fear extinguishes sufficiently either during or shortly after the CS to generate CI to that CS, particularly when there can be forthcoming US presentations, as in explicitly unpaired training. We would suggest that separate training sessions and especially pretraining sessions with just the US can dispose the animal to being acutely afraid of any abrupt and unpredictable change in that context. This could well result from "pseudoconditioning" (cf. Sheafor, 1975; Wickens & Wickens, 1942) or even from sensitization. For example, the startle reactions of rats subject to high arousal (as undoubtedly would be the case with unsignaled shocks) show sensitization, rather than habituation, with repeated presentations of a neutral stimulus (e.g., Davis, 1974; see also Groves & Thompson, 1970). Either or both of these effects should also operate with random as opposed to alternating presentations of an explicitly unpaired CS: It's abrupt and unpredictable occurrence can heighten both fear and reactivity to that CS, allowing it to become an inhibitor by virtue of its association with the decrement in fear that follows (cf. Wagner, Mazur, Donegan, & Pfautz, 1980). Viewed in this fashion, unsignaled USs do not so much establish CE to the context as they potentiate an acute fear reaction to the presentation of any stimulus in that context.[2]

[2]Although our comments on contextual mediation concern the effects of aversive USs, they suggest similar considerations regarding the effects of appetitive USs. For example, in appetitive conditioning, pretraining is customarily provided in the form of "magazine" training with the US, and that could well potentiate an excitatory reaction to a neutral CS presented in that context. Furthermore, such magazine training often implicates other stimuli (e.g., a light in the food hopper) that, if similar to the CS (e.g., a keylight), could enhance excitatory generalization to the CS. Hence, if unpaired with the US in subsequent training, that CS should develop as a conditioned inhibitor (see Kaplan, 1984).

Excitatory Variations for a Backward CS: An Unsignaled US, a Signaled US, or a Signal for the US (Experiment 4)

Our most recent assessment of CI procedures has returned to the backward-conditioning case to compare explicitly unpaired, differential, and Pavlovian procedures. In the present experiment, different groups of rats were exposed to a nonreinforced CS that always occurred in an immediate, backward relationship to (1) an unsignaled US, (2) a signaled US, or (3) a signal for the US that was reinforced only when presented alone. Note that condition (1) preceding is the standard, backward-conditioning arrangement of an explicitly unpaired procedure, condition (2) is a differential procedure, and condition (3) is a sequential Pavlovian procedure. All three procedures have a common basis, however, in that onset of the putative inhibitor (viz., the backward CS) occurs coincidentally with the termination of an excitatory stimulus, i.e., an unsignaled or a signaled US or just a signal for that US.

Our primary purpose in using these variations of backward conditioning was to assess factors controlling the development of CI. Experiment 1 had indicated that a backward CS acquired CI by being nonreinforced in the presence of US rehearsal. It followed, therefore, that training conditions affecting the "impact" of the US (i.e., the strength and duration of its rehearsal) would modulate CI development to a backward CS. One way in which this could be accomplished, without altering the parameters of the US or its temporal relationship to the backward CS, was by signaling the US. It was well known, for example, that the effect of a US, as indicated by the strength of its unconditioned response (UR), was appreciably decremented by providing a consistent signal for that US (e.g., Kimble & Ost, 1961; Kimmel, 1966). That phenomenon ("conditioned diminution of the UR") plus others, such as the fact that rats typically prefer signaled as opposed to unsignaled shocks (e.g., Badia, Harsh, & Abbott, 1979; Fanselow, 1979), implied that CI would develop less well to a CS that followed a signaled rather than an unsignaled US. In addition, Wagner and Terry (1975) had shown that the initial development of excitatory conditioning to a backward CS was impaired when the US was always preceded by an excitatory CS. According to them (see also Wagner, 1976, 1978), the excitatory CS primed a US representation in short-term memory that attenuated rehearsal of the actual US during occurrence of the backward CS. By our similar way of thinking (see Experiment 1), priming of a US memory representation by the excitatory CS would result not only in poorer CE to the backward CS, but also in poorer CI to that CS.

Our way of thinking was not shared by everyone. Schull (1979, p. 81) had specifically predicted that CI to a backward CS would be stronger with a signaled than with an unsignaled US. According to Schull, the backward CS in the signaled-US procedure would be associated with a strong opponent reaction that was occasioned by the signal and was assumed to be the primary basis by which "anti" (cf. inhibitory) properties developed to a CS contiguous with that reaction. Furthermore, the position advanced by Solomon (1980; Solomon & Corbit, 1974) implied that it did not much matter whether the US was signaled or not, because the opponent process, by which CI could develop, was posited to be primarily a reaction to the affective process generated by the US itself. There were others, though, who agreed with our prediction. Rescorla and Wagner's (1972; Wagner & Rescorla, 1972) position also implied that CI to a backward CS would be better with an unsignaled than with a signaled US, but for decidedly different reasons. Their position argued that the unsignaled-US procedure would be better because the backward CS of that procedure would occur in the presence of excitatory contextual cues whereas, with a signaled US, CE to the context would be overshadowed by the signal for the US.

These different positions could be further distinguished by including a sequential Pavlovian procedure in which the nonreinforced CS occurred immediately after a signal for the US that was reinforced only when not followed by the backward CS. For example, by Rescorla and Wagner's (1972) interpretation, this Pavlovian procedure would produce moderate CI to the backward CS because of moderate excitatory conditioning to the context. That would occur because the signal for the US would not be reinforced when accompanied by the backward CS; hence, it would partially extinguish and would not block CE to the context as well as would a signal that was consistently reinforced. From our perspective, though, the partially reinforced signal of the Pavlovian procedure could only generate a US representation that was weaker than that provided by the unsignaled US itself. For that reason, and those preceding concerning conditioned diminution of the UR, etc., we were merely predicting that an unsignaled US would produce stronger CI to a backward CS than would either a signaled US or just the signal for that US.

In the present experiment, the subjects were given baseline training in both the X and Y chambers on alternate days (with licking responses recorded only in X) and then CS habituation in X. Inhibitory conditioning was conducted entirely in Y, and consisted of a daily 8-min drinking session during which there were 4 presentations of a .5-sec, 1.25-mA shock US on a 2-min VT schedule (range = 50–190 sec). For subjects trained with an unsignaled US (UNS), a 10-sec, white-noise (N) CS always occurred immediately upon termination of the US. The same temporal arrangement of the US and N prevailed for subjects trained with a signaled US (SIG), but, in addition, the US always occurred upon termination of a 10-sec, overhead flashing-light (L) CS. The same arrangement of L and the US prevailed for subjects trained with a Pavlovian procedure (PAV), but, in addition, these subjects received an equal number of L presentations that terminated with N instead of the US. For these PAV subjects, the sequential-compound (LN) presentations were randomly interspersed with the reinforced L presentations but with the restriction that LN occupy a variable mid-locus position, as defined in Experiment 3.

Because we were uncertain of the rate at which CI would develop to N in the three training conditions and wanted to avoid a "ceiling" effect that would obscure differences among them, we assigned the subjects of each training condition to two subgroups that received a total of either 28 or 56 US presentations in inhibitory conditioning. Conditioning for subgroups receiving 28 US presentations was delayed so that it terminated in conjunction with the end of conditioning for subgroups receiving 56 US presentations. To provide an assessment of CI to N that would not be biased by whether the subjects had received N following a US and/or a signal for the US (as might result with a test that summated N with a signal for the US), we employed only a retardation test in X following inhibitory conditioning in Y. However, because that test allowed the possibility of differential US-preexposure effects for the groups (cf. Baker & Mackintosh, 1979; Baker, Mercier, Gabel, & Baker, 1981; LoLordo & Randich, 1981; Randich & LoLordo, 1979), we employed two control groups that did not receive any presentations of N during inhibitory conditioning. One control received presentations of the US alone (USa), and the other received presentations of the US consistently signaled by L (CS+). Like the experimental groups, though, both controls were partitioned into subgroups that received a total of either 28 or 56 US presentations in inhibitory conditioning.

The left panel of Fig. 4.5 presents the pooled data for subgroups of each training condition in 4-trial blocks over the course of the retardation test in X, when N was reinforced on a 50% basis, and on a pretest (PRE) day when N was nonreinforced. The data for trial-blocks 2–4, wherein group differences emerged, are presented in the right panel of Fig. 4.5, as a function of the number of US presentations received during training. As indicated in the left panel, there was reliably less (i.e., retarded) suppression for the UNS subjects as compared with the SIG and PAV subjects, which did not differ; and, as shown in the right panel, both of these effects were independent of the number of US presentations received during training. The right panel also shows that there was less suppression for the USa controls than for the CS+ controls in the 56-US training

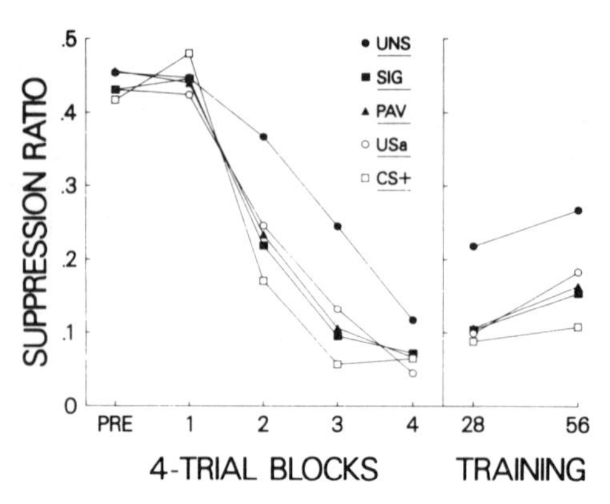

FIG. 4.5. Mean suppression to N for the UNS, SIG, PAV, USa, and CS+ groups during the pretest (PRE) and the retardation test (left panel); and over trial blocks 2–4, as function of the number of US presentations received in training (right panel). Pooled *SE* for N is .021 for the pretest, .022 for Trial-blocks 1–4, and .028 for Trial-blocks 2–4 (cf. right panel).

condition. However, this effect did not alter the overall outcome favoring the UNS subjects. The difference between the UNS subjects and their USa controls was reliably greater than that between the pooled SIG and PAV subjects and their CS+ controls. Furthermore, in the 28-US training condition, where there was no evidence of a differential US-preexposure effect, the UNS subjects showed significantly less suppression than all the other subjects.[3]

The observed differences among the backward-conditioning treatments are apparently not influenced by the type of CI test employed. Overmier and Patterson (1983) have recently reported a similar finding favoring UNS over SIG subjects in a summation test where the backward CS was presented on a Sidman baseline. They suggested that whereas their finding was not amenable to either the Schull (1979) or Solomon (1980) interpretations, it was compatible with the Rescorla–Wagner (1972) position, which predicts that CI will be greater for UNS than for SIG subjects because of excitatory contextual cues that accompany the backward CS in the UNS procedure. However, Overmier and Patterson did not include a PAV procedure in their study, and the present outcome, showing comparably weak CI development with the PAV and SIG procedures, is at odds with the Rescorla–Wagner interpretation. By that interpretation, CI should have been stronger in the PAV than in the SIG procedure, because CE to the context can be sustained at a moderate value in the PAV procedure due to occasional nonreinforcement of the signal. Indeed, the spirit, if not the letter, of the Rescorla–Wagner interpretation implies that CI will be exceedingly strong in the PAV procedure, because the backward CS occurs at the exact moment when the US is predicted to occur and yet does not. Even if one dismisses that consideration and also allows that there may not be much of a difference in the strength of CE to contextual cues in the PAV and SIG procedures, the Rescorla–Wagner interpretation can still be challenged on the basis of the argument given in Experiment 3: Whatever the excitatory strength of contextual cues accompanying a backward CS, those same cues follow that CS virtually unabated in excitatory strength.

We believe that the present findings are more amenable to the analysis that we previously offered. A CS that is backward to an unsignaled US will be subject to the full impact and rehearsal of that US and therefore, over the course of conditioning, strongly reinforced as an inhibitor by the large decrement in US processing that consistently follows it. In contrast, the impact and rehearsal of a signaled US will be considerably tempered because the priming of a US representation by the signal (cf. Wagner, 1976, 1978) will make rehearsal of the actual US redun-

[3]For assorted explanations of the US-preexposure effect, see the references cited earlier on this subject. Note, however, that one interpretation should be added to their list: that US-alone sessions, through sensitization and/or pseudoconditioning, can potentiate an acute fear reaction to the abrupt and unpredictable occurrence of any stimulus; and, in turn, that mediated excitatory effect can block conditioning to the stimulus during a retardation test.

dant. Indeed, from an evolutionary standpoint, it does not make much sense for an animal to waste time processing redundant information, especially when not processing new information can be detrimental to the animal's well-being. Thus, the backward CS of this procedure will be subject to weak inhibitory reinforcement with the decay of that weak US processing.

The same weak CI effect should prevail with a PAV procedure. A signal for the US that is only partially reinforced should produce a far weaker US representation than is provided by the unsignaled US itself, at least with moderate amounts of training, and therefore the backward CS of this procedure should be subject to weak inhibitory reinforcement with the decay of that weak US processing. On the other hand, if a signal for the US were first consistently reinforced on its own, and only then occasionally nonreinforced in compound with a backward CS, the evoked US representation would be substantial, perhaps even as great as that provided by the unsignaled US itself. In this case, the backward CS would be subject to substantial inhibitory reinforcement as, indeed, has been observed (e.g., Grelle & James, 1981). That is undoubtedly why one can develop robust CI to a novel CS that is paired with an excitatory CS during the latter's extinction (e.g., Rescorla, 1979), and also why the lore of the lab touts the Pavolvian procedure as the best. Everyone typically provides A+ training prior to concomitant A+ and AB− training.

The findings of the present and preceding experiments can be assimilated fairly easily because they all point to the same conclusion: The development of CI is dependent on the putative inhibitory CS being processed in both the presence of excitation (i.e., a representation of the US) and the subsequent decay of that excitation. As the different experiments indicate, however, that excitation can be mediated in a variety of ways: by rehearsal of the US, as in a backward-conditioning procedure; by rehearsal of the CS, as in a trace-conditioning procedure; by the presence of a discrete (and reliable) signal for the US, as in a Pavlovian CI procedure; and by contextual cues that potentiate acute fear reactions through sensitization and/or pseudoconditioning effects. Most importantly, though, as Experiment 4 indicates, it is the vividness or strength of that US representation that determines the strength of inhibitory conditioning; and that is where the differences in CI procedures lie. Let us attempt now to support and extend this description.

FACTORS AFFECTING THE MAINTENANCE AND EXTINCTION OF CI

Despite our contesting of the Rescorla–Wagner (1972) interpretation in the preceding section, we have always viewed it as one of the more comprehensive and elegant accounts of conditioning phenomena. Much of our thinking for the prior experiments was influenced by that interpretation; and, indeed, the findings of our experiments suggest a principle of CI development that is not fundamentally different from that expressed by their model: Conditioned inhibition to a stimulus depends on a "negative discrepancy" between the excitation that is present on a trial and that which follows (cf. Rescorla and Wagner's "net associative strength" of concurrent CS elements and "lambda" of the US). The primary difference in interpretation lies in our emphasis on US memory processing as the underlying form of excitation, and even that is not appreciably different from the emphasis developed in subsequent extensions of their model (see Wagner, 1976, 1978, 1981). For these reasons, we have been especially sensitive to the successes and, particularly, the failures of the model as they concern changes in the associative value of an inhibitory CS.

One apparent failure that has plagued the Rescorla–Wagner (1972) theory virtually from its inception (see Wagner & Rescorla, 1972) is its prediction that an inhibitory CS (CS^-), like an excitatory CS (CS^+), will extinguish when the US is omitted and the CS is presented alone. This deduction follows from the requirement of the theory that the negative value of a CS^- generate a "positive discrepancy" by comparison with the zero value of reinforcement that follows it. Accordingly, the negative value of the CS^- should be incremented to zero. Repeated assessments of this prediction, however, have failed to provide any evidence of a loss in the associative value of a CS^- that is presented with the reinforcer omitted (e.g., DeVito, 1980; Rescorla, 1982; Wagner & Rescorla, 1972; Witcher, 1978; Zimmer–Hart & Rescorla, 1974). Furthermore, there is no reliable evidence to support a corollary of the model; namely, that a neutral stimulus in compound with the CS^- will acquire excitatory value as a result of the positive discrepancy that is produced by nonreinforcing the inhibitory compound (Baker, 1974; see also Rescorla, 1971). These failures cannot be taken lightly because they challenge the basic structure of the model.

Possible Excitatory Cues That Can Block the Extinction of CI (Experiment 5)

It is conceivable, though, that all the prior assessments of inhibitory extinction were inadequate because they allowed the CS^- to occur in the presence of conditioned excitatory cues during the inhibitory-extinction phase. For example, all the cited studies appear to have used the same conditioning chambers throughout inhibitory conditioning and extinction, and therefore it is possible that excitatory contextual cues were present during presentations of the CS^- alone. Equally, if not more important, recent research with a Pavlovian CI procedure (e.g., Cunningham, 1981; Rescorla, 1981) has indicated that an association develops between the CS^- and the CS^+ by which the CS^- is established. That being so, presentations of a CS^- by itself could evoke a memory representation of the CS^+, such that the net associative value of the CS^- and the retrieved CS^+ (as well as any excitatory contextual cues) effectively approaches zero. Hence, with little, if any, positive discrepancy, extinction of the CS$^-$ would be blocked. This argument does not imply that the CS$^-$ will be ineffective when presented with the actual CS$^+$, because the excitatory representation evoked by the actual CS$^+$ would be expected to replace, and not summate with, that retrieved by the CS$^-$. Furthermore, the argument is not weakened by the possibility that CE to the retrieved CS$^+$ or the context extinguishes during nonreinforced presentations of the CS$^-$. Nonreinforced presentations of a CS$^-$ can block the extinction of a CS$^+$ that occurs in compound with the CS$^-$ (e.g., Chorazyna, 1962; Soltysik, Wolfe, Nicholas, Wilson, & Garcia–Sanchez, 1983).

In light of the foregoing possibilities, the purpose of our first investigation on this subject (Lysle & Fowler, in press, Experiment 1) was to determine whether presentations of a CS$^-$ by itself would extinguish that CS$^-$, if both the context and the memory representation of an associated CS$^+$ were neutralized prior to the inhibitory-extinction phase. To accomplish this, the following training conditions were imposed. After baseline training and CS habituation in chamber X, the subjects received on-baseline conditioned-suppression training to a 10-sec, clicker stimulus (A) that was consistently reinforced by a .5-sec, 1.2-mA shock US. Then, the subjects were given Pavlovian CI training in which there were 2 daily reinforced presentations of A and, for counterbalanced subjects within each group, 6 daily nonreinforced presentations of A in simultaneous compound with either a 10-sec flashing-light or a 10-sec white-noise CS, the putative inhibitor (B). Such training produced substantial CI: By the end of 20 sessions, the overall mean suppression ratio for A was .04, whereas that for the AB compound was .38. Thereafter, all subjects were given 96 nonreinforced presentations of A in X, over the course of 12 days, in order to extinguish CE to A and thereby neutralize A's memory representation. By the end of that training, mean suppression to A was .38. For 6 days thereafter, different sets of groups received just baseline sessions in the original X chamber or in a different, neutral chamber (Y) where the US and A had never been presented. This was intended to further extinguish any CE to X, as well as to provide a neutral Y context, in either of which the inhibitory CS could be extinguished.

For the inhibitory-extinction phase, which lasted 8 days, two experimental groups were given a total of 64 presentations of B alone while drinking in that context, X or Y, to which they had been exposed following A extinction; these two groups are designated B(X) and B(Y). To assess whether this training would extinguish inhibition, two control groups continued to receive just baseline sessions in their respective X and Y contexts; these two groups are designated (X) and (Y). In addition, to evaluate the extent of any inhibitory extinction for the experimental groups, two other control groups were trained and extinguished exactly like the experimental groups, except with a different inhibitory stimulus (C, a counterbalanced light or noise) used in place of B. Accordingly, these two C(X) and C(Y) control groups received a novel B stimulus in a subsequent retardation test of B's inhibitory property. Prior to that test, all groups were returned to the X context for 2 days of just baseline sessions, and then a single pretest session involving 4 nonreinforced presentations of B. In the retardation test, which also occurred in X, B was presented twice per day, for 6 days, and was reinforced on a 50% basis.

The results of both the pretest (PRE) and the retardation test on B are shown in Fig. 4.6. Although there were no differences among the groups in the pretest, suppression to B in the retardation test developed less rapidly for the B(X) and B(Y) experimental groups by comparison with the C(X) and C(Y) controls, for which B was a novel stimulus. That suggested that B was still an inhibitory CS for the experimental groups. However, there was an unexpected outcome: Suppression to the inhibitory B stimulus for the (X) and (Y) controls developed just as rapidly as that to the novel B stimulus for the C(X) and C(Y) controls. In other words, even though B had been trained as an inhibitory CS for the (X) and (Y) controls and had never been presented to these subjects during the inhibitory-extinction phase, it nonetheless extinguished! How could that have happened?

To gain further information, we gave all subjects additional baseline sessions in X following the retardation test and then reconditioned them to their extinguished A excitor by reinforcing 4 presentations of A on a 50% basis. That was followd by a 2-day summation test in which there were nonreinforced presentations of A alone, a simultaneous AC compound, and C alone. In this manner, we could at least determine whether C was still an inhibitory CS for the C(X) and C(Y) controls that had been trained (and presumably extinguished) with that stimulus. Figure 4.7 shows the results of that summation test, with the data for respective X and Y subgroups of each extinction condition pooled. As indicated, there was reliably less suppression to the AC compound for the C(X,Y) subjects, i.e., the pooled C(X) and C(Y) groups, as compared to the rest; and also reliably less suppression to C for these subjects. Because suppression to the novel C stimulus for the B(X,Y) and (X,Y) subjects was similar to their earlier levels of unconditioned suppression to that stimulus during CS habitua-

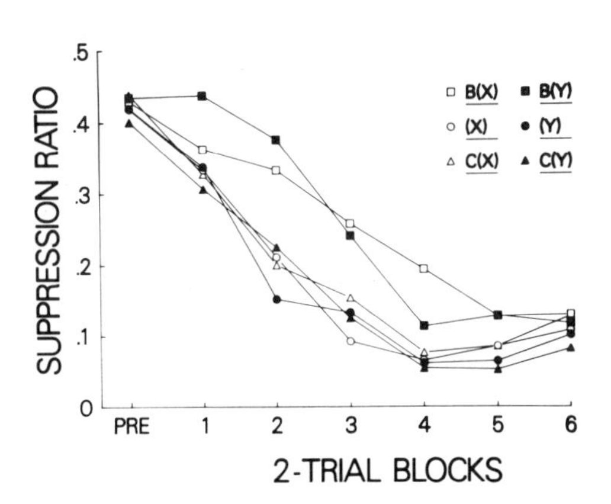

FIG. 4.6. Mean suppression to the B stimulus for the B(X), B(Y), (X), (Y), C(X), and C(Y) groups during the pretest (PRE) and the retardation test. Pooled *SE* for the B stimulus is .023 for the pretest, and .030 for Trial-blocks 1–6 of the retardation test. (From Lysle & Fowler, in press. © 1984 by the American Psychological Association. Reprinted by permission.)

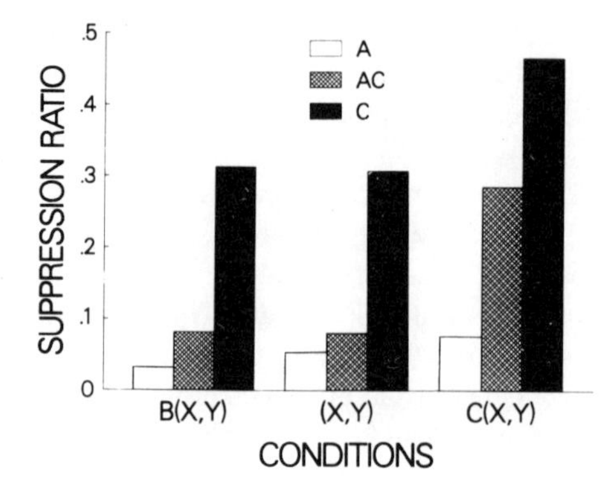

FIG. 4.7. Mean suppression to the A, AC compound, and C stimuli for the B(X,Y), (X,Y), and C(X,Y) groups during the summation test. Pooled *SE* for A, AC, and C is .035, .038, and .047, respectively. (From Lysle & Fowler, in press. © 1984 by the American Psychological Association. Reprinted by permission.)

tion, the reduced suppression to C for the C(X,Y) subjects can be taken to reflect inhibition of unconditioned suppression. That effect complements the inhibition of conditioned suppression evident for the C(X,Y) subjects in the AC compound.

It seems clear from the present results that presentations of a CS^- by itself do not extinguish inhibition to that CS even when a memory representation of its previously associated CS^+ has been neutralized by extinction of the CS^+ (see also Rescorla, 1982), and excitatory contextual cues have been extinguished or replaced with a neutral background. Although all animals received extensive extinction to their A excitor in X, followed by nonreinforced exposure to either X or Y, those subjects that subsequently received presentations of B alone showed an inhibitory effect to that stimulus in the retardation test, and those that subsequently received presentations of C alone showed an inhibitory effect to that stimulus in the summation test. Indeed, the only evidence of inhibitory extinction was that apparent for the (X,Y) subjects in the retardation test. But those subjects had never received extinction training to B.

Two explanations of the present results suggest themselves. One is that inhibition is a labile process that "decays" over an extended delay between inhibitory conditioning and testing (e.g., Henderson, 1978; but see Thomas, 1979). Several theorists have emphasized this unstable nature of the inhibitory process. For example, Pavlov (1927) accounted for spontaneous recovery from excitatory extinction by assuming that a delay would dissipate the inhibition that had accumulated during such extinction. Similarly, Solomon and Corbit (1974; Solomon, 1980) have postulated that an opponent (or inhibitory) reaction to the US is strengthened through use and weakened through disuse. Thus, it may have been the case that inhibition to B for the (X,Y) subjects was considerably weakened by the 28-day period that intervened between inhibitory conditioning and the retardation test. In contrast, the B(X,Y) and C(X,Y) subjects received presentations of their respective inhibitors toward the end of that period, and that experience may have been sufficient to maintain the inhibition to B and C that was apparent for these groups in the retardation and summation tests.

A second explanation of the present results is that inhibition is functionally dependent on (or a "slave" to the master process of) excitation and thus is inoperative when the excitatory process is not viable, i.e., has been extinguished. By this interpretation, the absence of an inhibitory effect of B for the (X,Y) subjects in the retardation test was due merely to prior extinction of the A excitor. That being the case, the inhibitory stimulus for the B(X,Y) subjects would also have been neutralized, but, as with a neutral stimulus, it would have been subject to a latent-inhibition effect due to the subsequent presentations of B alone (cf. Lubow, 1973; Reiss & Wagner, 1972). Neutralization or deactivation of the inhibitory stimulus for the C(X,Y) subjects would also have occurred with A^+ extinction, but due both to the excitatory conditioning of B during the retardation test and to the reconditioning of A following that test, C could have been reactivated as an inhibitor and thus rendered functional in the summation test. Note, in contrast, though, that the neutralized B stimulus for the (X,Y) subjects would not have been effectively reactivated by the shock US in the retardation test on B, because that US was contingent upon B. Hence, as a neutral stimulus, B would acquire an excitatory bias that would oppose reactivation and expression of its inhibitory property.

Decay of CI or Deactivation Through Excitatory Extinction? (Experiment 6)

Our purpose in the present experiment (Lysle & Fowler, in press, Experiment 2) was to evaluate the decay and slave interpretations suggested by the findings of Experiment 5. That was accomplished by giving the present subjects the same Pavlovian CI training as before (i.e., A+ and either AB− or AC−), but then differential excitatory extinction in which pairs of groups trained with B and C received one of the following treatments prior to a retardation test on B and a subsequent retardation test on C: (1) extensive extinction training to A^+ in the original X context; (2) comparable extinction training but just to X; and (3) no extinction training to either A^+ or X.

For those animals trained with B as an inhibitory CS, a slave interpretation predicts that inhibition in the B retardation test will be maintained without loss for the no-extinction subjects but will be appreciably decremented (i.e., deactivated) for subjects receiving A^+ extinction, and to a lesser extent for subjects receiving extinction training just to X. (The latter prediction stems from the finding that nonreinforced exposure to a training context can reduce CE to a CS^+ trained in that context; see Marlin, 1982). On the other hand, the same interpretation predicts that, in spite of the different excitatory–extinction treatments, inhibition in the C retardation test will be fully reactivated for all groups trained with C, because in the preceding retardation test all groups receive substantial excitatory conditioning to B. In contrast, a simple decay interpretation predicts that inhibition in both the B and C retardation tests will be comparably diminished for all groups trained with that particular inhibitor.

The training conditions in the present experiment were comparable to those employed in Experiment 5. Following baseline training and CS habituation in X, all subjects received on-baseline conditioned-suppression training to A (the 10-sec clicker), and then Pavlovian CI training involving reinforced presentations of A and, for different groups, nonreinforced presentations of A in simultaneous compound with either B or C. (Again, B and C were both the 10-sec white noise and flashing light, but counterbalanced within and across groups.) After 20 such sessions, the groups were given different excitatory–extinction treatments. One pair of groups trained with B and C received 96 nonreinforced presentations of A in X over the course of 12 days; these 2 groups are designated A(B) and A(C). A second pair of groups trained with B and C received the same number of sessions in X but without any presentations of A; these 2 groups are designated X(B) and X(C). A final pair of groups trained with B and C did not receive any sessions in X or presentations of A during the extinction phase; instead, these no-extinction groups, designated N(B) and N(C), were merely transported to the experimental room and given their daily 8-min drinking period in detention cages that were comparable to their home cages. Following the extinction phase, all subjects received a pretest in X, consisting of 4 nonreinforced presentations of B, and then a retardation test, during which B was presented twice per day and was reinforced on a 50% basis. Subsequent to that, all subjects were given a nonreinforced posttest on A, to ensure that group differences to A were maintained, and then a retardation test on C.

The results of both the pretest (PRE) and the retardation test on B are presented in Fig. 4.8. As shown, suppression to B developed comparably rapidly for the three groups that were trained with C but received a novel B stimulus in the retardation test. In contrast, suppression to B developed more slowly, and differentially, for the three groups that were both trained and tested with B. Among the

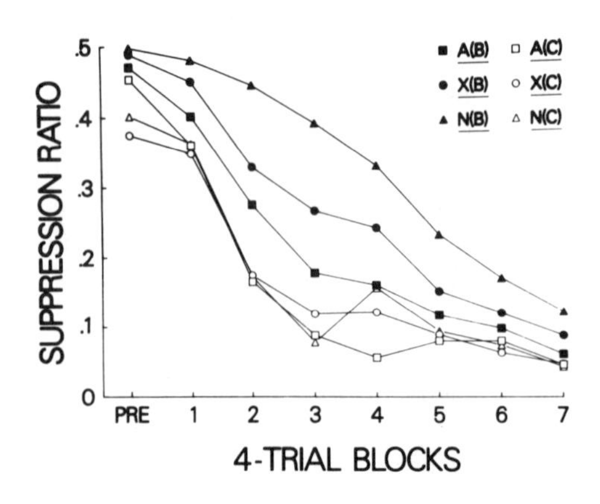

FIG. 4.8. Mean suppression to the B stimulus for the A(B), A(C), X(B), X(C), N(B), and N(C) groups during the pretest (PRE) and the retardation test. Pooled *SE* for the B stimulus is .027 for the pretest, and .021 for Trial-blocks 1–7 of the retardation test. (From Lysle & Fowler, in press. © 1984 by the American Psychological Association. Reprinted by permission.)

latter, the N(B) group developed suppression to B least rapidly, and the A(B) group, most rapidly; indeed, almost as rapidly as the controls that were tested with a novel B stimulus. It should be noted that the differences among the groups trained with B were reliably ordered across the excitatory–extinction treatments; i.e., suppression to B developed more rapidly with the A than X treatment, and more rapidly with the X than N treatment.

Following the B retardation test, all subjects received 3 sessions in X during which there were 4 nonreinforced presentations of A per day. These sessions served to extinguish any ''reinstated'' fear of A that could result for the A(B) and A(C) groups from the B retardation test. Following that, all subjects were given both a pretest and a retardation test on C identical to those administered for B. The left panel of Fig. 4.9 presents the results of the A posttest, and the right panel, those of the pretest (PRE) and the retardation test on C. As shown in the left panel, fear of A was appreciably reinstated for the A(B) and A(C) groups, but the difference in suppression to A between these two groups and the rest was reliable throughout, and was even amplified over, the course of A posttesting. The right panel shows that, despite the differences in suppression to A, suppression to C developed comparably slowly for the three groups that were both trained and tested with C, and comparably rapidly for the three groups that were trained with B but were tested with a novel C stimulus.

The present findings, in particular, those of the B retardation test, are not amenable to the interpretation that CI decays through disuse. The simplest version of that interpretation would predict equal decay of inhibition between inhibitory conditioning and testing for all groups, regardless of the type of excitatory–extinction treatment they received. On the other hand, a more intricate version of a decay interpretation (cf. Solomon, 1980; Solomon & Corbit, 1974) might postulate that the inhibitory (or opponent) process is activated during nonreinforced presentations of a CS$^+$, and therefore inhibition does not decay as rapidly

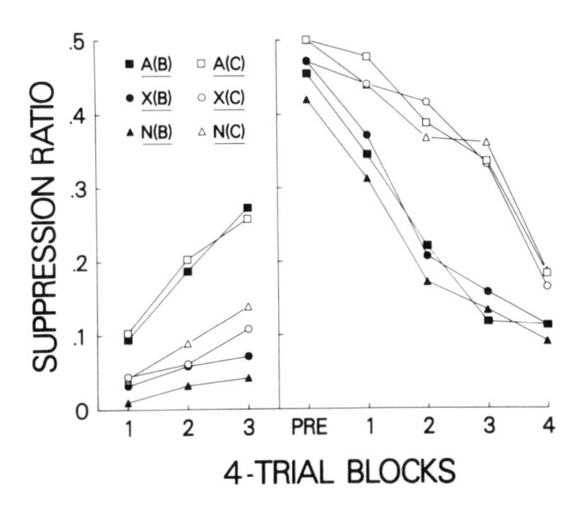

FIG. 4.9. Mean suppression for the A(B), A(C), X(B), X(C), N(B), and N(C) groups to the A stimulus during the posttest (left panel); and to the C stimulus during the subsequent pretest (PRE) and retardation test (right panel). Pooled *SE* for the A stimulus is .027 for Trial-blocks 1–3 of the posttest; and for the C stimulus, .017 for the pretest, and .011 for Trial-blocks 1–4 of the retardation test. (From Lysle & Fowler, in press. © 1984 by the American Psychological Association. Reprinted by permission.)

with this treatment as with a treatment involving no extinction of A. But contrary to either prediction, inhibition in the B retardation test was most pronounced for the N(B) group, and least pronounced for the A(B) group. Although these findings do not deny the possibility that an inhibitory process decays through disuse (see LoLordo & Randich, 1981), it is clear that such decay was not a determinant of the present outcome.

In contrast, the findings of the B retardation test are fully consistent with the interpretation that inhibition is functionally dependent on or a slave to excitation and operates only to the extent that the latter is viable. Accordingly, when conditioned excitation to A was maintained, as in the N(B) treatment, inhibition to B was retained apparently at full strength; but when A^+ and any fear of the context was extinguished, as in the A(B) treatment, inhibition to B was reduced virtually completely. Even though the reduction of inhibition to B in the A(B) treatment was not complete by comparison with those groups that were tested with a novel B stimulus, that outcome may have been due to residual unconditioned suppression to B for the novel-stimulus groups (see the pretest data in Fig. 4.8). Furthermore, it is known that a conditioned excitor retains some of its excitation even after prolonged extinction training (e.g., Reberg, 1972). Hence, to the extent that there was some residual excitation to A for the A(B) group, there should also have been some residual inhibition.

A slave interpretation can also accommodate the partial loss of inhibition to B for the X(B) group that received exposure to only the X training context during the excitatory-extinction phase. Although some theories of conditioning (e.g., Mackintosh, 1975; Rescorla & Wagner, 1972) argue that there should not be sustained conditioned excitation to contextual stimuli when the US is signaled by a discrete CS, other accounts (e.g., Rescorla, 1980; Wagner, 1978) do allow for the development of second-order conditioned excitation to contextual stimuli as signals for the CS^+. Furthermore, other theorists (e.g., Nadel & Willner, 1980) have argued that background stimuli are more than additional stimuli that act like nominal CSs; they suggest a hierarchical relationship in which the context becomes excitatory because it contains and predicts excitatory CSs. The common implication of these interpretations is that some excitation (i.e., fear) should develop to the context, and therefore nonreinforced exposure to the context alone can reduce that fear presumably by devaluing the context's relationship with the specific predictor (e.g., A^+) of an aversive US (see also Marlin, 1982). By a slave interpretation, the reduction of that fear should also reduce the strength of the inhibitory reaction.

The findings of the C retardation test showed comparably strong inhibition for all groups that were trained and tested with C. That outcome is also consistent with a slave interpretation because this interpretation argues that the effective strength, and hence the performance, of an inhibitory response is dependent on the viability of the excitatory process. Consequently, even though the original A excitor was differentially extinguished for the A, X, and N groups immediately prior to the C retardation test, a fully viable excitatory process was still present for these groups because B had been established as a strong conditioned excitor for all animals in the prior retardation test. The outcome of the C retardation test indicates, therefore, that the maintenance or reactivation of a conditioned inhibitor is not restricted to that excitor used to establish the inhibitor but extends as well to other, independently trained excitors. Such a view is fully consistent with the finding that a CS^- can function in an inhibitory capacity with an independently trained excitor, even though the original excitor, by which the CS^- was established, has been extinguished (e.g., Rescorla, 1982; Rescorla & Holland, 1977).

The Maintenance of CI: Context-Specific Versus a Generic Form of Excitation (Experiments 7 and 8)

Although the preceding experiment provides strong evidence that inhibition operates only to the extent that excitation is maintained, it does not establish whether that excitation (or fear) must be specifically associated with the context in which the inhibitor is trained and tested, or merely present "in" the animal in the form of some generic, memorial representation of the CS^+ and/or the US. This is an important question because it focuses on the mechanism by which inhibition is functionally dependent on excitation. For example, if fear of a CS^+ or of the US *in the test context* is the means by which a CS^- is functionally maintained in that context, it could be argued that an associative mechanism is at play whereby the context potentiates fear and, in that manner, activates an inhibitory reaction to the CS^-. However, if fear need only be present *in the animal* in order for inhibition to operate, then a nonassociative mechanism could be opted for whereby inhibition is motivated and thus maintained by an excitatory representation of generic form.

The purpose of the present experiment (Lysle & Fowler, in press, Experiment 3) was to assess whether inhibition would be maintained when conditioned excitation was extinguished in the original context (X) but was established for the same animals in a different context (Y). To effect this assessment, the present subjects were given baseline training and CS habituation in both the X and Y chambers on separate days. Then, on-baseline in X, they received conditioned-suppression training to A (the 10-sec clicker), followed by Pavlovian CI training in which, as before, there were reinforced presentations of A and, for different groups, nonreinforced presentations of A in compound with either B or C. (Again, B and C were both the 10-sec white noise and flashing light but counterbalanced within and across groups.) Thereafter, the subjects were given 16 days of conditional-discrimination training in which they received two 8-min sessions per day, one in X and one in Y, but with water available in only one context on any day, and with the order of presentation of X and Y counterbalanced within each group. During this phase, one pair of B and C groups received an average of

2 daily reinforced presentations of A in X (range = 1–3, randomly varied across days), and 6 daily nonreinforced presentations of A in Y. These two groups are designated B:A+/A and C:A+/A to indicate their inhibitory training with either B or C, and their A treatments in X and Y, respectively. In reverse fashion, a second pair of B and C groups received 6 daily nonreinforced presentations of A in X and an average of 2 daily reinforced presentations of A in Y; accordingly, these two groups are designated B:A/A+ and C:A/A+. Such conditional-discrimination training was highly effective. By the end of that training, mean suppression to A in X for the A+/A subjects was .07, whereas that for the A/A+ subjects was .47. (Responding in Y was not recorded.)

A fifth group, also trained with B as the inhibitory CS, was used to evaluate any general effects that conditional-discrimination training might exert on B's inhibitory property in a subsequent retardation test in X. During the conditional phase, this fifth group did not receive any presentations of A or of the US; instead, it received the no-extinction treatment of Experiment 5 (i.e., daily placements in detention cages instead of X) along with daily exposures to Y that matched the sequence of exposures given the conditional subjects. This fifth group is designated B:N/--, to indicate no extinction of any kind in X and merely exposure to Y.

The left panel of Fig. 4.10 presents the results of the pretest (PRE) and the retardation test on B that followed a single baseline session in X for all groups. As shown, suppression to B developed comparably rapidly for the two groups that were trained with C as an inhibitory CS but were tested with a novel B stimulus in the retardation test. However, it developed comparably slowly for all three groups that were both trained and tested with an inhibitory B stimulus. These results indicated that the inhibitory effect of B depended not on whether the animals feared A in X, but only that they feared A (in either X or some other context). As such, the findings of the retardation test argued that inhibition was

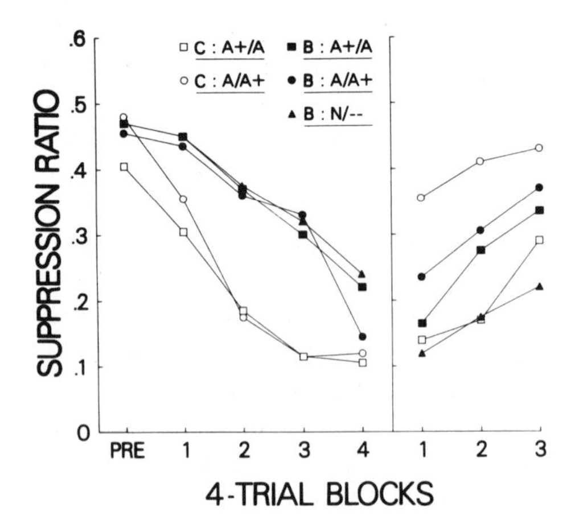

FIG. 4.10. Mean suppression for the C:A+/A, C:A/A+, B:A+/A, B:A/A+, and B:N/-- groups to the B stimulus during the pretest (PRE) and the retardation test (left panel); and to the A stimulus during the subsequent posttest (right panel). Pooled *SE* for the B stimulus is .022 for the pretest, and .029 for Trial-blocks 1–4 of the retardation test; and for the A stimulus, .032 for Trial-blocks 1–3 of the posttest. (From Lysle & Fowler, in press. © 1984 by the American Psychological Association. Reprinted by permission.)

maintained by some generic form of excitation that the animal carried with it into a context. To check this conclusion, we subsequently gave all subjects a 2-day posttest in X that involved a total of 12 nonreinforced presentations of A. The results of that posttest on A are shown in the right panel of Fig. 4.10.

As expected, suppression to A in X was minimal for the C:A/A+ group that had received nonreinforced presentations of A in X during the conditional phase, and it was substantial for both the C:A+/A and B:N/-- groups that had received reinforced presentations of A in X during either the conditional phase or just the earlier excitatory- and inhibitory-conditioning phases. By comparison, though, suppression to A in X was intermediate for both the B:A+/A and B:A/A+ groups, despite their differential treatments with A in X during the conditional phase. Evidently, fear of A in X had been transformed to a moderate value for these two groups during the retardation test. We are not certain how this happened. Our guess is that for these two groups, presentations of B in that test evoked a memory of A being both reinforced and nonreinforced during the conditional phase, and that recall, together with the 50% reinforcement in the retardation test, resulted in A's reprocessing as a moderate excitor. Whether A's reprocessing occurred in this manner or not, it is apparent that both groups were moderately afraid of A in X by the end of the retardation test, and therefore one can argue that B's operation as an inhibitor was not independent of fear of A in that context.

That argument, though, overlooks a pertinent consideration. In the previous study (Experiment 6), the group that was trained with B and extinguished to A in X prior to the B retardation test, i.e., the A(B) group, also showed a substantial reinstatement of fear to A in X in the following posttest (see Fig. 4.9, left panel). However, that A(B) group exhibited virtually no inhibition to B in the retardation test, in contrast to the robust inhibition observed for the B:A/A+ group of the present experiment. Given that difference, one can hardly attribute B's inhibitory effect to a reinstatement of fear to A in X. Rather, the difference in inhibitory effects for these two groups implicates the role of their different treatments prior to the retardation test, in particular, the development of excitation to A in Y for the B:A/A+ group, but not for the A(B) group.

Because of the seeming complications provided by the results of the A posttest in the present experiment, we conducted a final study (Lysle & Fowler, in press, Experiment 4) that also assessed whether inhibition would be maintained when excitation was extinguished in the original context (X) but was established in a different context (Y). However, the tact employed was slightly different. After baseline training and CS habituation in just X, and then on-baseline conditioned-suppression training to A (the 10-sec clicker), all subjects were given the same Pavlovian CI training as before but with just A+ and AB−. (Again, B was either the 10-sec white noise or flashing light, counterbalanced for subjects within each group.) Thereafter, the subjects received 16 days of differential conditioning across the X and Y contexts that involved a daily 8-min session in each context but with water available only in X. Four groups received 6 daily nonreinforced presentations of A in X, along with one of the following treatments in Y: (1) exposure only (designated A/--, to indicate the respective X and Y treatments); (2) US presentations only (A/+); (3) US presentations explicitly unpaired with A, with A occupying a variable mid-locus position, as earlier defined (A/+A); and (4) US presentations contingent upon a novel stimulus, C (A/C+). For subjects of the A/C+ group trained with B as either a noise or a light, C was a light and tone (10 sec, 80 dB, 1500 Hz), respectively. For groups that received the US in Y, there were, on the average, 2 such presentations per day (range = 1–3, randomly varied across days), as well as a corresponding number of A and C presentations for the A/+A and A/C+ groups, respectively. Again, a fifth group served as a no-extinction control and did not receive any presentations of A or the US, or exposures to X, during the differential phase. Instead, it received two 8-min sessions per day, one in the detention cage, where water was available, and the other in Y. This group is designated N/--, to indicate no extinction to A or X and only exposure to Y.

The differential phase was followed by a single baseline session in X for all subjects, a B pretest session, and then a B retardation test in X. The results of the pretest (PRE) and the retardation test are shown in the left panel of Fig. 4.11. As expected, suppression to B developed very rapidly for the A/-- group that had received just excitatory extinction to A in X, and reliably less slowly for the no-extinction control, N/--. This difference replicates the effect found in Experiment 6 between the A(B) and N(B) groups (cf. Fig. 4.8), although in the present experiment, conditioned suppression to B developed more rapidly. The more interesting result, however, was that for the three US-in-Y groups (A/+, A/+A, and A/C+). Even though these groups had received extinction training to A in X identical to that for the A/-- group (and with equal loss of suppression to A), they

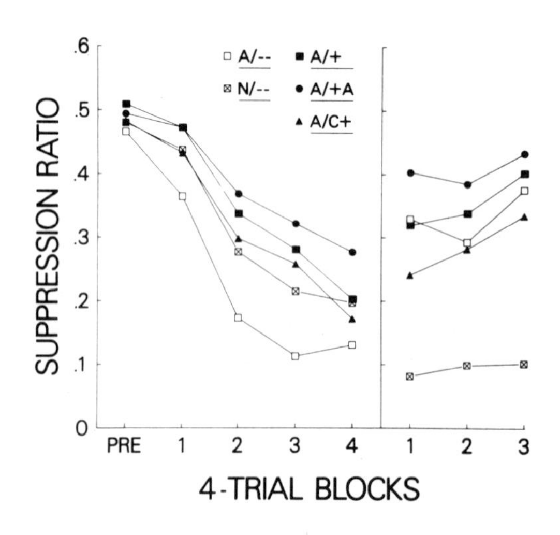

FIG. 4.11. Mean suppression for the A/--, N/--, A/+, A/+A, and A/C+ groups to the B stimulus during the pretest (PRE) and the retardation test (left panel); and to the A stimulus during the subsequent posttest (right panel). Pooled *SE* for the B stimulus is .013 for the pretest, and .028 for Trial-blocks 1–4 of the retardation test; and for the A stimulus, .024 for Trial-blocks 1–3 of the posttest. (From Lysle & Fowler, in press. © 1984 by the American Psychological Association. Reprinted by permission.)

showed retarded suppression to B that was comparable to (and not significantly different from) that for the no-extinction control, N/--. Hence, the present results, like those of the preceding experiment, indicated that the maintenance of inhibition depended not on fear of A in X, but upon some general excitatory representation that the animal carried with it into that context. There still was the question, though, of whether fear of A in X had remained extinguished for the three US-in-Y groups during the retardation test.

To answer that question, all subjects were subsequently given a posttest on A in X that was identical in all respects to that for the preceding experiment. The right panel of Fig. 4.11 presents the results of the posttest. As shown, suppression to A in X for the no-extinction (N/--) control was substantial, whereas it was appreciably and comparably reduced for the remaining groups, all of which had received extinction training to A in X during the differential phase. (Note that the seemingly weakest suppression of all, that for the A/+A group, can be attributed to the additional presentations of A in Y that were negatively correlated with the US for this group during the differential phase.) Because the reduced suppression to A in X for the three US-in-Y groups was comparable to that for the A/-- group, the difference between these groups in the B retardation test cannot be ascribed to differential fear of A in X. Hence, the present results argue clearly that the comparable inhibitory effect of B shown in the retardation test by the three US-in-Y groups and the no-extinction control was due merely to a viable excitatory representation for these groups, one that was generic in nature and not specifically associated with A and/or X.

SUMMARY AND CONCLUSIONS

The main findings of the preceding set of experiments (5–8) can be summarized fairly succinctly. When a conditioned inhibitor is established via a Pavlovian procedure involving reinforced presentations of one stimulus (A) and nonreinforced presentations of that stimulus in simultaneous compound with another (e.g., B), extinction of the A excitor will deactivate the inhibitory property of B, despite the lack of any manipulation of the latter (Experiments 5, 6, and 8). However, if, in conjunction with A extinction in the original context, the animal receives US presentations in a different context, either alone or for a novel CS, or even positively or negatively correlated with A, the inhibitory property of B will be maintained without apparent loss (Experiments 7 and 8). Furthermore, if, following A extinction in the original context, the animal receives US presentations for the same or a different CS in that context, the inhibitory property of B (cf. C) will be restored apparently to full strength (Experiments 5 and 6). Collectively, these findings indicate that conditioned inhibition is a slave process. It is functionally dependent on excitation in the form of some generic representation of the CS^+ and/or the US, for it operates only to the extent that an excitatory representation is maintained.

We can readily assimilate the findings of our earlier set of experiments (1–4) because they also speak to the manner in which inhibition is dependent on excitation. As earlier described, the results of those experiments argued essentially for two tenets: (1) The development of CI to a CS occurs only when that CS is processed together with the occurrence and decay of an excitatory representation in short-term memory. (2) Although such a representation can be mediated either by conditioned excitatory cues (of a discrete or ambient nature) or by rehearsal of the US itself, the initial strength and subsequent decay of that representation determines the strength of the conditioned inhibitory response. We would now propose that whereas such initial processing occupies short-term memory, the inhibitory reaction to the CS^- becomes motivationally dependent on a generic representation of the US and/or the CS^+ that is registered in the animal's long-term memory store. That description of the development of CI accords well with the maintenance of CI because Experiments 5–8 argue that, to the extent that an excitatory representation is eliminated from the animal's long-term memory store, the CS^- loses its functional property, only to be reactivated when an excitatory representation is restored.

ACKNOWLEDGMENTS

The research reported in this chapter was supported in part by both a National Institute of Health grant (MH24115) and a University of Pittsburgh Research Development Award to Harry Fowler. Experiments 1 and 2 and Experiments 5 and 6 were, respectively, submitted by Morris C. Kleiman and Donald T. Lysle in partial fulfillment of the requirements for the Master of Science degree at the University of Pittsburgh.

REFERENCES

Badia, P., Harsh, J., & Abbott, B. (1979). Choosing between predictable and unpredictable shock conditions: Data and theory. *Psychological Bulletin, 86,* 1107–1131.

Baker, A. G. (1974). Conditioned inhibition is not the symmetrical opposite of conditioned excitation: A test of the Rescorla–Wagner model. *Learning and Motivation, 5,* 396–379.

Baker, A. G. (1977). Conditioned inhibition arising from between-sessions negative correlation. *Journal of Experimental Psychology: Animal Behavior Processes, 3,* 144–155.

Baker, A. G., & Mackintosh, N. J. (1979). Preexposure to the CS alone, US alone, or CS and US uncorrelated: Latent inhibition, blocking by context or learned irrelevance? *Learning and Motivation, 10,* 278–294.

Baker, A. G., Mercier, P., Gabel, J., & Baker, P. A. (1981). Contextual conditioning and the US preexposure effect in conditioning fear. *Journal of Experimental Psychology: Animal Behavior Processes, 7,* 109–128.

Bolles, R. C., Collier, A. C., Bouton, M. E., & Marlin, N. E. (1978). Some tricks for ameliorating the trace-conditioning deficit. *Bulletin of the Psychonomic Society, 11,* 403–406.

Burkhardt, P. E. (1980). One-trial backward fear conditioning in rats as a function of US intensity. *Bulletin of the Psychonomic Society, 15,* 9–11.

Chorazyna, H. (1962). Some properties of conditioned inhibition. *Acta Biologiae Experimentalis, 22,* 5–13.

Cunningham, C. L. (1981). Association between the elements of a bivalent stimulus. *Journal of Experimental Psychology: Animal Behavior Processes, 7,* 425–436.

Davis, M. (1974). Sensitization of the rat startle response by noise. *Journal of Comparative and Physiological Psychology, 87,* 571–581.

Desmond, J. E., Romano, A. G., & Moore, J. W. (1980). Amplitude of the rabbit's unconditioned nictitating membrane response in the presence of a conditioned inhibitor. *Animal Learning and Behavior, 8,* 225–230.

DeVito, P. L. (1980). *The extinction of a conditioned fear inhibitor and conditioned fear excitor.* Unpublished doctoral dissertation, University of Pittsburgh.

Fanselow, M. S. (1979). Naloxone attenuates rat's preference for signaled shock. *Physiological Psychology, 7,* 70–74.

Fowler, H., & Lysle, D. T. (1982). (*Attenuation of a conditioned inhibitor by transfer to a novel context*). Unpublished raw data.

Grelle, M. J., & James, J. H. (1981). Conditioned inhibition of fear: Evidence for a competing response mechanism. *Learning and Motivation, 12,* 300–320.

Groves, P. M., & Thompson, R. F. (1970). Habituation: A dual process theory. *Psychological Review, 77,* 419–450.

Henderson, R. W. (1978). Forgetting of conditioned fear inhibition. *Learning and Motivation, 9,* 16–30.

Heth, C. D. (1976). Simultaneous and backward fear conditioning as a function of number of CS–US pairings. *Journal of Experimental Psychology: Animal Behavior Processes, 2,* 117–129.

Heth, C. D., & Rescorla, R. A. (1973). Simultaneous and backward fear conditioning in the rat. *Journal of Comparative and Physiological Psychology, 82,* 434–443.

Hinson, R. E., & Siegel, S. (1980). Trace conditioning as an inhibitory procedure. *Animal Learning and Behavior, 8,* 60–66.

Jacobs, W. J., & LoLordo, V. M. (1980). Constraints on Pavlovian aversive conditioning: Implications for avoidance learning in the rat. *Learning and Motivation, 11,* 427–455.

Kaplan, P. S. (1984). Importance of relative temporal parameters in trace autoshaping: From excitation to inhibition. *Journal of Experimental Psychology: Animal Behavior Processes, 10,* 113–126.

Keith–Lucas, T., & Guttman, N. (1975). Robust-single-trial backward conditioning. *Journal of Comparative and Physiological Psychology, 88,* 468–476.

Kimble, G. A., & Ost, J. W. P. (1961). A conditioned inhibitory process in eyelid conditioning. *Journal of Experimental Psychology, 61,* 150–156.

Kimmel, H. D. (1966). Inhibition of the unconditioned response in classical conditioning. *Psychological Review, 73,* 232–240.

Kleiman, M. C., & Fowler, H. (1984). Variations in explicitly unpaired training are differentially effective in producing conditioned inhibition. *Learning and Motivation, 15,* 127–155.

LoLordo, V. M., & Randich, A. (1981). Effects of experience of electric shock upon subsequent conditioning of an emotional response: Associative and non-associative accounts. In P. Harzem & M. H. Zeiler (Eds.), *Advances in analysis of behavior: Predictability, correlation, and contiguity* (Vol. 2, pp. 247–285). Sussex, England: Wiley.

Lubow, R. E. (1973). Latent inhibition. *Psychological Bulletin, 79,* 398–407.

Lysle, D. T. (1983). *Nonassociative extinction of a conditioned inhibitor.* Unpublished Master's thesis, University of Pittsburgh.

Lysle, D. T., & Fowler, H. (in press). Inhibition as a "slave" process: Deactivation of conditioned inhibition through extinction of conditioned excitation. *Journal of Experimental Psychology: Animal Behavior Processes.*

Mackintosh, N. J. (1973). Stimulus selection: Learning to ignore stimuli that predict no change in reinforcement. In R. A. Hinde & J. Stevenson-Hinde (Eds.), *Constraints on learning* (pp. 75–100). London: Academic Press.

Mackintosh, N. J. (1974). *The psychology of animal learning.* London: Academic Press.

Mackintosh, N. J. (1975). A theory of attention: Variation in the associability of stimuli with reinforcement. *Psychological Review, 82,* 276–298.

Mahoney, W. J., & Ayres, J. J. B. (1976). One-trial simultaneous and backward fear conditioning as reflected in conditioned suppression of licking in rats. *Animal Learning and Behavior, 4,* 357–362.

Mahoney, W. J., Kwaterski, S. E., & Moore, J. W. (1975). Conditioned inhibition of the rabbit nictitating membrane response as a function of CS–US interval. *Bulletin of the Psychonomic Society, 5,* 177–179.

Maier, S. F., Rapaport, P., & Wheatley, K. L. (1976). Conditioned inhibition and the UCS–CS interval. *Animal Learning and Behavior, 4,* 217–220.

Maier, S. F., Seligman, M. E. P., & Solomon, R. L. (1969). Pavlovian fear conditioning and learned helplessness: Effects on escape and avoidance behavior of (a) the CS–US contingency and (b) the independence of the US and voluntary responding. In B. A. Campbell & R. M. Church (Eds.), *Punishment and aversive behavior* (pp. 299–342). New York: Appleton–Century–Crofts.

Marchant, H. G., III, & Moore, J. W. (1974). Below-zero conditioned inhibition of the rabbit's nictitating membrane response. *Journal of Experimental Psychology, 102,* 350–352.

Marlin, N. A. (1982). Within-compound associations between the context and the conditioned stimulus. *Learning and Motivation, 13,* 526–541.

Moscovitch, A., & LoLordo, V. M. (1967, April). *Backward conditioning and cessation conditioning produce inhibition.* Paper presented at the meeting of the Eastern Psychological Association, Boston.

Moscovitch, A., & LoLordo, V. M. (1968). Role of safety in the Pavlovian backward fear conditioning procedure. *Journal of Comparative and Physiological Psychology, 66,* 673–678.

Nadel, L., & Willner, J. (1980). Context and conditioning: A place for space. *Physiological Psychology, 8,* 218–228.

Overmier, J. B., & Patterson, J. (1983, May). *A transfer of control assessment of Solomon's opponent-process model.* Paper presented at the meeting of the Midwestern Psychological Association, Minneapolis.

Pavlov, I. P. (1927). *Conditioned reflexes.* Oxford: Oxford University Press.

Plotkin, H. C., & Oakley, D. A. (1975). Backward conditioning in the rabbit (*oryctolagus cuniculus*). *Journal of Comparative and Physiological Psychology, 88,* 586–590.

Randich, A., & LoLordo, V. M. (1979). Preconditioning exposure to the unconditioned stimulus affects the acquisition of a conditioned emotional response. *Learning and Motivation, 10,* 245–277.

Reberg, D. (1972). Compound tests for excitation in early acquisition and after prolonged extinction of conditioned suppression. *Learning and Motivation, 3,* 246–258.

Reiss, S., & Wagner, A. R. (1972). CS habituation produces a "latent inhibition effect" but no active "conditioned inhibition." *Learning and Motivation, 3,* 237–245.

Rescorla, R. A. (1966). Predictibility and number of pairings in Pavlovian fear conditioning. *Psychonomic Science, 4,* 383–384.

Rescorla, R. A. (1967). Pavlovian conditioning and its proper control procedures. *Psychological Review, 74,* 71–80.

Rescorla, R. A. (1968). Pavlovian conditioned fear in Sidman avoidance learning. *Journal of Comparative and Physiological Psychology, 65,* 55–60.

Rescorla, R. A. (1969a). Conditioned inhibition of fear resulting from negative CS–US contingencies. *Journal of Comparative and Physiological Psychology, 67,* 504–509.

Rescorla, R. A. (1969b). Pavlovian conditioned inhibition. *Psychological Bulletin, 72,* 77–94.

Rescorla, R. A. (1971). Variations in the effectiveness of reinforcement and nonreinforcement following prior inhibitory conditioning. *Learning and Motivation, 2,* 113–123.

Rescorla, R. A. (1976). Second-order conditioned inhibition. *Learning and Motivation, 7,* 161–172.

Rescorla, R. A. (1979). Conditioned inhibition and extinction. In A. Dickinson & R. A. Boakes (Eds.), *Mechanisms of learning and motivation: A memorial volume to Jerzy Konorski* (pp. 83–110). Hillsdale, NJ: Lawrence Erlbaum Associates.

Rescorla, R. A. (1980). *Pavlovian second-order conditioning: Studies in associative learning.* Hillsdale, NJ: Lawrence Erlbaum Associates.

Rescorla, R. A. (1981). Within-signal learning in autoshaping. *Animal Learning and Behavior, 9,* 245–252.

Rescorla, R. A. (1982). Some consequences of associations between the excitor and the inhibitor in a conditioned inhibition paradigm. *Journal of Experimental Psychology: Animal Behavior Processes, 8,* 288–298.

Rescorla, R. A., & Holland, P. C. (1977). Associations in Pavlovian Conditioned Inhibition. *Learning and Motivation, 8,* 429–447.

Rescorla, R. A., & LoLordo, V. M. (1965). Inhibition of avoidance behavior. *Journal of Comparative and Physiological Psychology, 59,* 406–412.

Rescorla, R. A., & Wagner, A. R. (1972). A theory of Pavlovian conditioning: Variations in the effectiveness of reinforcement and nonreinforcement. In A. H. Black & W. H. Prokasy (Eds.), *Classical conditioning II: Current research and theory* (pp. 64–99). New York: Appleton–Century–Crofts.

Schull, J. (1979). A conditioned opponent theory of Pavlovian conditioning and habituation. In G. H. Bower (Ed.), *The psychology of learning and motivation* (Vol. 13, pp. 57–90). New York: Academic Press.

Sheafor, P. J. (1975). "Pseudoconditioned" jaw movements of the rabbit reflect associations conditioned to the contextual background cues. *Journal of Experiment Psychology: Animal Behavior Processes, 104,* 245–260.

Shurtleff, D., & Ayres, J. J. B. (1981). One-trial backward excitatory fear conditioning in rats: Acquisition, retention, extinction, and spontaneous recovery. *Animal Learning and Behavior, 9,* 65–74.

Siegel, S., & Domjan, M. (1971). Backward conditioning as an inhibitory procedure. *Learning and Motivation, 2,* 1–11.

Smith, M. C., Coleman, S. R., & Gormezano, I. (1969). Classical conditioning of the rabbit's nictitating membrane response at backward, simultaneous and forward CS–US intervals. *Journal of Comparative and Physiological Psychology, 69,* 223–231.

Solomon, R. L. (1980). The opponent-process theory of motivation: The costs of pleasure and the benefits of pain. *American Psychologist, 35,* 691–712.

Solomon, R. L., & Corbit, J. D. (1974). An opponent-process theory of motivation. *Psychological Review, 81,* 119–145.

Soltysik, S. S., Wolfe, G. E., Nicholas, T., Wilson, W. J., & Garcia–Sanchez, J. L. (1983). Blocking of inhibitory conditioning within a serial conditioned inhibitor compound: Maintenance of acquired behavior without an unconditioned stimulus. *Learning and Motivation, 14,* 1–29.

Thomas, D. A. (1979). Retention of conditioned inhibition in a bar-press suppression paradigm. *Learning and Motivation, 10,* 161–177.

Wagner, A. R. (1976). Priming in STM: An information-processing mechanism for self-generated or retrieval-generated depression in performance. In T. J. Tighe & R. N. Leaton (Eds.), *Habituation: Perspectives from child development, animal behavior, and neuropsychology* (pp. 95–128). Hillsdale, NJ: Lawrence Erlbaum Associates.

Wagner, A. R. (1978). Expectancies and the priming of STM. In S. H. Hulse, H. Fowler, & W. K. Honig (Eds.), *Cognitive processes in animal behavior* (pp. 177–209). Hillsdale, NJ: Lawrence Erlbaum Associates.

Wagner, A. R. (1981). SOP: A model of automatic memory processing in animal behavior. In N. E. Spear & R. R. Miller (Eds.), *Information processing in animals: Memory mechanisms* (pp. 5–47). Hillsdale, NJ: Lawrence Erlbaum Associates.

Wagner, A. R., Mazur, J. E., Donegan, N. H., & Pfautz, P. L. (1980). Evaluation of blocking and conditioned inhibition to a CS signaling a decrease in US intensity. *Journal of Experimental Psychology: Animal Behavior Processes, 6,* 376–385.

Wagner, A. R., & Rescorla, R. A. (1972). Inhibition in Pavlovian conditioning: Application of a theory. In R. A. Boakes & M. S. Halliday (Eds.), *Inhibition and learning* (pp. 301–336). London: Academic Press.

Wagner, A. R., & Terry, W. S. (1975). Backward conditioning to a CS following an expected vs a surprising UCS. *Animal Learning and Behavior, 3,* 370–374.

Weisman, R. G., & Litner, J. S. (1969). The course of Pavlovian excitation and inhibition of fear in rats. *Journal of Comparative and Physiological Psychology, 69,* 667–672.

Weisman, R. G., & Litner, J. S. (1971). Role of the intertrial interval in Pavlovian differential conditioning of fear in rats. *Journal of Comparative and Physiological Psychology, 74,* 211–218.

Wickens, D. D., & Wickens, C. (1942). Some factors related to pseudoconditioning. *Journal of Experimental Psychology, 31,* 518–526.

Witcher, E. S. (1978). *Extinction of Pavlovian conditioned inhibition.* Unpublished doctoral dissertation, University of Massachusetts.

Zimmer–Hart, C. L., & Rescorla, R. A. (1974). Extinction of Pavlovian conditioned inhibition. *Journal of Comparative and Physiological Psychology, 86,* 837–845.

5 Does Inhibition Differ From Excitation? Proactive Interference, Contextual Conditioning, and Extinction

A. G. Baker and Patricia A. Baker
McGill University

What appears to us to be the seminal issue in the study of conditioned inhibition is its relation to conditioned excitation. In the past, conditioned inhibitory mechanisms have been claimed to be symmetrically opposite to or even totally different from those of conditioned excitation. For instance, Wagner and Rescorla (1972) have claimed inhibition arises from negative habit strength, whereas Rescorla (1979, see also Konorski, 1948 and Pavlov, 1927, for similar accounts) has separately argued that conditioned inhibition is produced by changes in the threshold for the manifestation of the conditioned response. One would think, considering the proliferation of such claims, that there must exist a strong body of empirical evidence showing fundamental differences between excitatory and inhibitory conditioning.

Surprisingly there is not much evidence that even equivocally separates the two mechanisms. Our strategy in discussing the "differences" between inhibition and excitation involves briefly describing a recent theoretical conceptualization of inhibition, we then discuss some of our own and other research that we consider to be relevant to the issue and, finally, we claim that inhibition and excitation arise from similar learning mechanisms. A consequence of our analysis is that much of the "difference" between inhibition and excitation may be due more to the associative models designed to characterize them than to any real differences in the processes themselves.

One fairly consistent, although rarely discussed, difference between inhibitory and excitatory conditioning is the relative ease with which they are obtained. In the conditioned emotional response preparation, for instance, strong conditioning is usually obtained following three or four trials. On the other hand, inhibitory conditioning takes many more trials to develop (e.g., Baker, 1974,

Experiment 1). On the face of it, it would appear from this rather simple difference that there may be a fundamental difference in the mechanisms by which inhibition and excitation are conditioned.

There is, however, a fundamental logical problem with such comparisons that dates back, at least, to Thorndike's analysis of the negative law of effect. Thorndike wished to determine whether annoyers were as effective in stamping out behavior as satisfiers were at stamping it in. The design of the experiments was elegantly simple. Thorndike directly compared satisfiers and annoyers in order to determine which had the most profound effect upon behavior. Unfortunately, whereas this is simple in principle, it is obvious that for such a comparison to be meaningful the reinforcers must be equally salient and equally motivating. In Thorndike's case the satisfier needed to be just as "good" as the annoyer was "bad."

In an often-cited example of one of these experiments Thorndike (1932) compared warmth, food, light, and social stimulation as a satisfier, with restraint in a small, cool, dark box as an annoyer for chicks. He found the satisfier to be much more effective in stamping in behavior than the annoyer was in stamping it out. He used those results, among others, to repeal the negative law of effect and, indirectly, to minimize the role of punishment in North American education and child rearing. An alternative explanation of his results, however, was that, rather than comparing two separate learning mechanisms, he was simply comparing the effectiveness of two different reinforcers. It is easy to think up a case in which he might have found annoyers much more effective than satisfiers. For instance he might have compared a small amount of bird seed as a satisfier, with a compound of a strong electric shock, a loud noise, the silhouette of a flying predator, and a cat as the annoyer. If this comparison had been made Thorndike likely would have been forced to repeal the positive law of effect and our educational system might have been ruled by the rod.

This example shows how a theoretical bias can easily influence the interpretation and the perceived importance of data. We argue that a more parsimonious view of Thorndike's results is that events influence behavior depending on their salience, motivational significance, predictability, and controllability. In his experiment, Thorndike simply demonstrated that one reinforcer was more effective than the other.

To return from this digression we want to point out that the US is the reinforcer for excitatory conditioning and its absence is the reinforcer for inhibitory conditioning. The effectiveness of a reinforcer is determined by its salience, motivational significance, predictability, and controllability. Based on both salience and significance a US, say a shock, is likely to be learned about rather more rapidly than its absence.

Prior to the first conditioning trial the shock is novel to the animal so it is attention getting; it is painful so it is motivating. On the other hand at the beginning of inhibitory training the animal has had a lifetime of experience with the absence of shock, thus fairly heroic efforts must be made to make this absence salient. Furthermore, the absence of shock signals no change in the animal's usual state of affairs so it is not highly motivating. The animal must be made to expect something of importance before its absence is motivating.

The example of speed of acquisition that we have set up is clearly a "straw man." To our knowledge, no popular conceptualization of inhibition takes the observed differences in acquisition as crucial. The point that this example strives to make is that, before one seriously considers a difference between the inhibitory and excitatory mechanisms, it is important to analyze the design of the experiment; not only should the salience and motivational significance of the reinforcer be considered but also the information available to the animals. We come back to this very important point later when we discuss the problem of extinction of inhibition.

What are the Requisites for Acquisition of Inhibition?

Our interest in the necessary conditions for the development of inhibition came from our early considerations of Rescorla and Wagner's (1972) provocative model of conditioning. Associative models of conditioning have usually claimed that learning occurs when the two potential elements of the association occur contiguously. Inhibitory learning requires that animals learn that under certain circumstances the US will not occur. If learning involves associations between contiguous events, just what are the events that are contiguous? A conditioned excitor might predict or be associated with a shock, but a conditioned inhibitor must predict or be associated with the absence of shock.

Rescorla's (1966) experiments on the negative correlation procedure best illustrate this problem. A number of CSs and USs were presented at irregular intervals with the only constraint being that the two events could not occur close together in time. Functionally, the CS predicted a period of time during which the US could not occur.

This negative correlation procedure leads to strong conditioned inhibition, yet it appears to violate the principle of temporal contiguity. The animals seem to learn specifically that the CS is not contiguous with the US. Rescorla and Wagner (1972) solved this problem by claiming that this procedure was a special case of Pavlov's (1927) conditioned inhibition procedure involving contextual conditioning. Before elaborating and describing some tests of this explanation, we briefly discuss an experiment illustrating the role and potential power of contextual conditioning in excitatory conditioning.

The Role of the Context in Conditioning

We leave the problem of inhibition temporarily to discuss an experiment that shows the powerful integrative role of contextual conditioning in associative theory. It is a common finding (cf. Randich & LoLordo, 1979a) that if animals are exposed to the US prior to conditioning they will condition slower than nonpreexposed controls. The explanation of this proactive interference effect that is of most relevance here is an associative one (cf. Tomie 1976a, b). According to this context blocking account, a US preexposure experiment is seen as analogous to Kamin's (1969) blocking experiment. The analogy between the two procedures is shown in Table 5.1.

In Kamin's blocking procedure two crucial groups are exposed to a number of reinforced trials of a compound stimulus (AB). The important difference between the groups is that in the experimental group one of the stimuli (A) was previously paired with shock. In the final test phase of the experiment Kamin found that B came to control less conditioned responding in the experimental group than in the control group(s). Prior training to A blocked later conditioning to B. Kamin (1969) explained this result by claiming that conditioning of the CS would occur only if the US was surprising. In the blocked group A came to predict shock perfectly well during the reinforced A trials, so the shock was not reinforcing during the subsequent AB trials and not much was learned about B. This principle of surprise was formalized and extended by Rescorla and Wagner (1972).

The associative explanation of the US preexposure effect requires only that we assume that the experimental context is conditionable and acts in much the same manner as other, more discrete, stimuli, and there is much evidence to

TABLE 5.1
Analogy Between Kamin's Blocking and the US Preexposure Effect. A and B Are Discrete Stimuli and C Is the Experimental Context

Kamin's Blocking			
Group	*Phase 1*	*Phase 2*	*Test*
Blocked	16 A-Shock	8 AB-Shock	B
Control	—	8 AB-Shock	B
US Preexposure			
Group	*Phase 1*	*Phase 2*	*Test*
Exposed	C-Shock	CB-Shock	B[a]
Control	C	CB-Shock	B[a]

[a]The test in the US preexposure effect is often confounded by being carried out in the training context and is thus a compound of the dicrete stimulus and the context (CB).

support this assumption (cf. Baker, Mercier, Gabel, & Baker, 1981; Odling–Smee, 1975a, b, & 1978). As Table 5.1 indicates, during exposure to the US the context (C) becomes conditioned. The subsequent conditioning trials to the discrete CS (B) are viewed as being conditioning trials to the compound stimulus CB. During these trials the animal already expects shocks in the presence of the context; thus there is little surprise on the CB trials and conditioning to B is blocked.

This rather elegant and parsimonious analysis of the US preexposure effect predicts that the interference found following exposure to the US is a direct consequence of the amount of contextual conditioning during exposure. Another consequence of this analysis is that, if the experimental context can block or overshadow discrete stimuli, then under some circumstances discrete stimuli should be able to block or overshadow the context. Baker and Mackintosh (1979) took advantage of this in an experiment designed to test the blocking by context explanation of the US preexposure effect. The design of this experiment is shown in Table 5.2.

The experiment, as well as most of the subsequent ones, used the conditioned emotional response paradigm (CER), in which fear conditioned to a CS is monitored by assessing the CS's ability to suppress appetitive behavior. In our case the appetitive behavior involved lever pressing for food on a variable interval (VI) reinforcement schedule. In all these CER experiments the animals were trained to bar press and received eight or more 50-minute sessions of practice on the VI schedule prior to any experimental manipulations. The present experiment included three groups of animals. One group (Group Unsignaled) was exposed to unsignaled shocks presumably causing the context to become excitatory and thus block future conditioning to a discrete stimulus (B). The second grroup (Group Signaled was exposed to the same sequence of shocks as Group Unsignaled except that each shock was preceded by a discrete stimulus. The rationale for the use of this group was that the salient and contiguous stimulus A should compete with the context C for conditioning and thus reduce contextual conditioning.

TABLE 5.2
Design of Baker and Mackintosh's (1979) Experiment. C Represents the Contextual Stimuli. A and B Represent Discrete Stimuli (Light and Clicker, Respectively). The Experiment Used the CER Preparation

Group	*Exposure Phase*	*Test Phase*
Unsignaled	C-Shock	CB-Shock
Signaled	AC-Shock	CB-Shock
Control	C	CB-Shock

Later, in the test phase during which the second CS (B) was conditioned, the context would be expected to control less conditioning in Group Signaled than in Group Unsignaled. Thus, because interference with conditioning to B was assumed to be a function of contextual conditioning, signaling the shocks during exposure would be expected to reduce interference. The results of this experiment were quite clear-cut. Not only did signaling the shocks reduce interference with conditioning compared to the unsignaled condition, but Group Signaled was virtually indistinguishable from the nonexposed controls (cf. Baker & Mackintosh, 1979 Fig. 1).

This result, which strongly supports the blocking by context explanation, is quite robust. We have replicated the basic finding several times in the CER preparation (Baker et al., 1981; see also Randich, 1981). In addition Tomie (1976a, b) and Durlach (1983) have reported similar findings in autoshaping.

Contextual Conditioning and Conditioned Inhibition

We alluded earlier to the fact that Rescorla and Wagner (1972) used contextual conditioning to explain how inhibition could accrue following exposure to Rescorla's negative correlation procedure. According to them, the mechanism of inhibitory conditioning in this procedure was analogous to that occurring in Pavlov's conditioned inhibition procedure. Just as in their explanation of blocking, they used a formalized and extended version of Kamin's concept of surprise to explain how the principle of temporal contiguity was not violated in these inhibitory procedures. In Pavlov's procedure an initially neutral stimulus (B) becomes inhibitory when an excitatory stimulus is paired in nonreinforced compound (AB) with an excitatory stimulus. The argument is similar to the one for blocking. The animals came to expect the US in the presence of A. When the AB compound is not reinforced, the animals are "surprised" by the absence of the shock. This "surprise" (more formally it is called a negative discrepancy in associative strength) is contiguous with B, and the inhibitory association involves these two elements.

The explanation for the negative correlation procedure is similar. When animals are exposed to the shocks the context (C) becomes conditioned. When the CI, B, is presented and not reinforced, this is viewed as a CB compound. Because the animal expects shocks in the presence of C but no shocks occur during CB, the animal is "surprised." This "surprise" is contiguous with B and B becomes an inhibitor. Our use of surprise as an explanatory shorthand for the more formal negative discrepancy stretches a bit thin here, but it still represents the logical essence of the theory quite well.

Just as the US preexposure effect was assumed to be a function of the amount of contextual conditioning, the amount of inhibition arising from a negative correlation procedure should also be a function of contextual conditioning. Again it follows from the Rescorla–Wagner model that if something is done to reduce conditioning of the context, such as blocking or overshadowing it with a discrete CS, inhibitory conditioning should be reduced. Baker (1977) reported two experiments that used a design and rationale that was very similar to the US preexposure experiment we have just described. Basically the experiment was carried out to demonstrate that if the shocks in a negative correlation procedure are signaled, thereby blocking contextual conditioning, the amount of inhibition should be reduced. Both a summation and a retardation test for inhibition were used to measure this reduction (cf. Rescorla, 1969).

The two experiments used a variant of the negative correlation procedure called the "between days" procedure. This procedure involves presenting the stimulus that is to become a CI, and the US on alternate days. If this sequence is repeated several times, the stimulus becomes a strong CI. The design of the training phase of both experiments is shown in Table 5.3.

One important feature of these experiments is the contextual conditioning phase that preceded the inhibitory phase of the experiments. The rationale of these experiments requires that there be no excitatory contextual associations during inhibitory training in the signaled shock groups. For if there were, some inhibition might accrue, due to these associations, on the early inhibitory training days, and thus the effectiveness of signaling the shocks in reducing inhibition might be attenuated. The signal (A) was provided to block or overshadow contextual conditioning by becoming conditioned itself. It is, however, a common observation in our laboratory that early in conditioning before the CS becomes conditioned there is some general suppression of baseline responding. This suppression, which can be considered a measure of contextual conditioning, goes away after a few days. In order to prevent this contextual conditioning from being present on inhibitory conditioning days, we always give several days of conditioning to the signal before the inhibitory conditioning begins.

TABLE 5.3
Design of Baker's (1979) Between-Days Inhibition Experiments. For the Retardation Test Experiment A was a Light and the Test Involved Reinforcing B (the Clicker). For the Summation Test A was a Pulsed Tone and the Test Involved Reinforcing the Light then Pairing it with B (the Clicker).

Group	*Context Conditioning (4 Days)*	*Inhibitory Training (10 Days)*		*Test*
		Odd Days	*Even Days*	
Negative Correlation	2 Sh	6 Sh	6 B	Summation of Retardation of B
Signaled Shock	2 A-Sh	6 A-Sh	6 B	"
Control	—	—	6 B	"

The first experiment used a retardation test (cf. Rescorla, 1969) in which inhibitory conditioning is assessed by the tendency of the CI to resist the acquisition of excitatory conditioning. The animals in the negative correlation group received clickers and shocks on alternating days. The animals in the signaled shock group received the same sequence of clickers and shocks except that each shock was signaled by a discrete light stimulus. The control group was exposed to only the clickers. The between-days inhibition phase of this and subsequent experiments was carried out off-baseline; that is the levers were removed from the boxes and the food reinforcement schedule was discontinued. This was done because the shock exposure schedule was quite dense (6 or 8 shocks per session), and we wished to avoid strong baseline suppression of lever pressing and any possible response–shock associations.

Immediately following inhibitory training the levers were returned to the boxes and the animals were given one lever-press recovery day, during which the VI food reinforcement schedule was in effect but no stimuli were presented. Following this came the retardation test for conditioned inhibition to the tone. This test lasted for 5 days on each of which there were two shock-reinforced clicker trials.

The results of this experiment were quite straightforward. The animals that had been exposed to alternating days of unsignaled shocks and clickers conditioned more slowly than the control group, as would be expected if the clicker had became inhibitory (cf. Baker, 1977, Fig. 3). The finding of main interest concerned the signaled shock group. This group conditioned at about the same rate as the control group. Thus the results of this experiment provide quite strong support for the notion that conditioning of the context mediates the inhibitory conditioning that accrues in the negative correlation procedure.

There is a problem with the interpretation of the retardation test in this experiment. Exposure to unsignaled shocks interferes with conditioning (the US preexposure effect). We have demonstrated that signaling the shocks during exposure can eliminate interference. It is also well known that exposing animals to the CS prior to conditioning retards conditioning (cf. Lubow, 1973). This phenomenon is called latent inhibition, and we have reason to discuss it more thoroughly later. Any negative correlation procedure involves exposing animals to the CS and to unsignaled shocks. We have described this procedure from the viewpoint of inhibitory conditioning, but it could equally be described as a proactive interference experiment that combines latent inhibition and the US preexposure effect. Thus the interpretation of many experiments using retardation tests is confounded by the possible summed effects of latent inhibition and US preexposure. We discuss this problem more fully later.

One conclusion that must be drawn from the preceding analysis is that it is necessary to use a summation test (cf. Rescorla, 1969) in order to have any confidence that manipulations of contextual conditioning influence conditioned inhibition. Baker (1977) reported a second experiment that included a summation test for inhibition rather than a retardation test. The experiment also used the inhibitory training design shown in Table 5.3. The only important procedural difference between the experiments was that the signal in this experiment was a pulsed tone rather than a light.

Following inhibitory training came the summation test that lasted for 6 days. On each of the first 3 days the animals received two shock-reinforced presentations of the light, designed to turn the light into an excitor. The next 3 days were the summation test proper. On each of these days the animals received one presentation of the light and one presentation of the clicker–light compound stimulus. The compound stimulus and the first light stimulus were not reinforced, whereas the last two light stimuli were.

The results of this experiment were again clear-cut. In Group Negative Correlation there was evidence of strong inhibition to the clicker; the clicker reliably reduced suppression to the light (cf. Baker, 1977, see Fig. 5 for details). There was little evidence of any tendency for the clicker to reduce suppression to the light in either the control group or in the group of animals that were exposed to signaled shocks. Just as in the previous experiment that used a retardation test, there was strong evidence in favor of the argument that signaling the shocks during exposure to the negative correlation blocked contextual conditioning, which thereby reduced conditioned inhibition.

When we completed these experiments, we considered them to provide fairly unequivocal evidence for the role of contextual conditioning. However, we have already discussed some problems with the retardation test experiment, and there are at least two problems with our summation test experiment. First, in order to avoid a summation test that would use two auditory stimuli, we used an auditory stimulus for the signal (a pulsed tone) and for the CI (a clicker) and used the light as the excitor in the summation test. It is thus possible that equal inhibition may have accrued to the clicker in the signaled and unsignaled groups. If in addition to this some of the excitation to the signal generalized to the CI, the CI's ability to reduce the behavioral effect of the excitor in a summation test would be compromised. Further, the excitatory conditioning to the light was carried out after the inhibitory training so that during this training the animals had had differential exposure to the US. To the extent that this differential exposure influenced excitatory conditioning, and we shortly describe an experiment that documents such an effect, the summation test might again be biased. Finally, signaling shocks not only reduces contextual conditioning but also changes the predictability of the shocks within a session. We have recently provided evidence that these changes in predictability, independent of any changes in contextual conditioning, influence the acquisition of excitatory conditioning (cf. Baker et al., 1981). Thus it is possible that signaling shocks during the negative correlation procedure influences the acquisition of inhibition because of these changes in predictability instead of, or as well as, its effects on contextual conditioning.

Fortunately there is a way around these problems. Contextual conditioning may be manipulated in ways other than signaling the shocks with a discrete CS. In our US preexposure experiments we have described several other treatments that have reduced contextual conditioning (cf. Baker et al., 1981). One feature of these treatments is that, whereas they do influence the predictability of the US, they do so differentially. To increase our confidence that manipulations of the context actually influence the acquisition of inhibition through their effects on contextual conditioning rather than by introducing differences in shock predictability, we carried out two experiments, one using a retardation test and the other a summation test, in each of which signaling the shocks was compared with another procedure for reducing contextual conditioning.

The first experiment used a summation test to assess inhibition and compared exposure to shocks in another context with signaling shocks as methods of reducing the acquisition of inhibition. The design of this experiment is shown in Table 5.4. After the visual VI training the animals all received excitatory conditioning trials to the pulsed tone, which would be the signal during inhibitory training for the signaled shock group, and to the clicker, which would later be used in the summation test. The off-baseline inhibitory training phase followed and included a control group plus three groups that received alternating days of shocks and the light stimulus that was to become the CI. Group Negative Correlation received only these lights and shocks. For Group Signaled Shock the shocks were all signaled by the pulsing tone. The novel groups in this experiment were the control group, which received uncorrelated presentations of the clicker and shocks on odd days and were left in their home cages on even days (Rescorla's, 1967, truly random control procedure) and Group O.B. (Other Box), which received the light in the conditioning chambers but which received the shocks in different chambers. These chambers have been described elsewhere (Baker et al., 1981). They were modified lever boxes but they differed from the training boxes in size, shape, type of grid floor, light level, and type of soundproof enclosure. We have independently demonstrated that this context is dis-

TABLE 5.4
Design of the CER Experiment That Used a Summation Test to Compare the Efficacy of Exposure in Another Context (Other Box) and Signaling the Shocks as Means of Reducing Inhibition in the Between-Days Inhibition Procedure. Cl = Clicker, T = Pulsed Tone, L = Overhead Light, Sh = Shock.

Group	*Pretrain (3 Days)*	*Inhibition Train (10 Days)*		*Summation Test*
		Odd Days	*Even Days*	
Negative Correlation	2 T-Sh, 2 Cl-Sh	8 L	8 Sh	Cl vs Cl + L
Signaled Shock	"	8 L	8 T-Sh	"
Other Box	"	8 L	8 Sh in Other Box	"
Control	"	8 L, 8 Sh	Home Cage	"

criminably different from the usual training context, and that there is little generalization of fear of the context acquired in these chambers to the training chambers. Thus the rationale for inclusion of this group was that any contextual conditioning produced by the unsignaled shocks should be controlled by the other context and thus not be present on the alternate days in which the lights were presented, thereby reducing the acquisition of inhibition by the light.

The design of this experiment meets two of the procedural objections to Baker's (1979) experiment. In this experiment the discrete stimulus that was designed to block contextual conditioning was an auditory stimulus, whereas the inhibitory stimulus was visual. Thus the likelihood that the animals would generalize between the two stimuli on the summation test was reduced. In addition, conditioning to the clicker that was to be used in the summation test was carried out before inhibitory training. This reduced the likelihood that our associative manipulations of the context would influence conditioning of the excitor and thereby reduce the reliability of the summation test. The results of the summation test are shown in Fig. 5.1. This figure shows the levels of responding to the clicker and to the compound of the clicker and the light over the 3 test days. The measure of conditioning in this and all our CER experiments is the suppression ratio (Annau & Kamin, 1961). Ratios of .5 indicate no relative suppression during the CS and hence no conditioned response, whereas ratios of 0 represent strong suppression and hence a strong conditioned response. These data show quite clearly that the "between-days" negative correlation produced quite strong inhibition as compared to the control group that had received uncorrelated presentations of the light and shocks. The animals that received their shocks in a discriminably different context from the context in which they received the light also showed little evidence that the light became inhibitory.

There is also evidence that signaling the shocks reduced conditioned inhibition just as it had in our earlier experiment, but this evidence is somewhat

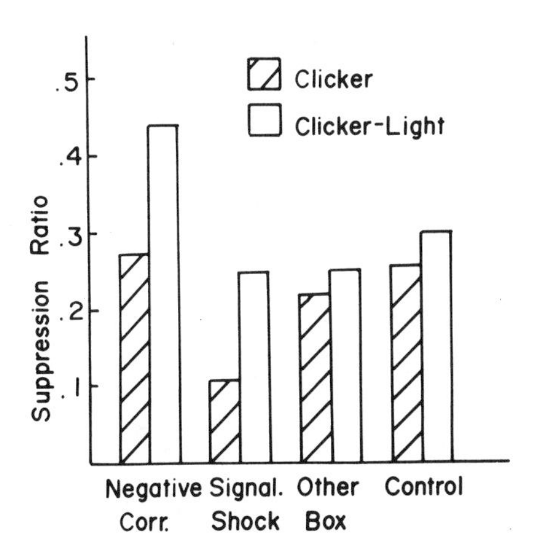

FIG. 5.1. Overall results of the summation test in the experiment (see Table 5.4) designed to assess the effectiveness of signaling the shocks with a pulsed tone or presenting them in another context on reducing the conditioned inhibition arising from the between-days conditioned inhibition procedure.

equivocal. The signaled shock group was somewhat more suppressed to the clicker than the other groups. It is possible that this difference in suppression, which is perhaps due to generalization with the tone signaling stimulus, might influence the summation test. Although the light did appear to have inhibitory properties in the signaled shock group, a closer analysis of the test indicates that this inhibition is more transient than that in the negative correlation group. Figure 5.2 shows daily difference scores for inhibition that were calculated by subtracting each animal's suppression to the clicker from its suppression to the clicker–light compound. This figure shows clearly that the inhibitory effect of the light increased throughout the test in the negative correlation group, whereas the signaled shock group showed no marked increase in difference scores over this period, and on the third day their scores did not differ from those of the animals that had been exposed in the other context.

We were thus able to conclude from this experiment that reducing contextual conditioning in the presence of the negatively correlated stimulus by giving the US in a discriminably different context reduced inhibition to that stimulus. Also, this experiment when taken with Baker's (1979) earlier demonstration shows that blocking contextual conditioning with a discrete CS may also reduce inhibitory conditioning.

It should be pointed out that an implication of these present data is that signaling the US may not always completely reduce inhibition. This is not surprising for at least two reasons. First, we independently assessed the effectiveness of signaling shocks and exposing animals to shocks in another context as a means of reducing contextual conditioning (cf. Baker & Mercier, in press,

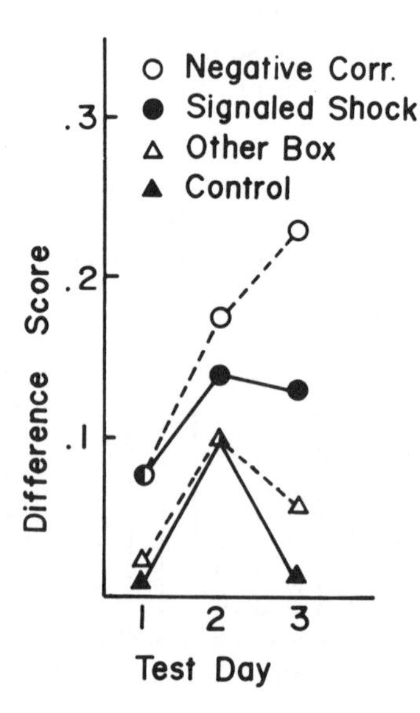

FIG. 5.2. Difference scores (clicker–light minus clicker suppression ratios) for the 3 days of the summation test in the experiment (see Table 5.4) designed to assess the effectiveness of signaling the shocks or presenting them in another context on reducing the inhibition arising from the between-days inhibition procedure.

Baker et al., 1981), and we found that signaling was a less effective means of reducing contextual conditioning. Thus it is not surprising that exposing the animals in another context might reduce inhibitory conditioning more effectively. Second, the signaled shock procedure is actually a Pavlovian discrimination and, although this procedure is often used as a control procedure in the CER (e.g., Holland, Rescorla, this volume), we have sometimes found it to produce conditioned inhibition, and it is regularly used as an inhibitory procedure in autoshaping (e.g., Kaplan & Hearst, 1983).

Table 5.5 shows the design of a second experiment that was designed to test the effectiveness of another associative manipulation of the context as a means of reducing the acquisition of conditioned inhibition, this time using a retardation test. In this experiment we again included a negative correlation group, a signaled shock group and a control group. The novel group was a group for which the contextual stimuli were modified on shock exposure days by turning on the overhead light CS for the entire session. The light was turned off on the alternate days in which the animals were exposed to the clicker. If this change of contextual stimuli between the shock and the clicker days caused reduced contextual conditioning on clicker days, less inhibitory conditioning should accrue.

The procedural details of this experiment were similar to the previous three. The animals were initially trained to respond on a VI schedule for food. Next came the context and signal conditioning phase during which the negative correlation and the session-long light groups received unsignaled shocks. The overhead light was turned on for the session-long CS group. The signaled shock group received shocks signaled by the discrete light CS. The control group received no shocks. During the subsequent between-days inhibitory training phase, the three shock groups received shocks that were unsignaled, signaled by the discrete light, or presented in the presence of the overhead light. On

TABLE 5.5
Design of the CER Experiment That Used Retardation Test to Compare the Efficacy of Signaling the Shocks or a Session-Long CS as Means of Reducing Inhibitory Conditioning in the Between-Days Procedure. Cl = Clicker, L = Overhead Light, Sh = Shock.

Group	Pretrain (3 Days)	Inhibition Training (10 Days)		Retardation Test (7 Days)
		Odd Days	*Even Days*	
Negative Correlation	2 Sh	8 Sh	8 Cl	2 Cl-Sh
Signaled Shock	2 L-Sh	8 L-Sh	8 Cl	"
Session-Long CS	2 Shock Light On	8 Sh Light On	8 Cl	"
Control	—	—	8 Cl	"

alternate days all animals received clicker stimuli under the normal houselight conditions. Following the inhibitory training phase, which had been again carried out off-baseline, all animals received a lever-press recovery day followed by the retardation test for conditioned inhibition to the clicker. This test lasted for 7 days and on each of these days the animals received two shock reinforced clicker trials. The results of the retardation test are shown in Fig. 5.3. The negative correlation group and the signaled shock group were very similar to Baker's (1977) earlier results. Consistent with the notion that conditioned inhibitors develop excitation relatively slowly, Group Negative Correlation conditioned more slowly than the control group. In addition to this it is also quite clear that the animals that had received signaled shocks conditioned more rapidly than the negative correlation group and, in fact, behaved very similarly to the controls. These results are consistent with the notion that signaling shocks reduced contextual conditioning and thereby reduced the acquisition of inhibition.

The group that received the session-long CS also acquired excitation slightly more rapidly than the negative correlation group. The difference between this group and the inhibitory control group was not nearly as marked as it was for the signaled shock group. This result, however, is still on the face of it consistent with the argument that inhibitory conditioning is mediated by contextual conditioning. It must be mentioned again that this experiment used a retardation test, and such tests can confound latent inhibition and the US prexposure effect with inhibition. The implications of this problem are discussed later.

We have now described four experiments that tested the notion that manipulations of contextual conditioning would influence the amount of conditioned

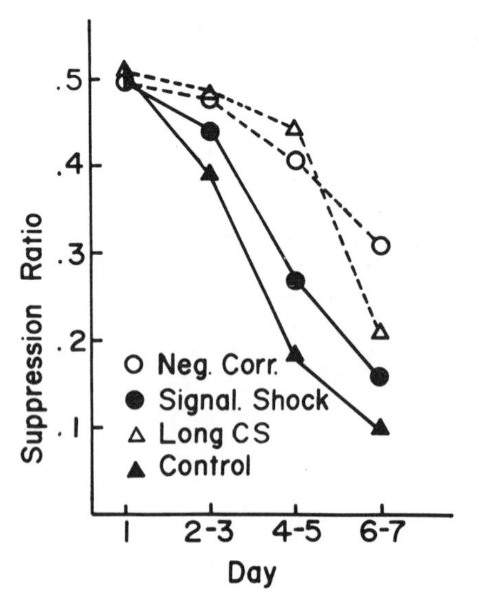

FIG. 5.3. Results of the retardation test in the experiment (see Table 5.5) designed to determine whether signaling the shocks with a discrete light CS or using a session-long light on the shock days would reduce inhibition accruing to a clicker in the between-days conditioned inhibition procedure. The animals received two shock-reinforced clicker presentations each test day.

inhibition accruing to a CS that is negatively correlated with shock. These results allow two major conclusions to be at least tentatively drawn. First, Kamin (1969) argued that a US only produces excitatory conditioning to a CS if that US is unexpected or surprising. These experiments show that this same notion of surprise can be used with very little modification to account for inhibitory conditioning. This certainly supports the argument that inhibitory conditioning may be very much like excitatory conditioning, at least with respect to the necessary conditions for its acquisition. Thus these experiments provide evidence of empirical similarities between inhibitory and excitatory conditioning.

A second conclusion would involve arguing that the experiments presented here were all derived from a model of conditioning that used temporal contiguity and associations as fundamental building blocks. To the extent that our experiments were "successes," this particular model and the general notion of association by temporal contiguity receive some support. However, just because experiments are generated by a theory and also support the theory does not necessarily imply that the theory best characterizes the results. We now briefly review some of our experiments on the US preexposure effect that extend the signaled shock experiment of Baker and Mackintosh (1979) in much the same way we extended Baker's experiments on conditioned inhibition. These experiments implied that excitatory and, indeed we now argue, inhibitory conditioning are most readily fitted in a framework incorporating the notion of a temporal cognitive map (e.g., Baker & Mercier, 1983).

Contextual Conditioning and the US Preexposure Effect

We were somewhat less than candid when we described Baker and Mackintosh's (1979) experiment as providing fairly unequivocal evidence for the blocking by context argument. That experiment had certain weaknesses. The first of these was that it only included one method of modifying contextual conditioning—signaling the shocks. As we have mentioned, this manipulation not only effects the level of contextual conditioning during exposure but the overall predictability of the shocks within the session as well. The second weakness was that we made little consistent effort to measure contextual conditioning in the experiment. Thus although signaling the shocks theoretically should have reduced contextual conditioning, the only indication that this was true was the fact that signaling the shocks reduced the US preexposure interference effect. Thus without independent evidence of contextual conditioning, it is very difficult to provide evidence against the context-blocking explanations of the US preexposure effect. Of course a similar argument could be made concerning our attempts to influence the acquisition of inhibitory conditioning through these manipulations.

In order to test more directly the context-blocking explanations of the US preexposure effect, we began a series of experiments that not only measured the effect of supposed manipulations of contextual conditioning on interference to conditioning to a discrete CS but also tried to assess the direct effect of these treatments on contextual conditioning itself (Baker et al., 1981; Baker & Mercier, 1982a). This scheme was quite simple in execution. Our experiments were CER experiments involving exposing animals to shocks. If animals become conditioned to the context during exposure to the shocks, this fear of the context should be measurable as suppression of responding for food in that context (baseline suppression), assuming that the conditioning achieved was above some threshold for responding. Although such baseline suppression is not always found (e.g., Randich & LoLordo, 1979b), we discovered that when sensitive measures were used, such as baseline suppression ratios (cf. Baker et al., 1981) or latency to the first response, we were able to obtain consistent and reliable evidence of contextual conditioning on the recovery days following exposure to shocks.

Because we were able to reliably measure contextual conditioning, we could also systematically manipulate this conditioning by blocking, overshadowing, or extinguishing the context. We tried all these treatments and found that contextual conditioning behaved in much the same manner as conditioning to a discrete stimulus (see Baker & Mercier, in press, for a summary). All the preceding manipulations reduced contextual conditioning to a varying extent. The specific treatments that we used included: signaling the shocks, administering the shocks in a different context, administering the shocks in the presence of a session-long stimulus and then removing the stimulus on the recovery days, and extinguishing the context by placing the animals in it for several days between shock sessions and the lever-press recovery days. The first three of these manipulations were included in our conditioned inhibition experiments. Thus some independent evidence is provided that these manipulations may actually have influenced contextual conditioning, as well as having influenced the course of inhibitory conditioning. The blocking by contextual conditioning explanation of the US preexposure effect claims that interference with conditioning is a direct function of the level of contextual conditioning following exposure to the US. There should be a monotonic relationship between contextual conditioning and the interference with conditioning to a discrete CS following exposure to the US. Following exposure to the US and all the preceding manipulations of contextual conditioning, we then assessed the rate of acquisition of suppression to a discrete CS. The results of these experiments, which are summarized in Table 5.6, were enlightening.

The most important conclusion that may be drawn from this tabulation is that, contrary to the predictions of Rescorla and Wagner, there was no monotonic relationship between contextual conditioning and interference with the acquisition of conditioned suppression. Indeed, signaling the shocks, which was the only treatment to virtually eliminate the US preexposure effect, was if anything the least effective method of reducing contextual conditioning with our param-

TABLE 5.6
Effectiveness of Several Methods of Manipulating Contextual Conditioning in Reducing Baseline Suppression and in Reducing the US Preexposure Effect (Baker et al., 1981; Baker & Mercier, 1982a; Summary Figures for the Data in This Table May be Found in Baker & Mercier, in Press, Fig. 1 & 2)

Treatment	*Contextual Conditioning*	*Interference with Conditioning*
Unsignaled Shocks	Strong	Strong
Signaled Shocks	Moderate	Weak
Session-long CS	Weak	Strong
Other Context	Weak	Strong
Extinction of Context	Weak	Moderate

eters. These data were condensed from seven published experiments and, thus, involve between-experiment comparisons (cf. Baker et al., 1981 and Baker & Mercier 1982a; the data are summarized in Fig. 1 & 2 in Baker & Mercier, in press). This does not pose a strong problem because, to make the point that there is no strong relation between contextual conditioning and interference, it is only necessary to note that, although the first three manipulations reduced context fear, only signaling the shocks reduced interference with conditioning. In any case the stronger point that, although signaling the shocks was perhaps the least effective method of reducing context conditioning, it was the only procedure that consistently and strongly reduced interference, may be made using a single within-experiment comparison (cf. Baker et al., 1981, Experiment 2). In this experiment exposing the rats in a different context strongly reduced context conditioning but did not reduce interference. Signaling the shocks reduced context conditioning less effectively but virtually eliminated interference. We are thus left with what to us seems to be the inescapable conclusion that, although signaling shocks reduces contextual conditioning, it is not this reduction that mediates the elimination of the US preexposure effect.

Our US preexposure experiments provided precious little evidence for a monotonic relation between contextual conditioning and the US preexposure effect. They did, however, provide fairly strong evidence that the context behaved in the same "lawful" manner as punctate CSs. The thrust of these data was that these laws were not the traditional laws of temporal associationism as embodied by such models as Rescorla and Wagner (1972) (or for that matter Mackintosh, 1975, or Pearce & Hall, 1980). Although other models of conditioning specifically rule out blocking by context (e.g., Scalar Expectancy Theory, Gibbon & Balsam, 1979), our findings that a discrete CS reduces the US preexposure effect and that signaling shocks reduces fear of the context are inconsistent with them (Baker et al., 1981, also see Durlach, 1983).

The ''laws'' that we felt best accounted for our data were more a set of empirical cognitive principles than laws. We have proposed that during conditioning animals are sensitive to, *and actually learn or represent,* the temporal correlation or contingency between events to which they are exposed. Animals come out of a conditioning or exposure session with not only associations between events but with a representation or temporal cognitive map of the order and time of occurrence of the events and of their relationship with and/or predictability by one another.

When animals are exposed to unsignaled shocks, they not only form associations between the shocks and the context, but they also learn that the shocks are unpredictable. Later during conditioning to a discrete CS the representation of these shocks as unpredictable proactively interferes with learning that they are predicted by a discrete CS. When the animals are exposed to shocks signaled with a discrete CS, this CS might block or overshadow associations between the context and shocks, but, as well, the signal should prevent the animal from characterizing the shocks as unpredictable. Signaling shocks should prevent future interference with conditioning arising from both context blocking and the unpredictability of the shocks. On the other hand, whereas the other treatments, such as exposing the animals to the shocks in another context or in the presence of a session-long stimulus, should reduce contextual conditioning, these treatments do not change the basic unpredictability of the shocks within a session and thus do very little to reduce the proactive interference with learning the subsequent predictability of shocks. This argument is analogous to that used in the ''learned helplessness'' paradigm, in which it is claimed that animals learn that responses cannot control consequences and thus instrumental learning is retarded (e.g., Maier & Seligman, 1976). We claim that if the animal has learned that a potential US is unpredictable, this will interfere with subsequent classical conditioning.

This framework, that at present is more empirical than theoretical, may be applied to inhibitory conditioning. Its main features are that animals can represent the predictability of important events, and that, further, the mechanism of conditioning involves searching for that predictability, contingency, or correlation. There is nothing new about characterizing conditioning as involving a ''search'' for contingency or predictability; this, after all, is what associative models were designed to explain. What is somewhat unusual about our view is the catholicism of the relations that may be learned. One consequence of models of temporal contiguity is that they only allow animals to learn positive correlations among stimuli. Explanations of inhibitory conditioning involve redefining the conditioning events so that the negative contingency between the events actually is seen as some form of contingency between the inhibitor and some sort of ''null'' event. For example the formalization of surprise or the negative discrepancy of associative strength that we have described is such a characterization. Accepting the temporal cognitive map avoids the need for such gymnastics to protect the "parsimony" of the associative system by simply stating that animals may represent positive relations between events in excitatory conditioning, negative relationships in inhibitory conditioning, and null relationships or the absence of relationships in learned irrelevance, learned helplessness, latent inhibition, or the US preexposure effect. Moreover, the fact that animals can learn that the experimental context predicts the US (contextual conditioning) but that the US is unpredictable within the session forces us to assume that the animal can maintain several representations of the same event simultaneously.

Our empirical framework easily accommodates the seeming paradox that our manipulations of contextual conditioning, such as exposing the animals in a separate context, had little effect on the US preexposure effect but seemed to effectively reduce the acquisition of inhibitory conditioning. We have already discussed the US preexposure effect and described how manipulations that involved the predictability of the shocks seemed to influence this phenomenon. Similarly, in inhibitory conditioning the animal may be characterized as searching for events that predict the presence and the absence of the US. In a pure negative correlation the US may occur anytime in the absence of the CI, with certain limitations, so that the CI is the only and hence the most valid predictor of the presence and the absence of the CS; more accurately it is its absence that is the best predictor of the CS. Because of this the CI is the most valid stimulus available. When the shocks are signaled the CI is no longer the best safety signal nor is its absence the best predictor of the US, because the added signal now efficiently predicts when the US will occur and that signal's absence outlines all the safe time. According to this analysis the potential CI is blocked by a more efficient predictor in much the same manner as signaling the shocks blocks contextual conditioning. Likewise with a session-long CS or exposure to the US in the other context, the absence of either the session-long cue or the other context are better predictors of safety than the CI.

It should be noted that taken in isolation this analysis may appear to be a somewhat radical temporal associationism that claims that there is little to conditioned inhibition as a phenomenon. We have claimed that a stimulus will only become an inhibitor when its absence is the most contiguous predictor of the US. It is not a large step to extend this analysis by claiming that there is only excitatory conditioning, and that this excitatory conditioning is strictly competitive along the lines of this contiguity. The CI only becomes important when its absence is the most reliable predictor of the US. The CI thus is not a safety signal but the only nonexcitatory stimulus. Such a model would require a redefinition of the mechanism of the summation test and would have trouble in paradigms in which the animal develops a measurable response to the CI (e.g., Kaplan & Hearst, 1983), but it would be very congenial to the early associationists (e.g., Guthrie, 1935). We, however, feel that the claim that animals actually represent the negative correlation is forced by the fact that this allows the inclusion of learned irrelevance, learned helplessness, and the US preexposure effect under the same empirical framework.

This discussion may seem to have strayed far from the original aim of determining whether or not there are fundamental differences between inhibition and excitation. The point is that the characterization of inhibition and excitation as fundamentally different is a consequence of temporal associationism, and that, if the empirical principles that we have described are used, both types of conditioning are consequences of the same information-searching system. We now briefly describe how the proactive interference phenomena called latent inhibition and learned irrelevance provide more evidence for similarities between inhibition and excitation. We then describe how the traditional summation and retardation tests that are recently coming under some pressure (e.g., Pearce, Nicholas, & Dickinson, 1982) can be reconceptualized. Finally, we discuss one of the main assymetries between excitation and inhibition: the resistance of inhibition to extinction (e.g., Zimmer–Hart & Rescorla, 1974).

Latent Inhibition and Learned Irrelevence: Blocking by Context

Exposing animals to the CS prior to conditioning retards the acquisition of conditioned responding. This interference does not occur because the CS becomes inhibitory, because in a summation test a preexposed stimulus shows little, if any, tendency to reduce excitation to a conditioned excitor and exposing the CS prior to conditioning retards conditioned inhibition as well as conditioned excitation (Rescorla, 1971). There is an associative explanation of this effect that claims that latent inhibition may be explained by a blocking-by-context argument that is very similar to that used for the US preexposure effect (cf. Wagner, 1978). This argument claims that during exposure to the US an association forms between the contextual stimuli and the CS. Later this association interferes with the formation of an association between the CS and US. The language used is somewhat different than in the Rescorla–Wagner model; Wagner speaks of priming in the short-term memory rather than associative strength, but the principles at our level of discussion are still the same.

Just as with the US preexposure effect, manipulations that should reduce the associations between the context and the CS should also reduce latent inhibition and there is some evidence for this claim (e.g., Anderson, O'Farrell, Formica, & Caponigri, 1969; Anderson, Wolf, & Sullivan, 1969; Mackintosh, 1973). Elsewhere, we have reported a series of experiments that are analagous to our investigations of conditioned inhibition and the US exposure effect (cf. Baker & Mercier 1982b, in press). In general, these results did not support the associative notion that manipulations of the context should reduce latent inhibition. We have argued that these experiments provide compelling evidence that the interference with conditioning in latent inhibition occurs because the animals learn that the

CS signals nothing of importance during exposure, and thus future conditioning is retarded (Baker & Mercier, in press).

Another proactive interference effect occurs when animals are exposed to uncorrelated presentations of both the CS and the US (cf. Mackintosh, 1973). This in itself is not particularly surprising because a zero correlation involves unsignaled USs and unsignaled CSs, two treatments that independently cause proactive interference. Baker and Mackintosh (1979) have argued that there is more to this interference than the simple sum of latent inhibition and US preexposure. The main evidence for this claim was the finding that signaling shocks, which reduced the US preexposure effect, did not reduce learned irrelevance. If learned irrelevance was simply the sum of latent inhibition and the US preexposure effect, then any treatment that reduced one of them should have reduced learned irrelevance as well. Baker and Mackintosh (1979) argued that animals exposed to a zero correlation should not only learn whatever is learned in latent inhibition and the US preexposure effect but should also, above and beyond this, learn that the CS and US are specifically uncorrelated. This specific *learned irrelevance* (a term coined by Mackintosh, 1973) should later retard conditioning when the animals were expected to learn a positive correlation between the CS and US.

Latent inhibition and learned irrelevance are robust phenomena in excitatory conditioning. Baker and Mackintosh (1977) reported two experiments that were designed to demonstrate that exposure to the CS and to uncorrelated CSs and USs has similar disruptive effects on both inhibitory and excitatory conditioning. This experiment used an appetitive conditioning preparation called conditioned licking, in which thirsty rats are placed in a small conditioning chamber that contains a fixed drinking tube through which water may be intermittently presented. The conditioning procedure consists of presenting 30-sec CSs that are overlapped in their last 20 sec with a water presentation. In this paradigm conditioning is assessed by measuring the anticipatory licking at the tube that occurs during the CS in the 10-sec period before the water presentation begins (CS period). This rate of licking is compared to the rate of licking in the 10 sec immediately prior to the tone presentation (Pre-CS period). The actual measure of conditioning used is a difference score that is calculated by subtracting the Pre-CS scores from the CS scores. The design of the experiments is shown in Table 5.7.

The subjects in all groups were first adapted to the apparatus and were then given 10 light–water pairings a day for 8 days. The light was to be the conditioned excitor in the later inhibitory training that was to use Pavlov's conditioned inhibition procedure. Following this conditioning phase the animals received one of four exposure treatments. The animals were exposed to no stimuli, to water presentations alone, to tone presentations alone, or to uncorrelated presentations of the tone and water (Rescorla's truly random control procedure, 1969). This phase of the experiment lasted for 4 days, and following it half the animals were given excitatory pairings of the tone and water. For the other half the tone was made inhibitory by being paired in nonreinforced compound with the light.

TABLE 5.7
Design of Baker and Mackintosh's (1977) Conditioned Licking Experiments, Which Compared Latent Inhibition and Learned Irrelevance in Appetitive Inhibitory and Excitatory Conditioning.

Group	*Excitatory Conditioning (10 Days)*	*Treatment (4 Days)*	*Test (8 Days)*
Learned Irrelevance	10 Light–Water	10 Tone, 10 Water	Tone–Water or Light–Water/Tone–Light
Latent Inhibition	"	10 Tone	"
US Preexposure	"	10 Water	"
Control	"	—	"

The results of the experiments were clear-cut. Exposing the animals to the tone retarded both excitatory and inhibitory conditioning (see Baker & Mackintosh, 1977, Fig. 1 & 2). Exposing the animals to uncorrelated presentations of the tone and water also retarded both excitatory and inhibitory conditioning and did so to a much greater and more permanent extent than exposure to either stimulus alone. Although these experiments emphasize the similarity between excitatory and inhibitory conditioning, they provide no further strong evidence that learned irrelevance is not the simple sum of latent inhibition and the US preexposure effect. Such a demonstration is no longer so crucial to the "temporal cognitive map," however, because we have already argued that learned irrelevance-like mechanisms operate in both of these independent interference effects.

These results are of relevance to our present discussion for two reasons. First, they represent another demonstration that excitatory and inhibitory conditioning behave in much the same way. In this instance both learned irrelevance and latent inhibition have similar effects on both types of conditioning. Second, because excitatory and inhibitory conditioning control opposite response tendencies, results such as those that we have just discussed rule out various competing response explanations of learned irrelevance and latent inhibition in excitatory conditioning. These explanations posit that during preexposure the animals develop some behavioral tendency, such as inactivity, which interferes with the acquisition or performance of the conditioned response. If some response that competed with the acquisition of excitatory conditioned licking was learned during exposure, it is difficult to imagine how the same response could interfere with inhibitory conditioning that involves the suppression of licking.

Retardation Tests, Latent Inhibition, and US Preexposure

Rescorla (1969) originally argued that in order to be convinced that a stimulus was a CI the stimulus must pass both a retardation and a summation test for conditioned inhibition. We have already described how, particularly in the case of the negative correlation procedure, the treatment of exposing the animals to unpaired CSs and USs has much in common with both latent inhibition and US preexposure. Pearce et al. (1982) have shown that nonreinforced exposure to the CI after inhibitory training increases retardation on a retardation test but has little effect on a summation test. It is thus possible that the specific results from the retardation tests are confounded by both latent inhibition and the US preexposure effect. For example, Fig. 5.3 shows a retardation test following exposure to shocks that were signaled, unsignaled, or presented in the presence of a session-long stimulus. On alternate days the animals were exposed to the clicker as a potential CI. The results are straightforward. The inhibitory group conditioned most slowly, whereas the signaled shock group and the control group conditioned most rapidly. Finally the session-long stimulus group conditioned slightly more rapidly than the inhibitory group. These results are certainly consistent with the context-blocking hypothesis, but they are also quite inextricably confounded by an alternative explanation.

During exposure all four groups were exposed on even days to the clicker. This should produce a marked and equal latent inhibition effect in all groups. On odd days three groups were exposed to shocks and the fourth group was not. Thus in addition latent inhibition these three groups would be expected to show a marked US preexposure effect. Table 5.6, which describes the results of our US preexposure experiments, indicates that with our parameters unsignaled shocks produced a large US preexposure effect, whereas signaling the shocks with a discrete stimulus largely eliminated it. Finally, this table indicates that signaling the shocks with a session-long stimulus, although an effective manipulation of contextual conditioning, did very little to reduce the US preexposure effect. It would seem therefore that an acceptable analysis of the results of this experiment is that they reflect the summed effects of latent inhibition and the US preexposure effect and not conditioned inhibition at all.

To make this point, perhaps rather pedantically, we carried out a simple two-group experiment to demonstrate the possible pitfalls of the use of retardation tests. This experiment was a simple replication of the negative correlation and signaled shock procedures that we have already described. The experiment used a design similar to that shown in Table 5.3. The animals were trained to respond on a VI schedule for food in a typical manner. After subsequent context and signal (a pulsed tone) conditioning days the animals were exposed to shocks and clickers on alternating days. In the signaled shock group all shocks were signaled by the tone, and for the negative correlation group all shocks were unsignaled. It will be readily recognized that this is the design of an experiment meant to test whether blocking conditioning to the context with a signal will reduce inhibitory conditioning to the clicker. To test for this using a retardation test it is necessary to follow inhibitory training with a number of reinforced clicker presentations and observe whether or not Group Negative Correlation shows more "retardation," that is, conditions more slowly than Group Signaled Shock.

Such a result would support the context-conditioning explanation of inhibitory conditioning. Alternatively one could claim that the slower learning in the negative correlation group occurred simply because both groups showed latent inhibition acquired during exposure, and the negative correlation group showed the US preexposure effect as well, but that this extra interference was blocked in the signaled shock group thereby producing faster conditioning in this group. To reiterate, one explanation claims that the differences in acquisition occur because of differences in the clicker's inhibitory properties, whereas the other claims that it is due to differences in the effectiveness of the US. To demonstrate the plausibility of the second argument we carried out a "retardation test," but rather than condition the clicker in this test we conditioned a novel stimulus, specifically a light. The results of this "retardation test" are shown in Fig. 5.4. It is clear that the animals in the signaled shock group conditioned more rapidly than the unsignaled group, just as they had in our earlier experiments that used the inhibitor rather than a novel stimulus (see also Miller, this volume). This result shows that a "successful" retardation test may well measure things other than inhibition.

It is not our intention to claim that all retardation tests are necessarily confounded. Under some circumstances signaling shocks reduces the US preexposure effect, whereas under others that are sometimes unspecifiably different (compare for instance Randich & LoLordo, 1979b and Randich, 1981) signaling the shocks does not reduce the US preexposure effect. In fact in Baker's (1977) original demonstration that signaling the shocks reduced interference with future conditioning to an inhibitor, the excitor for the summation test, a light, was conditioned after the inhibitory training phase. This is very similar to our present

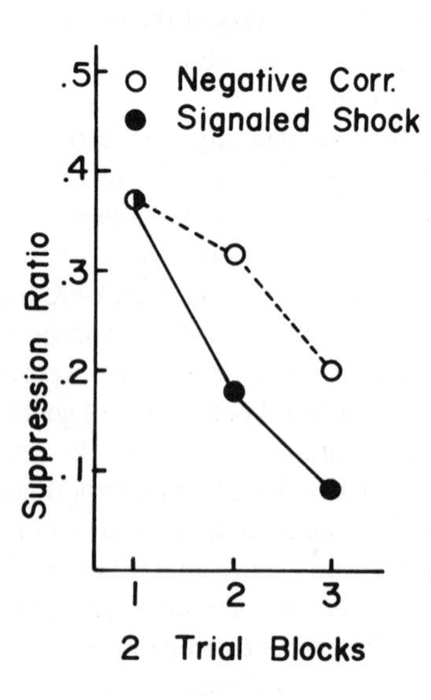

FIG. 5.4. Conditioning scores to the novel light following inhibitory training to the clicker (Negative Correlation) and the signaled shock procedure in the "mock" retardation test.

demonstration; however, there were no differences in the rate of conditioning to the light in the signaled and unsignaled shock groups, whereas on the retardation test with the inhibitory clicker there were marked differences in conditioning using those parameters (cf. Baker, 1979, Fig. 4 & 5).

The reason that Rescorla (1969) proposed that a stimulus must pass both tests in order to be considered a conditioned inhibitor was because each test could easily be confounded by effects such as those that we have described here. The argument goes, however, that the same confounding could not influence both tests, thus any stimulus that passes both tests must be inhibitory. Our experiment and that of Pearce et al. (1982) may be considered as demonstrations that his assumptions are true for the retardation test. It also should be easy to demonstrate a "successful" summation test in which a stimulus disrupts suppression to another that it is paired with. For example, Pavlov (1927) describes the external inhibition procedure in which a novel stimulus disrupts responding to a conditioned excitor.

The problem, however, is more profound than this. At the very least the retardation test is so confounded by interference phenomena, other than conditioned inhibition that might arise in the procedures that produce inhibition, as to be almost useless except as control for the novelty of the CS. Further it would appear that the logic of both tests seems to have its roots in the notions of associationism. A summation test is carried out because the negative habit strength of the inhibitor summates with that of the excitor and thus reduces the net associative strength of the compound. A retardation test is done because it is expected that it will take time for the reinforcement to overcome the negative associative strength of the inhibitor (cf. Wagner & Rescorla, 1972) or to overcome the increased threshold for associative strength (Rescorla, 1979).

Our analysis of conditioning handles these problems quite readily. Animals represent the events that they have encountered in a temporal cognitive map. If any treatment follows another that involves a different correlation between previously seen events or involves changes in the importance of these events, the previous experience will proactively interfere with forming a new cognitive map. Therefore it is not surprising that what is called a retardation test is "confounded" by all sorts of proactive interference effects. After all, such a test measures proactive interference and perhaps not anything that would be recognizable as "inhibition," at least in the sense of negative associative strength or thresholds for the manifestation of the response. The summation test is also straightforward; the animal has been trained to expect shock in the presence of one stimulus and its absence in the presence of another. When the two are presented together there is a conflict or a contradiction of information, and suppression to the excitor is disrupted.

Regardless of whether our arguments concerning the logical validity of the tests are accepted, the problems with the retardation test as an assay of inhibitory conditioning remain. Stimuli that are inhibitors are likely to be learned about slowly; however, if the intention of the experimenter is to compare several methods of producing inhibitory conditioning or compare several methods of reducing it in which a sensitive measure of inhibition is necessary, the retardation test does not fill the bill. As a test it is just too confounded with other proactive interference effects.

Extinction of Conditioned Inhibition

Wagner and Rescorla (1972) characterized inhibition as being the consequence of negative associative strength. According to their model conditioning occurred any time there was a discrepancy between the associative strength of a stimulus and that supportable by the reinforcement that followed it. Another not too surprising assumption was that nonreinforcement would support no associative strength. The conditioning that would occur on a trial would be in the direction of the discrepancy between the associative strength of the stimuli that were present on that trial and the associative strength that the reinforcer would support; that is, a positive discrepancy produced excitatory conditioning, whereas a negative discrepancy produced inhibition. A consequence of this point of view was the prediction that if an inhibitor was presented and nonreinforced there would be a positive discrepancy between nonreinforcement's ability to support no associative strength and the negative associative strength of the inhibitor. Thus if an inhibitor is presented by itself it will gradually extinguish because of the increments of positive associative strength produced by the nonreinforced presentations.

Zimmer–Hart and Rescorla (1974) investigated the effects of procedures designed to eliminate conditioned inhibition. They found that nonreinforced presentation of the CI was not an effective means of eliminating this inhibition. This result has been replicated by Pearce et al. (1982), who reported even more interference with conditioning on a retardation test following "extinction" than following inhibitory training. Zimmer–Hart and Rescorla (1974) found that reinforcing the compound of the CI and the excitor did prove to be an effective means of reducing inhibitory conditioning.

Rescorla (1979) correctly interpreted these results as being inconsistent with the notion of inhibition as negative associative strength. However, rather than question the notion that inhibition was fundamentally different from excitation, he reactivated one of Konorski's notions that inhibition acts not through negative associative strength but through changing the threshold for excitatory conditioning. An implication of this theoretical framework is that an inhibitor is assumed to be inactive in the absence of excitation. Thus it is not surprising that inhibition is not extinguished by presentation of the CI by itself because such procedures involve no excitation.

Kaplan and Hearst (1983) have recently reported a particularly elegant series of experiments that have attempted to answer the question of whether or not inhibition is active in the absence of excitation, or, in other words, whether inhibition is active in the absence of the expectation of the US. These experiments used the sign-tracking procedure with pigeons as subjects. This paradigm is particularly useful because inhibition is indexed by the overt response of conditioned withdrawal. They also attempted to answer the reciprocal question of whether or not excitation would be active in the presence of an excitatory background. In general their inhibitory experiments were consistent with Rescorla's position. Inhibitory stimuli did not seem to be active in the absence of contextual conditioning (i.e., if the context had been extinguished), but as soon as excitation was reinstated (by either unsignaled USs or excitatory CSs) the inhibitors regained their ability to elicit conditioned withdrawal.

Their experiments on conditioned excitation seemed to support the notion that excitation and inhibition are not symmetrical. In these experiments, making the context excitatory by presenting free food had little measurable effect on the conditioned approach to the excitatory CS. Thus one plausible conclusion from their experiments is that conditioned inhibition and excitation are fundamentally different. However, there is an alternative to this argument. The attempts to extinguish the experimental context involved exposing the animals to a food density that was exactly that predicted by the inhibitory stimulus—zero. It should also be mentioned that it took very little excitatory reinstatement to resurrect the inhibitor (10 food presentations or a few excitatory CSs). On the other hand the attempts to make the context excitatory in the conditioned excitation case never involved a food density during excitatory conditioning that was as high as that during the CS, and, in any case, the tests were generally carried out in contextual extinction (in the absence of food), a manipulation that is logically equivalent to the reinstatement procedures for inhibition. These procedural asymmetries thus weaken any conclusions concerning theoretical asymmetries between inhibition and excitation.

Kaplan and Hearst (1983) were not unaware of these points. In discussing their results and other data on the reinstatement of excitatory conditioning, they mentioned that the data were not only consistent with a formulation that emphasized the differences between excitation and inhibition, but also with an informal model. According to this model the reason that inhibition does not extinguish easily is that an inhibitor is a stimulus that predicts the absence of the US. The extinction procedure of presenting nonreinforced CIs presents no information that is inconsistent with this expectation. Further, their data on reinstatement are also consistent with this formulation. In the presence of a context in which there is no likelihood of food, why would a signal that predicts no food be particularly aversive and thus energize withdrawal?

Kaplan and Hearst's informal notion is at least a cousin of the temporal cognitive map that we have been championing here. It is clear that CEs may be extinguished at least temporarily by presenting them in the absence of reinforcement or, in other words, in a situation in which the expectation that the CE is followed by a US is violated. It would be most surprising if a CE would "extinguish" if it was presented while still paired with shock. Zimmer–Hart and Rescorla demonstrated that if the expectation of no-US is violated, by reinforcing the inhibitory compound, a CI will extinguish. It is also not surprising that if the no-US expectation is not violated during extinction, for example when the CI is presented and not reinforced, conditioned inhibition should be very resistant to extinction.

We have now discussed the problem of the extinction of CIs and CEs, and it is clear that when appropriate methods are used there is very little evidence for any qualitative differences between CIs and CEs. Moreover, there are procedural parallels between the proactive interference paradigms that we have described and various extinction procedures. Latent inhibition, the US preexposure effect, and learned irrelevance involve presenting the animal with the various conditioning events before conditioning. Various "extinction" procedures involve exposing the animals to these treatments after conditioning. We have discussed exposing the animals to the CI or CE alone, before, and after both inhibitory and excitatory conditioning. The US preexposure effect involves exposing the animals to the US prior to conditioning, and the various reinstatement paradigms involve exposing the animals to the US after conditioning. We have argued that none of these provide evidence for qualitative differences between CEs and CIs. Finally animals have been exposed to uncorrelated CSs and USs both before and after excitatory conditioning and before inhibitory conditioning. To our knowledge there are no reports of experiments exposing animals to uncorrelated CSs and USs after inhibitory conditioning. We have recently completed two unpublished experiments that have done just this.

In one experiment that involved a within-group comparison, we attempted to "extinguish" a CI by presenting the animals with uncorrelated CIs and USs following inhibitory conditioning. This CER experiment involved 1 group of 16 rats. After 7 days of VI training and 2 excitatory conditioning days to a light, the animals received 10 days of inhibitory training using the between-days negative correlation procedure. The only innovation in this experiment was that 2 auditory stimuli (the clicker and the pulsed tone) were presented on the nonshock days and thus became CIs. Following this came the 10-day "extinction" phase of the experiment, during which 1 auditory stimulus (counterbalanced) received continued inhibitory training; that is, it was presented on alternate days to shocks, whereas the other stimulus was presented uncorrelated with the shocks on the shock days. Under this uncorrelated procedure the probability of shock during the "extinguished" inhibitor was the same as in its absence (i.e., Rescorla's, 1967, truly random control procedure). On the alternate days these animals received no shocks under this procedure, so in fact this extinction procedure might be viewed as a between-days positive correlation procedure and, thus, a particularly good extinction procedure. Following this phase, the animals received a 3-day summation test on each day of which they were presented with single trials of the excitatory light, the noise–light, and the clicker–light compound. During the test none of the compound trials were shocked, but the light trials on the second and third day were followed by shock. The results of the summation test are shown in Fig. 5.5. This figure shows quite clearly that both the control stimulus and the "extinguished" stimulus quite effectively reduced suppression to the light stimulus. These results that are consistent with considerable data on conditioned excitation indicate that although exposure to uncorrelated CSs and USs retards the acquisition of conditioned inhibition, once conditioned, a CI is quite resistant to "extinction" by presentation of uncorrelated CSs and USs. This demonstration does not stand in isolation. We have also carried out a similar experiment using a between-group comparison that has also shown that inhibition is very resistant to "extinction" by uncorrelated presentations of CIs and USs.

These experiments report acceptances of the null hypothesis because they failed to reduce inhibition following a treatment. Such a comparison can never be considered definitive because it is always possible to argue that if a more dense or a prolonged treatment was used, inhibition would have been reduced. The parameters used here are, however, very similar to those that we have used before in learned irrelevance experiments, in which they cause large and consistent proactive effects. Thus at the very least it would seem that inhibition and, as others have shown, excitation are quite robust and once formed are very resistant to reduction unless strong measures are used. For instance, following inhibitory training with excitatory pairings of the CI with the US will reduce inhibition, but the level of uncorrelated presentations that we have used will not extinguish it.

Conclusions

One aim of the research that has been described here was to investigate factors that influence the acquisition of inhibitory conditioning and excitatory condition-

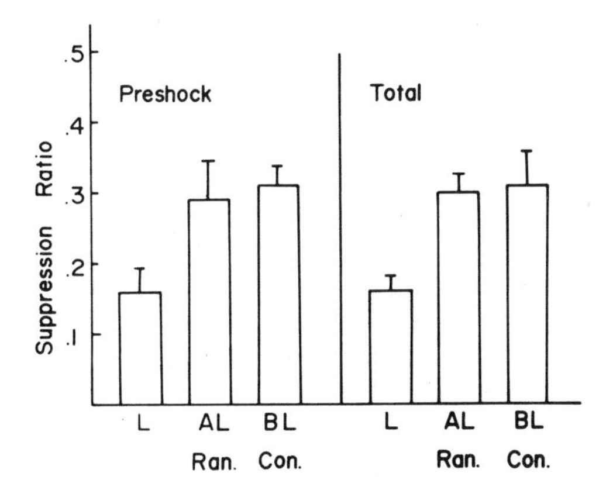

FIG. 5.5. Summation test documenting the failure to extinguish inhibition to Stimulus A by presenting it uncorrelated with shock (ran. = random) compared with B, which underwent continued inhibitory training. The left-hand panel represents the first day of the summation test during which there were no shocks. The right-hand represents the entire 3-day test during which L was followed by shock on the last 2 days. A and B were tone or clicker counterbalanced. L was the light. The error bars represent standard errors of the mean.

ing, and to determine whether we could find any qualitative differences between the two processes. More specifically, we investigated the role of conditioning of the experimental context in the acquisition of inhibitory conditioning. We found that manipulations of the experimental context that were designed to reduce the level of contextual conditioning extant when a stimulus was presented in a negative correlation with the US would reduce the tendency of that stimulus to become a conditioned inhibitor. These results were consistent with traditional associative notions concerning the role of the experimental context in the acquisition of excitatory conditioning (e.g., Tomie, 1976a, b). We argued, however, that the main organizing feature of these data and of our other data in the CER preparation (Baker et al., 1981) is not the role of contextual conditioning in mediating conditioning, although they do provide considerable evidence of the conditionability of the context, but the effect of these manipulations on the predictability of the US during exposure. The factors that seem to most reliably influence conditioning are the predictive relations between stimuli.

Another major feature of both inhibitory and excitatory conditioning is that the conditioning system "searches" for predictive relations between signaling events (CIs and CEs) and important events (USs). The evidence further seems to indicate that once the animal has represented a predictive relation between events this relation is quite resistant to change. This conclusion is supported by the findings that both excitatory and inhibitory conditioning are quite resistant to extinction, and that once a CI or CE is extinguished it is easy to reinstate the predictive representation. There are other examples of this "inertia" toward change in conditioning. For example, in Kamin's blocking, once a predictive relationship between one stimulus and the US is learned, a neutral stimulus that is added later will not readily come to control behavior.

This inertia toward change has theoretical implications. First, it brings the notion of independence of path into question; that is, it questions the notion that if two stimuli, for example, a novel stimulus and an extinguished stimulus, have the same behavioral effect, subsequent conditioning manipulations of them should have similar effects on the behavior that they control (cf. Rescorla & Wagner, 1972). Another implication is that these data tend in general to support theories that imply changes in the associability of stimuli during conditioning (e.g., Mackintosh, 1975, Pearce & Hall, 1981). Finally they suggest limits on the integration of the temporal cognitive map into behavior. The major weakness of "cognitive" positions such as ours is that they are consistent with everything and seem to exclude nothing. The present experiments strongly question traditional associative assumptions, but they also imply empirical limits to the flexibility of these cognitive representations. It seems that animals may represent all the events that occur. However, not all these representations seem to equally affect behavior. The animal integrates much information to make a "decision" that influences behavior. Once this decision is made it is difficult to change, although the animal may retain much information about other events that does not influence behavior much. This "decision" that we have referred to has many of the properties of "habit" and thus might be considered as involving traditional associations. Thus it is possible that the ultimate control of behavior in the conditioning system might involve associations, whereas behavioral changes such as the acquisition of conditioning might involve complicated cognitive representations. Such a dichotomy might be efficient and adaptive. Conditioning takes into account more of the information than traditional associationism would allow, but well-learned behavior is controlled by associative mechanisms. Thus associative systems might well be responsible for automatic processing as Wagner (1978) has suggested. Our claim that both associative mechanisms and the temporal cognitive map influence behavior might seem unparsimonious; however, such speculations may well begin to impose some limits on the cognitive map and thus bring such notions closer to being a testable model.

Regardless of whether the preceding theoretical considerations are taken seriously, the present data have more concrete implications. First, it is clear that the retardation test is a contaminated measure of inhibitory conditioning, and, therefore, experiments reporting only these tests should be greeted with considerable skepticism. Second, manipulations of contextual conditioning or US predictability influence the acquisition of inhibitory and excitatory conditioning strongly but have less effect once responding has been established. Third, we have found very little evidence for consistent differences between excitatory and inhibitory conditioning. However, it is clear that before this conclusion is accepted a much more systematic search for differences must be made, using within-experiment comparisons rather than between-experiment comparisons as we have done here. Finally, as Skinner (1938) argued, too much theory is a dangerous thing. Much important data can be gathered if experiments are generated by an empirical point of view that simply tries to outline the capabilities of the system regardless of theoretical nuisances. The experiments reported by Rescorla and Holland in this volume are good examples of this latter strategy.

REFERENCES

Anderson, D. C., O'Farrell, T., Formica, R., & Caponigri, V. (1969). Preconditioning CS exposure: Variation in place of conditioning and of presentation. *Psychonomic Science, 15*, 54–55.

Anderson, D. C., Wolf, D., & Sullivan, P. (1969). Preconditioning exposure to the CS: Variations in place of testing. *Psychonomic Science, 14*, 233–235.

Annau, Z., & Kamin, L. J. (1961). The conditioned emotional response as a function of the intensity of the US. *Journal of Comparative and Physiological Psychology, 54*, 428–432.

Baker, A. G. (1974). Conditioned inhibition is not the symmetrical opposite of conditioned excitation: A test of the Rescorla–Wagner model. *Learning and Motivation, 5*, 369–379.

Baker, A. G. (1977). Conditioned inhibition arising from a between session negative correlation. *Journal of Experimental Psychology: Animal Behavior Processes, 3*, 144–155.

Baker, A. G., & Mackintosh, N. J. (1979). Preexposure to the CS alone, the US alone or CS and US uncorrelated: Latent inhibition, blocking by context or learned irrelevance? *Learning and Motivation, 10*, 278–294.

Baker, A. G., & Mercier, P. (1982a). Manipulation of the apparatus and response context may reduce the US preexposure interference effect. *Quarterly Journal of Experimental Psychology, 34B,* 221–234.

Baker, A. G., & Mercier, P. (1982b). Extinction of the context and latent inhibition. *Learning and Motivation, 13,* 391–416.

Baker, A. G., & Mercier, P. (in press). Prior experience with the conditioning events: Evidence for a rich cognitive representation. In A. R. Wagner, R. Herrnstein, & Michael Commons (Eds.), *Quantitative analysis of behavior: Acquisition processes* (Vol. III). Ballinger: Cambridge, Mass.

Baker, A. G., Mercier, P., Gabel, J., & Baker, P. A. (1981). Contextual conditioning and the US preexposure effect in conditioned fear. *Journal of Experimental Psychology: Animal Behavior Processes, 7,* 109–128.

Durlach, P. J. (1983). Pavlovian learning and performance when CS and US are uncorrelated. In M. Commons, R. Herrnstein, & A. Wagner (Eds.), *Quantitative analyses of behavior:* (Vol. III) *Acquisition.* New York: Ballinger.

Gibbon, J., & Balsam, P. D. (1979). Spreading association in time. In C. M. Locurto, H. S. Terrace, & J. Gibbon (Eds.), *Autoshaping and conditioning theory.* New York: Academic Press.

Guthrie, E. R. (1935). *The Psychology of Learning.* New York: Harper.

Hearst, E. (1969). Extinction, inhibition and discrimination learning. In N. J. Mackintosh & W. K. Honig (Eds.), *Fundamental issues in associative learning. Delhousie Press: Halifax, N. S.*

Hull, C. L. (1943). Principles of behavior. New York: Appleton–Century–Crofts.

Kamin, L. J. (1969). Predictability, surprise, attention, and conditioning. In B. A. Campbell & R. M. Church (Eds.), *Punishment and aversive behavior.* New York: Appleton–Century–Crofts.

Kaplan, P. S., & Hearst, E. (in press). Contextual control and excitatory vs inhibitory learning: Studies of extinction, reinstatement, and interference. In P. D. Balsam & A. Tomie (Eds.). *Context and learning.* Hillsdale, NJ: Lawrence Erlbaum Associates.

Konorski, J. (1948). *Conditioned Reflexes and Neuron Organization.* England: Cambridge Press.

Lubow, R. E. (1973). Latent inhibition. *Psychological Bulletin, 79,* 398–407.

Mackintosh, N. J. (1973). Stimulus selection: Learning to ignore stimuli that predict no change in reinforcement. In R. A. Hinde & J. Stevenson–Hinde (Eds.), *Constraints on learning.* London/New York: Academic Press.

Mackintosh, N. J. (1974). *The psychology of animal learning.* London: Academic Press.

Mackintosh, N. J. (1975). A theory of attention: Variations in the associability of stimuli with reinforcement. *Psychological Review, 82,* 276–298.

Maier, S. F., & Seligman, M. E. P. (1976). Learned helplessness: Theory and evidence. *Journal of Experimental Psychology: General, 105,* 3–46.

Odling–Smee, F. J. (1975). The role of background stimuli during Pavlovian conditioning. *Quarterly Journal of Experimental Psychology, 27,* 201–209. (a)

Odling–Smee, F. J. (1975). Background stimuli and the interstimulus interval during Pavlovian conditioning. *Quarterly Journal of Experimental Psychology, 27,* 387–392. (b)

Odling–Smee, F. J. (1978). The overshadowing of background stimuli by an informative stimulus in aversive Pavlovian conditioning. *Animal Learning and Behavior, 6,* 43–51.

Pavlov, I. P. (1927). *Conditioned Reflexes.* London: Clarendon Press.

Pearce, J. M., & Hall, G. (1980). A model for Pavlovian learning: Variations in the effectiveness of conditioned but not unconditioned stimuli. *Psychological Review, 87,* 532–552.

Pearce J. M., Nicholas, D. J., & Dickinson, A. (1982). Loss of associability by a conditioned inhibitor. *Quarterly Journal of Experimental Psychology, 33B,* 149–162.

Randich, A. (1981). The US preexposure effect in the conditioning suppression paradigm: A role for conditioned situational stimuli. *Learning and Motivation.*

Randich, A., & LoLordo, V. M. (1979a). Associative and nonassociative theories of the UCS preexposure phenomenon: Implications for Pavlovian conditioning. *Psychological Bulletin, 86,* 523–548.

Randich, A., & LoLordo, V. M. (1979b). Preconditioning exposure to the unconditioned stimulus effects the acquisition of a conditioned emotional response. *Learning and Motivation, 10,* 245–277.

Rescorla, R. A. (1966). Predictability and number of pairings in Pavlovian fear conditioning. *Psychonomic Science, 4,* 383–384.

Rescorla, R. A. (1967). Pavlovian conditioning and its proper control procedures. *Psychological Review, 74,* 71–80.

Rescorla, R. A. (1969). Pavlovian conditioned inhibition. *Psychological Bulletin, 72,* 77–94.

Rescorla, R. A. (1971). Summation and retardation tests of latent inhibition. *Journal of Comparative and Physiological Psychology, 75,* 77–81.

Rescorla, R. A. (1979). Conditioned inhibition and extinction. In A. Dickinson and R. A. Boakes (Eds.) *Mechanisms of Learning and Motivation,* Lawrence Erlbaum Associates: Hillsdale New Jersey, 83–110.

Rescorla, R. A., & Wagner A. R. (1972). A theory of Pavlovian conditioning: Variations in the effectiveness of reinforcement and nonreinforcement. In A. H. Black & W. F. Prokasy (Eds.), *Classical conditioning II: Current research and theory* (pp. 64–99). New York: Appleton–Century–Crofts.

Skinner, B. F. (1938). *The Behavior of Organisms.* Appleton–Century–Crofts: New York.

Thorndike, E. L. (1932). Reward and punishment in animal learning. *Comparative Psychology Monographs, 8,* 1–39.

Tomie, A. (1976a). Interference with autoshaping by prior context conditioning. *Journal of Experimental Psychology: Animal Behavior Processes, 2,* 323–334.

Tomie, A. (1976b). Retardation of autoshaping: Control of contextual stimuli. *Science, 192,* 1244–1246.

Wagner, A. R. (1978). Expectancies and the priming of STM. In S. H. Hulse, H. Fowler, & W. K. Honig (Eds.), *Cognitive processes in animal behavior.* Hillsdale, NJ: Lawrence Erlbaum Associates.

Wagner, A. R., & Rescorla, R. A. (1972). Inhibition in Pavlovian conditioning: Application of a theory. In R. A. Boakes & M. S. Halliday (Eds.), *Inhibition and learning.* London: Academic Press.

Zimmer–Hart, C. L., & Rescorla, R. A. (1974). Extinction of Pavlovian conditioned inhibition. *Journal of Comparative and Physiological Psychology, 86,* 837–845.

6 Attention and Conditioned Inhibition

Geoffrey Hall
University of York

Helen Kaye, and John M. Pearce
University College, Cardiff

For at least 25 years (since, e.g., the work of Broadbent, 1958) students of information processing by human subjects have given a central role to the concept of "attention." It is no surprise, therefore, to find that a concept of attention has been intimately associated with the notion that the traditional procedures of conditioning can be profitably viewed as ones in which animals process information. The information-processing view of animal learning had its origins, at least in part, in the demonstration of "attention-like" processes in classical conditioning (Kamin, 1968); and several recent theories of conditioning have found a place for some sort of attentional mechanism (e.g., Frey & Sears, 1978; Mackintosh, 1975; Pearce & Hall, 1980; Wagner, 1978, 1979, 1981; see also the contribution by Moore to this volume).

The account proposed by Pearce and Hall (1980; see also Hall & Pearce, 1983; Pearce, Kaye, & Hall, 1983) is concerned primarily with the processing received by a conditioned stimulus (CS) during classical conditioning. Its starting point is the suggestion that a novel stimulus will be processed because it will be attended to by the mechanism responsible for associative learning. In other terminology, such a stimulus may be said to be high in associability. The central assertion of the theory is that, as conditioning proceeds, the associability of a stimulus will change in a manner determined by its informativeness. Associability will be maintained when the stimulus does not reliably predict a given outcome; associability will decline when the stimulus is a good predictor of its consequences. A stimulus that loses associability must still be attended to in some sense, because a well-trained CS will still evoke a conditioned response (CR). But, according to this account, such a CS will lose the power to command the attention of the associative-learning mechanism. This loss may perhaps be regarded as a shift from "controlled" to "automatic" processing (cf. Schneider & Shiffrin, 1977).

Evidence to support the theory proposed by Pearce and Hall (1980) has come almost exclusively from studies of excitatory classical conditioning but the mechanisms proposed should, in principle, apply just as readily to inhibitory conditioning. An inhibitory CS (which we will take to be a stimulus that signals the omission of a reinforcer) can, with appropriate training, become a reliable predictor of its consequences. It should, therefore, suffer a loss of associability as a result of this training. Our concern in this chapter is to provide experimental demonstrations of such an effect and to consider its implications for the interpretation of conditioned inhibition and attention more generally.

The chapter falls into four sections. First, we present experiments intended to establish that a well-trained conditioned inhibitor loses associability by showing that such a stimulus will form new associations only with difficulty. Second, we discuss experimental studies of an orienting response—studies that aim to provide a direct measure of the attention being paid to a stimulus during the course of inhibitory learning. Third, we discuss procedures that might restore lost associability and consider how the empirical findings can best be accommodated by a formal theory of the role of attention in conditioning. Finally, we consider the inplications of our findings for the interpretation of the transfer-of-training procedures that are routinely used in the assay of conditioned inhibition.

STUDIES OF THE TRANSFER OF TRAINING

The experiments described in this section comprise two main phases. In the first, a stimulus (B) is established as a conditioned inhibitor by discrimination training in which a second stimulus (A) is followed by reinforcement (by an unconditioned stimulus, US), whereas the simultaneous compound AB is presented without reinforcement. In the second phase the subject is required to learn further about stimulus B. If B loses associability during the first phase of training there should be negative transfer to the second phase, because any new association will be formed only with difficulty.

It will be apparent, however, that negative transfer might occur for reasons other than a loss of associability by B. In particular, B will acquire inhibitory associative strength as a result of Phase-1 training. This, in itself, would be enough to produce negative transfer to a second phase that attempted to establish B as an excitatory CS with a US of the same affective value—this, after all, is the principle underlying the retardation test for conditioned inhibition (Rescorla, 1969). Clearly it is necessary to devise a transfer test that allows us to distinguish between effects produced by changes in stimulus associability and those produced by the transfer of such associative strength as the stimulus may have acquired. We present next examples of two different techniques that permit this distinction to be made.

Transfer with an Associatively Neutral Compound Stimulus

Pearce and Hall (1979) reported a transfer-of-training experiment using rats as subjects and the conditioned suppression procedure. In the first phase of training, subjects in the Experimental condition received conditioned inhibition training with auditory stimuli that we may symbolize as A+/AB−. (The plus indicates that presentation of stimulus A was followed by electric shock; AB that two events were presented simultaneously as a compound.) Control subjects received simple discrimination training with quite different auditory stimuli (X+/Y−). At the end of this phase of training, the animals showed suppression of food-rewarded lever-press responding in the presence of their reinforced stimulus (A+ or X+) but responded normally in the presence of the nonreinforced stimulus (Y− or compound AB−). We assumed that, for the Experimental group, stimulus B had acquired inhibitory associative strength sufficient to neutralize the excitation governed by A (e.g., Wagner & Rescorla, 1972).

In Phase 2, subjects from both the Experimental and Control groups received conditioned inhibition training in which a new stimulus (a light) was followed by shock, whereas the compound AB plus the light was not reinforced. For Control subjects, therefore, a novel compound stimulus (AB) was trained as a conditioned inhibitor. If novel stimuli are highly associable, learning should proceed readily for these animals. For Experimental subjects, on the other hand, one element of this compound (B) had already acquired inhibitory associative strength. Nonetheless, we still hoped to see a retardation of Phase-2 learning. The presence of the excitatory stimulus A might be expected to counteract the inhibition governed by B, producing an AB compound that is both associatively neutral and also, because of Phase-1 training, low in associability.

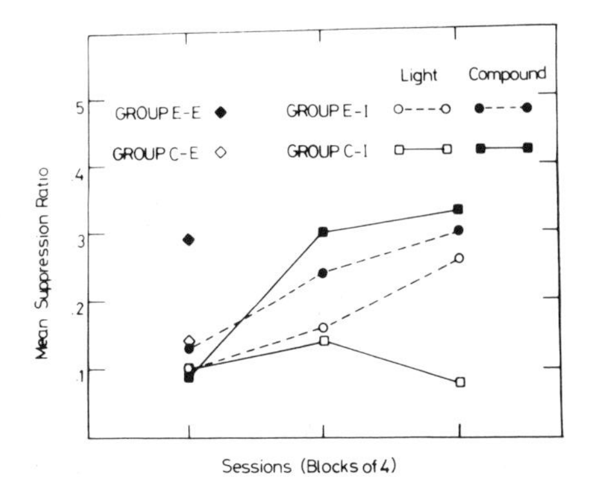

FIG. 6.1. Group mean suppression scores in a transfer test. Control groups (C–E and C–I) were pretrained on X+/Y−; Experimental groups (E–E and E–I) on A+/AB−. In the excitatory test (for groups C–E and E–E) the AB compound was reinforced. In the inhibitory test (for groups C–I and E–I) a discrimination between a reinforced light and a nonreinforced compound of AB plus the light was required. (From Pearce & Hall, 1979).

The course of Phase-2 training is shown in Fig. 6.1. Control subjects (group C-I for Control–Inhibitory) formed a clear discrimination, remaining suppressed in the presence of the light but losing suppression in the presence of the compound. Experimental subjects (group E-I) formed only a poor discrimination, a result to be expected if, for them, the AB compound had lost the ability to form new associations.

It might be objected that the results of Fig. 6.1 reflect not the lost associability of the AB compound in Group E–I but the fact that the compound was not truly associatively neutral for these animals. If AB retained some net excitatory strength from Phase I, this might be expected to produce poor learning in the Phase-2 task where AB signals the omission of the US. To investigate this possibility, further groups of Experimental and Control subjects were given Phase-2 training designed to turn the compound AB into an excitatory CS. The pooled results of eight reinforced (AB+) trials for these subjects are shown in Fig. 6.1. Experimental subjects (group E–E for Experimental–Excitatory) acquired much less suppression over these trials than did Control subjects (group C–E). The reverse outcome would be expected if AB were already somewhat excitatory for group E–E at the start of Phase 2. We conclude that, because the Experimental subjects showed poor learning when AB is trained both as an inhibitory CS and as an excitatory CS, the negative transfer observed in this experiment was not produced by the transfer of associative strength. Rather, the overall pattern of results suggests that Phase-1 training causes AB to lose associability.

Increasing the Magnitude of the Reinforcer

When a subject has been trained to asymptote with a CS predicting a given reinforcer, it may be possible to observe further learning if the magnitude of the reinforcer is increased in a second phase of training. Positive transfer from Phase 1 to Phase 2 of such an experiment is to be expected, in that the associative strength acquired by the CS in Phase 1 would reduce the amount of training required for the new asymptote of Phase 2 to be reached. But in a series of experiments using the conditioned suppression procedure we have found negative rather than positive transfer (Hall & Pearce, 1979, 1982). We argued that in these experiments some other source of transfer was outweighing the effect produced by the transfer of associative strength, and we identified this other source with the loss of associability that the CS is presumed to suffer when trained to asymptote in Phase 1. Overall negative transfer need not always occur in experiments of this type—with some parameters it is to be expected that the positive, associative transfer will outweigh the negative, attentional transfer. But when overall negative transfer is found, it provides rather good evidence for the suggestion that a well-trained CS loses associability.

Our previous experiments using this design have demonstrated negative transfer from excitatory phase-1 conditioning with a mild shock US to Phase 2 using the same CS and a more intense shock. We present now a formally equivalent study (hitherto unpublished) intended to reveal changes in associability in inhibitory conditioning. The experiment uses a different training procedure (autoshaping with pigeons as the subjects) and thus, as a preliminary, we report a study to show that the effect expected in the excitatory case can be found here too. The preliminary experiment may also be useful in establishing that the negative transfer effect found by Hall and Pearce (1979) is not confined to experiments using the conditioned suppression procedure.

Transfer in Excitatory Conditioning. In our autoshaping experiments the subjects were pigeons, the primary US was access to grain for 5 s, and CSs were provided by the illumination of a response-key for 10 s with lights that differed in color and white lines differing in orientation. Compound stimuli were produced by the superimposition of a white line on a colored background. The specific events used as CSs varied from one subject to another in a counterbalanced way, and they are referred to by arbitrary symbols henceforth.

The 8 pigeons in our preliminary experiment received 6 sessions of Phase-1 autoshaping with stimulus B presented 36 times per session, followed on each occasion by a small reward. In Phase 2 they received 2 sessions in which 18 presentations of B were intermixed with 18 presentations of a novel stimulus (N), each trial being followed by a larger reward. We hoped to be able to show more rapid conditioning to N that to B, in spite of the fact that B might be expected to have acquired some associative strength as a result of Phase-1 training. Stimulus N being a novel should be high in associability, whereas B should have lost associability. In order to vary size of reward we used the primary reinforcer in Phase 2, but in Phase 1, B was associated only with a secondary or conditioned reinforcer. Thus, the subjects were trained initially with a different stimulus (A) predicting food, and during their Phase-1 training they received 12 A-food presentations per session intermixed with the trials on which B was presented followed by A alone. This second-order conditioning procedure might be expected to generate some responding to B (Rescorla, 1980), but not so much as to prevent our seeing an increase in response rate as further learning occurs in Phase 2.

Figure 6.2 shows the percentage of the various types of trial upon which at least one response occurred. The isolated point at the left of the figure shows performance on the final session of preliminary training with stimulus A. For Phase-1 (second-order) conditioning, responding was maintained at a high level to this stimulus on those trials when it was immediately followed by food (A+). Performance on trials when A was presented following B is not shown in the figure, but responding was maintained on these trials too, there being a mean of

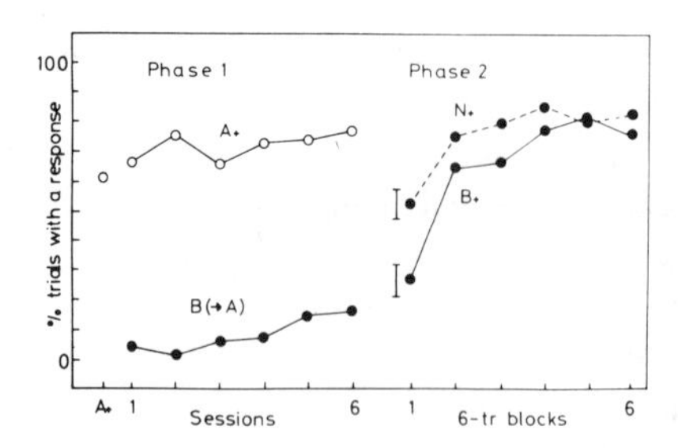

FIG. 6.2. Autoshaped responding in pigeons: percentage of trials of each type on which a response occurred (groups means). After initial reinforced training with stimulus A, all subjects received six sessions of Phase-1 conditioning consisting of reinforced A trials (A+) and presentations of B followed by A. Phase 2 consisted of reinforced presentations of B and of a novel stimulus, N. Vertical bars represent the pooled standard error of the mean scores to the N stimulus and to the B stimulus.

76% trials with a response on the final session of Phase 1. Some responding also occurred to B, and such responding increasing as training proceeded. All subjects responded more to B on the final session of phase one than they did on the first session, evidence that B had acquired some small measure of excitatory strength, and evidence, therefore, that our training procedure is distinct from the traditional latent inhibition paradigm.

Upon its first presentation in Phase 2, stimulus N evoked responding in only one subject; the pretrained stimulus evoked marginally more with three subjects responding on the first B trial. But, as Fig. 6.2 shows, the first few reinforced trials were sufficient to generate a relatively high level of responding to both stimuli. Phase-2 acquisition occurred very readily to stimulus N, suggesting that this stimulus was high in associability. Stimulus B took significantly longer ($p < .05$) to reach the same level of performance as stimulus N, suggesting that B had lost associability during Phase-1 training.

Transfer in Inhibitory Conditioning. We then applied this general technique to the investigation of attentional changes during inhibitory conditioning. We assumed the magnitude of an inhibitory reinforcer in appetitive conditioning to depend on the magnitude of the unfulfilled expectation of food (Pearce & Hall, 1980). In conditioned inhibition (A+/AB−) training, therefore, it will depend on the excitatory associative strength governed by A and, thus, on the magnitude of the US with which A is associated. In order to see further inhibitory learning after such training it is necessary to increase the size of the inhibitory reinforcer, and we may do this by increasing the magnitude of the US associated with A. In the autoshaping experiment described next, this was achieved by using a conditioned reinforcer in Phase 1 and immediate access to food on the excitatory reinforcer in Phase 2.

Ten pigeons were given Phase-1 training that consisted of trials of two types. There were 30 presentations per session of a simultaneous compound stimulus, BC, that were not reinforced. Intermixed were 10 presentations of stimulus C alone followed immediately by the presentation of a further stimulus, A, the offset of which was followed by food. It was hoped that this serial conditioning procedure would allow stimulus C to acquire somewhat more excitatory strength than would use of the second-order procedure employed in the previous experiment, but less than that generated by direct pairings of the stimulus with food. Stimulus C should, therefore, have acquired a moderate level of excitatory strength and B a moderate level of inhibitory strength. After seven sessions of Phase 1, all subjects received five sessions of training in each of which stimulus D was presented 18 times followed by immediate access to food. D should, therefore, become strongly excitatory.

Phase 2 comprised 4 sessions of discrimination training, in which reinforced D trials continued to occur (10 per session) intermixed with 15 nonreinforced trials to the BD compound and 15 nonreinforced trials to the ND compound (where N was a novel stimulus element). The number of trials of each type on which at least one response occurred was recorded as was the number of responses per trial. The results are shown in Fig. 6.3.

Both measures showed that, in Phase 1, responding to stimulus A rose to and stayed at a high level. Although performance did not achieve the same high level, some excitatory conditioning also occurred to C; thus our serial conditioning procedure appeared to be successful in endowing C with a moderate level of excitatory associative strength. Some responding occurred initially to the BC compound but disappeared by the end of Phase 1, suggesting that B had acquired inhibitory power. All subjects learned to respond at high levels to stimulus D (the figure shows performance on the last session of acquisition to D), and this responding was maintained throughout Phase 2. The addition of either B or N to stimulus D brought about a marked reduction in responding; but, as the right-hand panel of Fig. 6.3 shows, the reduction was initially greater on BD than on

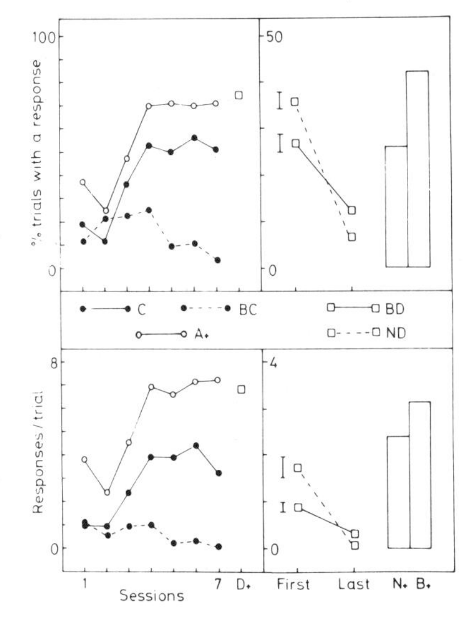

FIG. 6.3. Autoshaped responding in pigeons (group means): percentage of trials with a response (top panel) and number of responses per trial (lower panel). Phase 1 consisted of six sessions in which stimulus C was followed by a reinforced A trial and nonreinforced BC trials. After reinforced training to D a discrimination was required between D and the nonreinforced compounds BD and ND. Vertical bars represent the pooled standard errors of the mean scores to BD and to ND in Phase 2. The histograms show performance on a final session in which B and N were presented alone and reinforced.

ND trials. There was significantly ($p < .05$) less responding to BD than to ND on the first session of Phase 2, and we conclude that B acquired inhibitory properties during Phase 1. The BD compound acquired further inhibition only slowly. By the last session of this phase on which responding to the compound stimuli occurred (some animals failed to respond to these stimuli on session four), the difference between the conditions was reversed. Eight of the 10 subjects showed more responding to BD than to ND on the final session, and the interaction of session number and stimulus type was statistically reliable ($p < .05$).

Response rates were very low for the last session of Phase 2, and it seemed possible that the true size of the difference in inhibitory strength between N and B was being obscured by a "floor effect." We, therefore, concluded the experiment with a retardation test. This consisted of a single session in which N and B were each presented 10 times, each presentation being followed by immediate access to food. Responding to both stimuli was reestablished by this procedure; but, as Fig. 6.3 shows, stimulus B acquired excitatory strength more readily than did stimulus N. The histograms on the right of the figure show group mean performance on rewarded N and B trials. Performance was significantly superior ($p < .05$) to B on both measures. This outcome confirms the conclusion that N had acquired more inhibitory strength than B by the end of the preceding phase of training.

Conclusions

Conditioned inhibition (A+/AB−) training endows stimulus B with associative properties that can be revealed by the application of appropriate transfer tests. The first Phase-2 session of the experiment just described constitutes an example of a summation test and shows that the presumed conditioned inhibitor has the power to reduce in magnitude the CR normally elicited by an excitor. For the most part, however, we have concentrated upon variants of the retardation test. Although our experiments do nothing to invalidate the notion that conditioned inhibition can be revealed in the retardation of subsequent excitatory conditioning, they do serve to show that other forms of learning (in particular, further inhibitory conditioning itself) may suffer a retardation also. These results suggest to us that a conditioned inhibitor undergoes another change in the process of acquiring inhibitory associative strength—that it suffers a loss of associability.

In seeking further evidence relevant to this interpretation we now turned from these relatively complex transfer-of-training experiments to an observational procedure that, we hoped, would give us a direct measure of changes in attention during conditioning.

STUDIES OF AN ORIENTING RESPONSE

A naive rat in a dimly lit Skinner box will respond in a characteristic way to the illumination of a small, bright signal lamp, fixed to the wall of the chamber. The rat will stop its current activity (exploration, or grooming, or whatever), will turn toward the light, and may then rear in front of it making contact with its snout or paws. We call this pattern of behavior an orienting response (OR). The OR is well defined and independent observers have no difficulty in identifying it.

The occurrence of an OR means that the rat is attending to the novel light. Pearce, Kaye, and Hall (1983) reported a study in which the stimulus (a 10-s light) was presented six times a session over 14 sessions; there was an orderly decline in the frequency of occurrence of the OR, with none at all recorded on the final day of training. We concluded that the familiar stimulus was no longer capable of controlling attention in the way that it did when it was novel. It seems, then, that this observational technique can supply a direct measure of the attention that a rat is paying to a stimulus. Can the technique be used to demonstrate that a CS loses the ability to command attention as, during the course of conditioning, it becomes firmly associated with its consequences?; that is, is the OR to the light sensitive to variations in the predictive accuracy of that stimulus when it is used as a CS in a conditioning paradigm? That the OR will habituate is evidence that the light can lose the ability to attract attention in some sense. We have already pointed out, however, that there is more than one sort of attention. It remains to demonstrate that the OR can be used as an index of the form of attention that we are primarily concerned with—the mechanism that determines the readiness with which a stimulus enters into associative processing.

Evidence that the OR can be used as such an index comes from the other results reported by Pearce et al. (1983). They included a group of subjects that received training with the light as an excitatory CS, each presentation of the light being followed by access to food. Although it occurred more slowly, these subjects too, like those experiencing simple habituation training, showed a loss of the tendency to orient toward the light as training proceeded. We took this as preliminary evidence that the OR can be used as an index of the associability of the light, and that associability will be lost when the light is reliably followed by a given consequence, whether that be no event (as in habituation) or the presentation of food (as in excitatory conditioning). Support for this interpretation has recently been provided by a series of experiments by Kaye and Pearce (1984), which confirm the original findings and provide further tests allowing us to reject alternative interpretations.

The OR During Conditioned Inhibition Training

We have found that the frequency of occurrence of the OR will decline when the light consistently signals the delivery of a reinforcer. What will happen when the light is made to signal the omission of a reinforcer in conditioned inhibition training?

In an unpublished study Kaye (1983) trained a group of rats on a discrimination in which a 10-s tone was followed by the delivery of a food pellet, whereas a compound stimulus, consisting of the tone and the light presented simultaneously, was not. The subjects received 4 sessions of initial training designed to establish the tone as an excitor in each of which there were 6 reinforced tone presentations. There followed a pretest session containing 6 nonreinforced light trials (the illumination of a bulb 6 cm above floor level and 6 cm to the left of a food tray) on which the unconditioned effect of the light was observed. The occurrence of an OR was identified at 2 points during the 10-s stimulus (after 4s and 9s) by "freezing" a videotape recording of the rat's behavior. The subjects in the critical group (designated T+/LT−) received 14 sessions of discrimination training. On each session there were 6 nonreinforced presentations of the LT compound and 6 reinforced presentations of the tone alone. Two control groups were included. One (group C+/LT−) received identical training except that a different auditory stimulus (a clicker) was presented on reinforced trials. The second (group LT−) received reinforced pretraining with the clicker but presentations just of LT and no reinforced trials during the next stage of training. For these control groups the light would not be expected to acquire inhibitory properties, to the extent that inhibition is generated by the omission of a reinforcer signaled by a concurrently presented excitatory CS. (It is possible, however, that generalization from the reinforced clicker to the tone might allow some inhibitory learning, particularly in group C+/LT−).

The group mean results are shown in Fig. 6.4. The lower panel shows "magazine time"; that is, the length of time on each trial during which the rat pushed open a small flap that guarded the entrance to the food tray. Magazine time increased during excitatory pretraining and remained at a high level to the

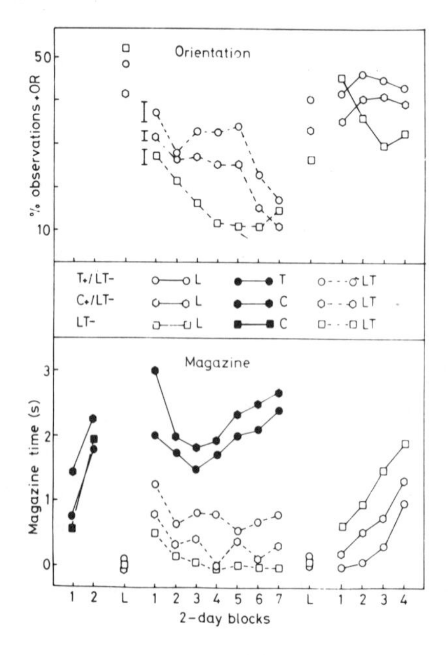

FIG. 6.4. Top panel: percentage of observations on which orienting to a light was scored for three groups of rats (group means). After an initial session with the light presented alone (L), all subjects received 14 sessions with a light–tone (LT) compound, a further L session, and 8 sessions with the light signaling food. Vertical bars represent, separately for each group, pooled standard errors of the mean scores for the stage of training with LT. Lower panel: group median time spent in magazine activity in the presence of the light and in the presence of a reinforced auditory stimulus, a tone (T) and a clicker (C).

reinforced auditory stimulus during discrimination training for groups T+/LT− and C+/LT−. No group showed much responding to the magazine during LT presentations, and what little there was tended to disappear as training progressed. The critical results for our present purpose are shown in the upper panel of the figure. When it was first presented, the light evoked an OR in all groups but the frequency of occurrence of the OR declined with repeated presentations. The decline was most rapid in group LT−, this group showing significantly fewer ORs than either of the other groups ($p < .05$). We suggest that for group T+/LT− (and perhaps also for group C+/LT−) the light is acquiring associative strength as a conditioned inhibitor; that such learning occurs over a number of trials but that as the association approaches its asymptotic strength, so the light loses associability.

After the discrimination phase had been completed, all subjects received a single test session with the light presented alone. There was an increase in the frequency of the OR in all groups ($p < .05$), a finding that is discussed later. There followed eight sessions of training in which the light was presented six times per session followed by food. This procedure constitutes a retardation test for conditioned inhibition, and we might expect the development of excitatory associative strength to occur slowly in those subjects for whom the light had acquired inhibitory properties. The lower panel of Fig. 6.4 shows that all groups started to exhibit magazine behavior in the presence of the light, and we take this behavior to reflect the acquisition of excitatory strength by the light. The groups differed on this measure ($p < .05$) with group LT− acquiring the behavior most rapidly and group T+/LT− most slowly, a finding consistent with the suggestion that the light was a conditioned inhibitor for the latter group at the start of this final phase of training.

The OR to a Stimulus Negatively Correlated with Reinforcement

The training procedure used in the preceding experiment is one of several that serve to produce inhibitory learning (see LoLordo & Fairless's contribution to this volume). What these procedures seem to hold in common is that they arrange, in one way or another, for there to be a negative correlation between the critical stimulus and a reinforcer. In the experiment described next we examined the effects of such a negative correlation on the OR to a light.

There were 3 groups of subjects. All received a pretest session with the light presented alone. Subjects in the negatively correlated group (NC in Fig. 6.5) received 10 sessions of training on each of which the light was presented 6 times and 6 food pellets were delivered. The light was presented at 4-min intervals but food was never delivered in its presence or within 1 min of its onset or offset. Subjects in a random control (RC) group received similar training except that for them the two events were uncorrelated, food being delivered according to an

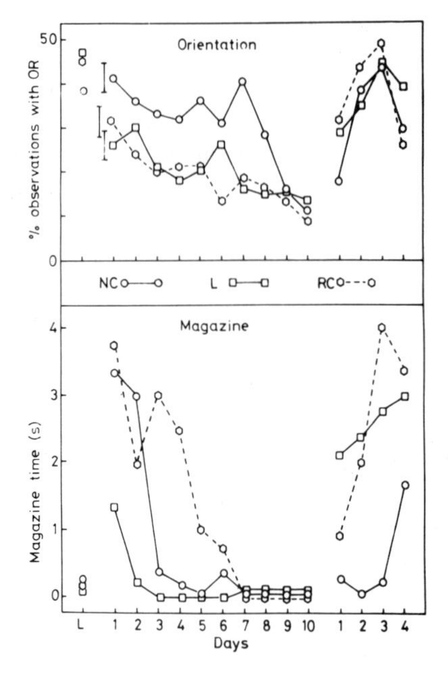

FIG. 6.5. Top panel: percentage of observation periods on which orienting to a light was scored for three groups of rats (group means). All subjects received an initial session with the light presented alone (L) followed by 10 sessions of light alone (group L), of light and food negatively correlated (NC), or of light and food randomly correlated (group RC). Vertical bars represent, separately for each group, pooled standard errors for the mean scores of this phase of training. For the final four sessions the light signaled food. Bottom panel: group median time spent in magazine activity in the presence of the light.

independent variable time 4-min schedule. Group L received presentations of the light but not of food.

The subjects (in groups NC and RC) that received food during training had, initially, a strong tendency to enter the food tray and they did so both in the absence of the light and (as the lower panel of Fig. 6.5 shows) in its presence. By session seven the magazine-time scores had declined to a low level, the decline occurring more readily in group NC (for which the light signaled that food would not be presented) than in group RC (for which food pellets were presented occasionally in the presence of the light). The upper panel of the figure shows the results for the OR. All groups had initially a fairly high level of responding, and in all groups there was a significant ($p < .05$) decline in response with repeated presentations of the light. This decline occurred most slowly in group NC (the group for which the light was an informative signal) but, by the end of training, group NC was showing the OR no more frequently than either of the other two groups.

We conclude that a stimulus of predictive value will command attention only while it is being learned about. In order to support this conclusion it is necessary to show that inhibitory learning did indeed occur to the light in group NC. Again we used a retardation test. In a final phase of training, all subjects received four six-trial sessions with the light predicting the immediate delivery of a food pellet. For all groups (as Fig. 6.5 shows) this change of conditions evoked a restoration of the OR that increased in frequency over the first three conditioning sessions before beginning a decline in the fourth. Magazine time was taken as an index of the acquisition of excitatory associative strength. All groups started to show magazine responding in the presence of the light (Fig. 6.5, lower panel), but the acquisition of this behavior was profoundly retarded in group NC that produced significantly lower scores ($p < .05$) than each of the other groups. We concluded that the negative correlation experienced by group NC endowed the light with inhibitory properties.

Conclusions

If the behavior of orienting to the light can be taken as an index of stimulus associability and magazine responding as an index of the associative strength of the light, the results presented previously suggest that, as inhibitory strength is acquired, so associability is lost.

The use of magazine responding as an index of associative strength is justified by the fact that it behaves as a CR might be expected to. In increases in frequency or duration in the presence of the light when that stimulus precedes food, and it disappears when the light is negatively correlated with food. Interpretation of the behavior shown to the light itself is more problematic in that others have regarded it too as being a CR. Thus Holland (e.g., 1977, 1980), in a series of observational studies of appetitive Pavlovian conditioning, has used a pattern of behavior, similar to that which we have called an OR, as an index of associative strength. In Holland's experiments the response did indeed have some of the properties of a CR in that he found it to be maintained throughout excitatory training, whereas we have repeatedly observed its decline even when the CS signaled the presentation of food. There are certain procedural differences between Holland's experiments and our own that might account for the discrepancy. They are discussed by Kaye and Pearce (1984) and we do not repeat them here, but rather we expand upon our reasons for thinking that the response generated by our experimental procedure is more appropriately regarded as an OR than a CR.

They are, first, that the response in question appears unconditionally when the light is first presented and that it declines with repeated presentation. Further, we found that subjects experiencing a negative correlation between the light and food showed a greater tendency to approach the light than did control subjects—the reverse of what might be expected if the behavior we have called an OR were in fact an autoshaped CR (cf. Wasserman, Franklin, & Hearst, 1974).

We should emphasize that our interpretation can be applied only to certain training situations. ''Sign tracking'' cannot always be safely used as an index of the associability of a CS. In some situations (autoshaping with pigeons for example and perhaps for rats with procedures different from ours) the response of approaching the stimulus is strengthened by reinforcement and has many of the properties of a CR. We have thus, in our experiments on autoshaped responding in pigeons, been willing to use keypecking as an index of associative strength and have assessed the associability of the key-light stimulus by indirect, transfer of training techniques.

More generally, however, our observations of the OR and our transfer-of-training studies point toward the same conclusion. A stimulus that supplies information that a given reinforcer will not be presented will lose associability as inhibitory learning proceeds. We turn next to the theoretical interpretation and implications of this conclusion.

INTERPRETATION

Although the results described previously are perhaps compatible with other theories of stimulus associability (e.g., Wagner, 1978, 1979, 1981), they had their origins in the model of conditioning proposed by Pearce and Hall (1980). We concentrate, therefore, on the latter theory. First we present a more formal statement of the way in which this model accounts for the results. We then move on to a discussion of the problems raised for this theory by experiments concerned with procedures that might restore lost stimulus associability.

Outline of a Theory

The model proposed by Pearce and Hall (1980) starts with the assumption that in conditioning the central representation of an effective stimulus forms an association with the central representation of the consequences of that stimulus. The strength of the association increases each time the two representations are conjointly activated until an asymptote is reached. In excitatory conditioning the consequence of the CS is the occurrence of a US, and a CS–US association is formed. In conditioned inhibition (A+/AB−) training, the consequences of stimulus B will include the state of frustration or of relief evoked by the nonoccurrence of the expected US. The resulting association can be summarized as being between the CS and a "no-US" representation or center (cf. Konorski, 1967).

When associations have been formed CSs have the power to evoke activity in the US centers. The level of activity in the no-US center will depend on V_I, the aggregate associative strength of such inhibitory CSs as are present. The level of activity in the US center will depend, in part, on V_E, the aggregate excitatory associative strength. But, in addition, we assume that activation of the no-US center will reduce the readiness with which the US center can be activated by an excitatory CS (cf. Rescorla, 1979).

The increase in associative strength produced by an inhibitory learning trial will depend on the magnitude of the inhibitory reinforcer (R_I) and on the associability of the CS (α). The former will depend on the extent to which the omitted US was expected, that is, upon the level of activity in the US center. Initially this will be given directly by V_E, but as inhibitory learning proceeds the magnitude of the inhibitory reinforcer will be reduced according to:

$$R_I = (V_E - V_I) \qquad 1$$

In accord with our assumption that the effect of an inhibitor is to reduce the ability of the US center to be activated, we assume that the value of $(V_E - V_I)$ cannot go negative (see Pearce, Nicholas, & Dickinson, 1982).

Omitting certain complications not relevant to our present concerns, the growth of inhibitory associative strength is now given by:

$$\Delta V_I = \alpha R_I \qquad 2$$

where ΔV_I refers to the increment produced by a training trial. The value of α can change, becoming reduced as learning proceeds. We suggest that the associability of a CS on trial n + 1 will be determined (in part, see Pearce et al., 1983) by what occurred on the previous trial, specifically by the discrepancy between what actually happened and what the associative strengths of the CSs predicted. Thus:

$$\alpha^{n+1} = |\lambda - (V_E - V_I)|^n \qquad 3$$

where λ varies with US magnitude. For the compound trials of conditioned inhibition training, λ will be zero.

Applying these equations shows that A+/AB− training will endow stimulus B with inhibitory strength that will grow to neutralize the excitatory strength governed by A. Learning will then stop, both because R_I falls to zero and because the associability of the stimuli will have declined to zero (Equation 3). Stimulus B will be effective in a summation test because of its inhibitory effect on the US center. And in a retardation test an excitatory CR will be slow to develop for two reasons. First, the existence of inhibitory associative strength will limit the ability of V_E, as it grows, to evoke a CR. Second, because α will be low, it will take at least one trial for a discrepancy between λ and $(V_E - V_I)$ to restore associability. This latter factor allows us to predict that a fully trained inhibitor will be learned about slowly not only in the traditional retardation test but also in other tests of the transfer of training.

Restoration of Lost Associability

Although the theory just presented can accommodate most of the results reported so far, there is one exception that requires immediate comment. The results of the OR study presented in Fig. 6.4 suggest a procedure whereby the lost associability of an inhibitory CS might be restored. They show that, for a group of subjects given A+/AB− training, the OR to B declines as the discrimination is formed, but that it reappears when B is presented alone. The possibility that B-alone trials can restore associability is not allowed by the equations given earlier.

If A+/AB− training produces a B stimulus having inhibitory strength (V_I) to match the V_E governed by A and an α-value of zero, then subsequent presentations of B alone will have no effect on either of these properties. In Equations 2 and 3, R_I, λ, and ($V_E - V_I$) will all take the value of zero; there will be no change in associative strength and α will remain at zero.

A Transfer-of-Training Experiment. Resolution of the divergence between findings and theory is complicated by the fact that in other experiments we have found no evidence that B-alone trials can restore associability. Pearce et al. (1982) trained rats by the conditioned suppression procedure on the discrimination A+/AB−. One group of subjects (A+/AB−, B−) then received a series of presentations of B alone; a control group was placed in the apparatus and no CSs were presented (group A+/AB−,O). The associative strength governed by B was then assessed by means of a summation test. The left-hand panel of Fig. 6.6 shows that the groups did not differ on this measure, confirming the finding of Zimmer–Hart and Rescorla (1974) that B-alone trials produce no change in the inhibitor strength of B.

It remains possible that the associability of B was changed by this treatment. In order to assess this, Pearce et al. gave two further groups of subjects a retardation test in which B was presented six times followed by the shock US. If B-alone trials leave associative strength unchanged but restore associability, we might expect to find fairly rapid acquisition in group A+/AB−. B−, The results in the right-hand panel of Fig. 6.6 show that the reverse was the case; it seems that B-alone trials produced not a restoration but a further loss of associability.

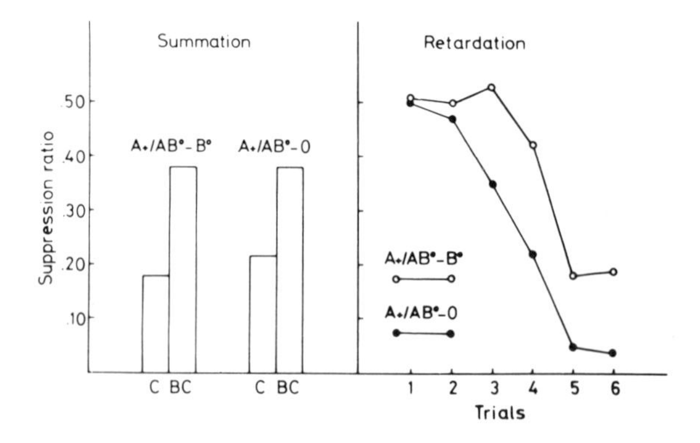

FIG. 6.6 Group mean suppression ratios on summation and retardation tests for rats given conditioned inhibition training (A+/AB−) followed by exposure to B (B−) or exposure to the training context in the absence of CSs (0). The summation test compares performance to a separately established excitor (C) with that to the BC compound; the retardation test consisted of six reinforced trials with B. (Based on data reported by Pearce et al., 1982).

Pearce et al. (1982) consider a number of possible explanations for their findings, but we consider here only the way in which they can be accommodated by the model outlined earlier. We have said that with R_I λ, and (V_E -V_I) all at zero, B-alone trials will produce changes neither in associative strength nor in associability. It is possible, however, that the associability of B could decline further if it was not yet a zero by the end of A+/AB− training; that is, if inhibitory conditioning has not reached asymptote so that V_E still exceeds V_I, Equation 3 yields an α with a value greater than zero. Presentations of B alone will produce no changes in associative strength, but, because λ and (V_E -V_I) will both take the value of zero on these trials, Equation 3 predicts that B will lose such small measure of associability as it might possess.

Restoration of the OR. If the account given previously is accepted, it still remains necessary to explain the apparent increase in α that occurred during the posttraining L-alone trials shown in Fig. 6.4.

The first point to make is that, although the increase in the frequency of the OR upon omission of T from the LT compound was greatest in the group for which L was a conditioned inhibitor, an increase was observed in the other groups too. Another way of looking at this is in terms of the significance of the omitted stimulus. For the group given conditioned inhibition training the tone was motivationally significant having been paired with food. For the two control groups the tone would have only generalized excitatory strength produced by the reinforced training they had experienced with another auditory stimulus, the clicker. Perhaps the associability of the light is governed not only by how well it predicts its consequence (the omission of food) but also by the associations it forms with other events presented concurrently (cf. Rescorla, 1982; Rescorla & Durlach, 1981). The omission of such an event, at least when it is motivationally significant, may be enough to restore associability in the same way as the omission of an orthodox reinforcer can (Hall & Pearce, 1982).

We investigated this matter further by training another group of rats (group CI) by the same basic procedures used for group T+/LT− of Fig. 6.4. (The procedure differed only in that more extensive preliminary T+ training was given.) After 15 sessions of conditioned inhibition training they received five sessions in which the light was presented alone. In order to minimize changes in the general context, T+ trials continued to occur. Control subjects (group RC) received identical treatment except that for them the delivery of food was not correlated with the presentation of any other stimulus.

The results for the discrimination and test phases are shown in Fig. 6.7. The general picture looks much like that presented by Fig. 6.4. Group CI showed a clear discrimination between T and LT in magazine responding, whereas group RC showed little tendency to enter the magazine during either of the stimuli. In group RC the OR to the light showed a steady decline. In group CI the OR

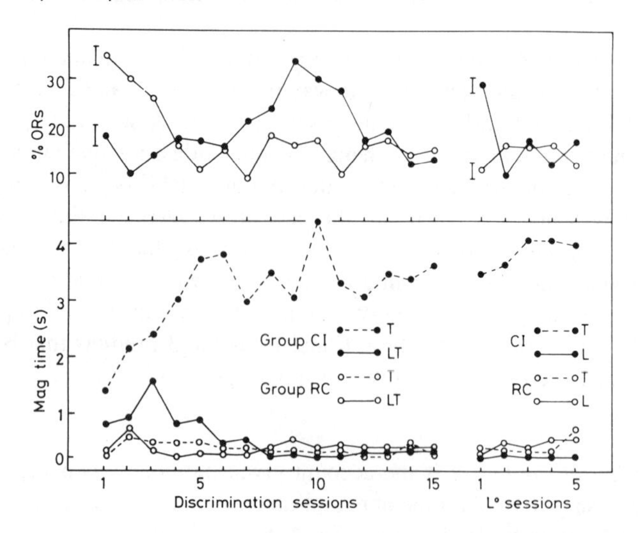

FIG. 6.7. Top panel: percentage of observations on which orienting to a light was scored for two groups of rats (group means). During the discrimination phase group CI received reinforced presentations of a tone (T) and nonreinforced light–tone (LT) trials; for group RC the delivery of food was uncorrelated with the stimuli. On L sessions the light was presented in the absence of the tone. Vertical bars represent, separately for each group, the pooled standard error of the mean scores shown for each phase of training. Lower panel: group median time spent in magazine activity in the presence of the stimuli.

showed an initial increase (on early sessions there was a strong tendency to enter the magazine during LT trials) but thereafter declined as the discrimination was formed. The groups differed in their behavior to the light on only the first of the L-alone sessions. The omission of the tone (a stimulus lacking motivational significance) produced no effect in group RC. Group CI, on the other hand, showed a marked increase on the OR score on the first session, thus replicating the finding shown in Fig. 6.4. Thereafter the OR rapidly declined and was at a low level by the end of training. These results, therefore, do not contradict those reported by Pearce et al. (1982). Presentation of an inhibitor alone can restore associability (as measured by the OR), but the effect is transient and should have no effect on a retardation test given after a long series of such trials.

These results mean that the theory proposed by Pearce and Hall (1980) requires extension. In particular, that theory took account only of associations between a CS and the US (or inhibitory reinforcer) that followed it. We proposed that the omission of a reinforcer, once a CS-reinforcer association had been formed, would lead to a restoration of associability (Hall & Pearce, 1982). What we must now acknowledge is that associations can be formed between events presented concurrently, and that the omission of one of these may also bring about a restoration of associability that will endure until the subject has fully learned about the new state of its world. To do this will make necessary some elaboration of the equations in which the formal model is expressed, but the basic notion is fully in accord with the psychological principles that underlie the model.

IMPLICATIONS

Conditioned inhibition training produces a CS that is, often, "behaviorally silent." Special transfer tests, such as the summation and retardation procedures, are required to demonstrate the CS's conditioned properties. The conditioned inhibitor is not unique in this respect. Latent inhibition training, for instance, will produce a stimulus with changed properties that are revealed only in some subsequent transfer test—for example, in the retardation of subsequent excitatory or inhibitory conditioning (Rescorla, 1971a).

The retardation test results for a latently inhibited stimulus have been taken as showing that exposure causes the stimulus to lose associability. Our results reported previously amount to a demonstration that latent inhibition of a sort can occur during the course of conditioning itself. A CS will lose associability as it approaches an asymptotic level of associative strength. This conclusion carries with it a number of implications for our understanding both of conditioned and of latent inhibition.

Interpreting Retardation Tests

First, our conclusion means that we must be cautious about the interpretation of retardation tests for conditioned inhibition. Learning will be retarded by any treatment that causes associability to decline, even when that treatment produces no change in the associative strength of the stimulus. The point is developed by Baker in his contribution to this volume, but it is worth mentioning that the experiment by Pearce et al. (1982, see Fig. 6) provides a particular clear-cut example of the problem. In that experiment, the results of the retardation test might lead one to believe that the presentation of a conditioned inhibitor in isolation can enhance its inhibitory strength; but the outcome of the summation test makes this conclusion untenable. We conclude, rather, that the results of the retardation test reflect a change in associability that occurred in the absence of any change in associative strength.

This is not to deny that conditioned inhibition can produce a retardation of subsequent excitatory learning. But because the rate at which conditioning occurs will depend on the associability of the stimulus and because the associability of the stimulus will decline during conditioned inhibition training, the retardation test must be regarded as, at best, an imprecise measure of inhibition.

Conditioned Inhibition and Latent Inhibition

Next, our results prompt a reconsideration of the learning that goes on during latent inhibition. According to Pearce and Hall (1980) among others, this learning is not associative in nature; rather, latent inhibition is interpreted as a form of "perceptual learning" (cf. Hall, 1980), expressed in terms of a decline in the value of the learning-rate parameter that determines associability.

Now, the latent inhibition procedure has much in common with conditioned inhibition (the critical stimulus is presented in the absence of a US), and in the latter we do allow that an association (CS-no US) is formed, in addition to the associability change that occurs. Our reasons for supposing that latent inhibition influences only the associability of the stimulus are found in its effects on subsequent inhibitory conditioning. Not only does latent inhibition produce a stimulus slow at acquiring excitatory strength (that much is true of a conditioned inhibitor too), but it is also retarded in acquiring inhibitory strength (Reiss & Wagner, 1972; Rescorla, 1971a). This latter result has been taken as evidence that a latently inhibited stimulus cannot have acquired "true" inhibition. But the experiments described earlier have established that even a true conditioned inhibitor can be poor at acquiring further inhibitory strength—if the initial inhibitory strength of the stimulus is slight, the advantage it bestows on further inhibitory learning will be outweighed by the loss of associability that the stimulus will have suffered. The fact that latent inhibition retards later inhibitory conditioning can no longer be taken to show that no true (associative) inhibition is acquired during nonreinforced preexposure to a stimulus.

In consequence we must now allow consideration of the possibility that some component of the associative learning that goes on during conditioned inhibition training also occurs during latent inhibition training. Clearly it cannot be that latent inhibition allows the formation of an association between the stimulus and the emotional consequences of the omission of an expected reinforcer in the way that conditioned inhibition training does. But in both procedures it is possible for the subject to learn that the stimulus is followed by no event at all (see the chapter by Mackintosh). To the extent that this learning forms a part of the complex association that we have summarized as CS-no-US, we might say that true inhibition is acquired during latent inhibition training.

It is not easy to provide evidence in support of this speculation. We cannot use a retardation test after latent inhibition, because the effects seen here can be put down to the reduced associability of the stimulus (which will indeed play an important role). The results of summation tests in which a preexposed stimulus is superimposed upon an established excitor have proved varied and difficult to interpret. Thus Kremer (1972) has found a latently inhibited CS to attenuate the CR produced by an excitor, whereas Rescorla (1971a) using similar experimental procedures found no effect; and the effect found by Reiss and Wagner (1972) can be explained away in terms of external inhibition (generalization decrement). The issue remains unresolved for the time being.

But, in spite of the prominence they have been given, the retardation and summation procedures are not the only tests for inhibition. We could, for instance, investigate the effects of a latently inhibited stimulus (B) on the acquisition of associative strength by a second stimulus (A) with which it is presented in compound. There will be little learning about B because this stimulus will be low in associability, but any associative strength it might possess would influence the learning that occurs to A (cf. Rescorla & Wagner, 1972). In particular, if B is an inhibitor, reinforcing the AB compound would produce "supernormal" conditioning to A (Rescorla, 1971b); and an inhibitory B stimulus would block learning about A if the AB compound were made into an inhibitor (Suiter & LoLordo, 1971).

We have recently begun experiments with rats and the conditioned suppression procedure designed to examine these possibilities. It would be premature to do more than give a brief summary of one such study. In our first experiment investigating the blocking of inhibitory learning we gave experimental subjects 32 nonreinforced preexposures to stimulus B. Control subjects received 32 presentations of a different stimulus. All subjects then learned a discrimination, C+/ABC−, designed to endow both A and B with inhibitory properties. The inhibitory power gained by A was then assessed by means of a retardation test. We found that experimental subjects learned somewhat more readily in the retardation test than did control subjects, suggesting that A had acquired rather less inhibitory strength in the experimental group. One interpretation of this outcome is that preexposure to B gave this stimulus inhibitory properties that blocked the acquisition of inhibition by A.

Conclusions

If further research can confirm the preliminary results just outlined, we may be compelled to acknowledge that latent inhibition and conditioned inhibition have more in common than has previously been supposed. First, the intention of the experiments reported in the first two sections of this chapter was to demonstrate that CS-associability declines during inhibitory learning. It has been accepted for some time (since, at least, Wagner & Rescorla, 1972) that a loss of associability characterizes latent inhibition training. It seems likely, therefore, that attentional learning occurs in both latent and conditioned inhibition.

Next, it is possible that conditioned and latent inhibition involve similar sorts of associative learning. We have suggested that latent inhibition might allow the formation of a CS-no-event association. The third section of this chapter developed the notion that inhibitory learning consists in the formation of a CS-no-US association. This latter association (or complex of associations) is assumed to allow the CS to evoke certain emotional states; but to the extent that it also embodies the knowledge that no external event will follow the CS it has a component in common with latent inhibition.

We acknowledge that incorporating these suggestions would make it necessary to recast several of our formal theories of conditioning, including, it should quickly be added, that proposed by Pearce and Hall (1980).

ACKNOWLEDGMENTS

The experimental work reported here was supported by grants from the United Kingdom Science and Engineering, and Medical Research Councils.

REFERENCES

Broadbent, D. E. (1958). *Perception and communication.* Oxford: Pergamon.

Frey, P. W., & Sears, R. J. (1978). Model of conditioning incorporating the Rescorla–Wagner associative axiom, a dynamic attention process, and a catastrophe rule. *Psychological Review, 85,* 321–340.

Hall, G. (1980). Exposure learning in animals. *Psychological Bulletin, 88,* 535–550.

Hall, G., & Pearce, J. M. (1979). Latent inhibition of a CS during CS–US pairings. *Journal of Experimental Psychology: Animal Behavior Processes, 5,* 31–42.

Hall, G., & Pearce, J. M. (1982). Restoring the associability of a preexposed CS by a surprising event. *Quarterly Journal of Experimental Psychology, 34B,* 127–140.

Hall, G., & Pearce, J. M. (1983). Changes in stimulus associability during acquisition: Implications for theories of acquisition. In M. L. Commons, R. J. Herrnstein, & A. R. Wagner (Eds.), *Quantitative analysis of behavior* (Vol. III). Cambridge, Mass.: Ballinger.

Holland, P. C. (1977). Conditioned stimulus as a determinant of the form of the Pavlovian conditioned response. *Journal of Experimental Psychology: Animal Behavior Processes, 3,* 77–104.

Holland, P. C. (1980). Influence of visual conditioned stimulus characteristics on the form of Pavlovian appetitive conditioned responding in rats. *Journal of Experimental Psychology: Animal Behavior Processes, 6,* 81–97.

Kamin, L. J. (1968). "Attention-like" processes in classical conditioning. In M. R. Jones (Ed.), *Miami Symposium on the Prediction of Behavior: Aversive Stimulation.* Miami: University of Miami Press.

Kaye, H. (1983). *The influence of Pavlovian conditioning on the orienting response in the rat.* Unpublished doctoral dissertation, University College, Cardiff.

Kaye, H., & Pearce, J. M. (1984). The strength of the orienting response during Pavlovian conditioning. *Journal of Experimental Psychology: Animal Behavior Processes, 10,* 90–109.

Konorski, J. (1967). *Integrative activity of the brain.* Chicago: University of Chicago Press.

Kremer, E. F. (1972). Properties of a preexposed stimulus. *Psychonomic Science, 27,* 45–47.

Mackintosh, N. J. (1975). A theory of attention: Variations in the associability of stimuli with reinforcement. *Psychological Review, 82,* 276–298.

Pearce, J. M., & Hall, G. (1979). Loss of associability by a compound stimulus comprising excitatory and inhibitory elements. *Journal of Experimental Psychology: Animal Behavior Processes, 5,* 19–30.

Pearce, J. M., & Hall, G. (1980). A model for Pavlovian learning: Variations in the effectiveness of conditioned but not of unconditioned stimuli. *Psychological Review, 87,* 532–552.

Pearce, J. M., Kaye, H., & Hall, G. (1983). Predictive accuracy and stimulus associability: Development of a model of Pavlovian learning. In M. L. Commons, R. J. Herrnstein, & A. R. Wagner (Eds.), *Quantitative analyses of behavior* (Vol. III). Cambridge, Mass.: Ballinger.

Pearce, J. M., Nicholas, D. J., & Dickinson, A. (1982). Loss of associability by a conditioned inhibitor. *Quarterly Journal of Experimental Psychology, 33B,* 149–162.

Reiss, S., & Wagner, A. R. (1972). CS habituation produces a "latent inhibition effect" but no active "conditioned inhibition." *Learning and Motivation, 3,* 237–245.

Rescorla, R. A. (1969). Pavlovian conditioned inhibition. *Psychological Bulletin, 72,* 77–94.

Rescorla, R. A. (1971a). Summation and retardation tests of latent inhibition. *Journal of Comparative and Physiological Psychology, 75,* 77–81.

Rescorla, R. A. (1971b). Variation in the effectiveness of reinforcement and nonreinforcement following prior inhibitory conditioning. *Learning and Motivation, 2,* 113–123.

Rescorla, R. A. (1979). Conditioned inhibition and extinction. In A. Dickinson & R. A. Boakes (Eds.), *Mechanisms of learning and motivation.* Hillsdale, NJ: Lawrence Erlbaum Associates.

Rescorla, R. A. (1980). *Pavlovian second-order conditioning.* Hillsdale, NJ: Lawrence Erlbaum Associates.

Rescorla, R. A. (1982). Some consequences of associations between the excitor and the inhibitor in a conditioned inhibition paradigm. *Journal of Experimental Psychology: Animal Behavior Processes, 8,* 288–298.

Rescorla, R. A., & Durlach, P. J. (1981). Within-event learning in Pavlovian conditioning. In N. E. Spear & R. R. Miller (Eds.), *Information processing in animals: Memory mechanisms.* Hillsdale, NJ: Lawrence Erlbaum Associates.

Rescorla, R. A., & Wagner, A. R. (1972). A theory of Pavlovian conditioning: Variations in the effectiveness of reinforcement and nonreinforcement. In A. H. Black & W. F. Prokasy (Eds.), *Classical conditioning II: Current research and theory.* New York: Appleton–Century–Crofts.

Schneider, W., & Shiffrin, R. M. (1977). Controlled and automatic human information processing: I. Detection, search, and attention. *Psychological Review, 84,* 1–66.

Suiter, R. D., & LoLordo, V. M. (1971). Blocking of inhibitory Pavlovian conditioning in the conditioned emotional response procedure. *Journal of Comparative and Physiological Psychology, 76,* 137–144.

Wagner, A. R. (1978). Expectancies and the priming of STM. In S. H. Hulse, H. Fowler, & W. K. Honig (Eds.), *Cognitive processes in animal behavior.* Hillsdale, NJ: Lawrence Erlbaum Associates.

Wagner, A. R. (1979). Habituation and memory. In A. Dickinson & R. A. Boakes (Eds.), *Mechanisms of learning and motivation.* Hillsdale, NJ: Lawrence Erlbaum Associates.

Wagner, A. R. (1981). SOP: A model of automatic memory processing in animal behavior. In N. E. Spear & R. R. Miller (Eds.), *Information processing in animals: Memory mechanisms.* Hillsdale, NJ: Lawrence Erlbaum Associates.

Wagner, A. R., & Rescorla, R. A. (1972). Inhibition in Pavlovian conditioning: Application of a theory. In M. S. Halliday & R. A. Boakes (Eds.), *Inhibition and learning.* London: Academic Press.

Wasserman, E. A., Franklin, S. R., & Hearst, E. (1974). Pavlovian appetitive contingencies and approach versus withdrawal to conditioned stimuli in pigeons. *Journal of Comparative and Physiological Psychology, 86,* 616–627.

Zimmer–Hart, C. L., & Rescorla, R. A. (1974). Extinction of Pavlovian conditioned inhibition. *Journal of Comparative and Physiological Psychology, 86,* 837–845.

7 Antiassociations: Conditioned Inhibition in Attentional—Associative Networks

John W. Moore and Kenneth J. Stickney
University of Massachusetts—Amherst

BACKGROUND

Inhibition has been one of the most elusive and poorly understood topics in learning theory. For a long time theorists were unclear as to precisely what set of phenomena to include under this heading. Some experts spoke of different types of inhibition arising from procedures involving the withholding of a US and/or responses: habituation, extinction, differential conditioning, long CS–US intervals, and the like. Other experts questioned the very existence of inhibition in these or any other behavioral paradigm. It is only recently that inhibition has gained wide acceptance as a legitimate topic for study by learning theorists. Apart from Hull, Spence, and a few of their disciples, behaviorists (e.g., Guthrie and Skinner) and neobehaviorists (e.g., Tolman) scarcely mentioned the topic. The term does not appear in the index of Keller and Schoenfeld's (1950) classic text, *Principles of Psychology*. To many, the concept of inhibition was superfluous. The events to which the term was applied by Hullians and Pavlovians could presumably be more parsimoniously explained by "reduced excitation" than by "increased inhibition."

In this chapter, the term conditioned inhibition (CI) refers to the familiar paradigm in which one CS (A) is paired with a US (+), whereas the same CS when compounded with another (B) is not reinforced (−). Renewed interest by Western theorists in CI was prompted by experimental research of the mid-1960s. A key publication was Rescorla and Lolordo's (1965) demonstration of CI in the A+/AB− paradigm. They demonstrated that, following A+/AB− "fear" conditioning, B had the effect of lowering the rate of Sidman avoidance shuttling in dogs; A had the expected effect of increasing this rate. That same year, 1965, H. M. Jenkins published a chapter in the context of the pigeon stimulus-generalization literature stating some of the criteria that inhibitors must satisfy to be so classified (Jenkins, 1965). Shortly thereafter, Brown and Jenkins (1967) employed the summation test to demonstrate the conditioned inhibitory properties of an S-delta for not responding in a pigeon-operant task.

In 1969, we reported our studies of auditory stimulus generalization in rabbits (Moore, 1972) at the McMaster conference on classical conditioning organized by A. H. Black and W. F. Prokasy. These studies were concerned in part with whether Spence's notion of interacting gradients of excitation and inhibition was applicable to rabbit NMR conditioning. We were following the lead of E. Hearst, who had developed rather precise and sophisticated tools for assessing this theory in the context of the pigeon operant-stimulus generalization literature (e.g., Hearst, 1969).

It was at the McMaster conference that R. G. Weisman suggested we employ the "feature negative" paradigm to assess inhibitory stimulus generalization gradients in our preparation (Hearst, 1978; Jenkins & Sainsbury, 1969). In one study (Mis, Lumia, & Moore, 1972) we demonstrated inhibitory generalization gradients following A+/AB− training. In that experiment Stimulus A was a noise burst and Stimulus B a tone of 1200 Hz; in another study (Mis, Norman, Hurley, Lohr, & Moore, 1974) A was electrical brain stimulation and B was a 1200 Hz tone. Generalization testing in extinction consisted of a series of test tones of 400, 800, 1200, 1600, and 2000 Hz compounded with A. In both studies, shallow but reliable V-shaped gradients were obtained about the 1200-Hz tone. These results compared favorable with similar gradients reported in the pigeon-operant paradigm.

Parametric Features of CI

Marchant, Mis, and Moore (1972), using a nictatating membrane preparation in rabbits, reported evidence consistent with the criteria for conditioned inhibition. This experiment entailed four stages of training and/or testing. There were two groups, a conditioned inhibition group and pseudo-CI group. All animals were conditioned to three CSs in Stage 1—light (L+), noise (N+), and tone (T+). In stage 2, CI rabbits received L+/LT− training, T being the potential inhibitor. At this time the pseudo-CI-rabbits received a like sequence of L and LT trials, but both trial types were reinforced on a 50%-random basis. In Stage 3, each CS was presented alone repeatedly, and the tone was presented in combination with noise, over the course of three extinction sessions. This was the summation test. Finally, in Stage 4 each CS was paired repeatedly with the US in Stage 1. This reacquisition training constituted the retardation test.

The results were straightforward. Following CI training, but not pseudo-CI training, conditioned responding in Stage 3 was lower to the NT compound CS than N alone. For both groups, however, the LN compound CS elicited more CRs than N alone. The first result is an instance of inhibitory summation. The second result is an instance of excitatory summation. The reacquisition data followed suit. Reacquisition to T was slower in the CI group than in the pseudo-CI group, and reacquisition to L and N was faster than T in both groups.

Encouraged by these results, we next sought to establish contact with theories of conditioned inhibition, such as the Rescorla–Wagner model and Spence's familiar theories of discrimination learning. To establish this link, Marchant and Moore (1974) examined conditioned inhibition from the perspective of these theories. We recast the criteria for conditioned inhibition in the following way. In order to assert that a given training experience (e.g., A+/AB−) results in a CS with inhibitory properties, we argued it must have *negative* associative strength. This simple statement implies that our test must: (1) establish a zero point for associative strength; (2) demonstrate that CR strength is below this zero point for conditioned inhibitors; and (3) demonstrate the CR strength is not below zero for other stimuli, i.e., the below-zero or negative effect of training is the result of learning rather than "nonassociative" factors such as attention decrements. The initial rate of conditioning by naive rabbits was selected as a zero-point. Such animals required 51 T+ trials on the average to attain a criterion of three consecutive CRs to T. Rabbits previously given L+/LT− with no preliminary conditioning to either CS required 134 T+ trials on the average to attain the same criterion. This profound retardation was specific to CI training to T: Other rabbits given L+/LN− training or N+/LT− training showed facilitated conditioning to T, averaging only 20 trials to criterion. The rapid conditioning in these two cases was due to a general transfer factor; once rabbits are conditioned to one CS, conditioning to a second CS (or of the contralateral eye) proceeds rapidly (e.g., Stickney & Donahoe, 1983). In light of this factor, the retarded conditioning to T seems all the more striking.

Marchant and others working in our laboratory performed a number of purely parametric studies in which summation and retardation tests were used to assess the strength of conditioned inhibition. Most of these studies employed the design used by Marchant et al. (1972) in which T, N, and L were conditioned prior to the introduction of L+/LT− training (T+/N+/L+). N served as a summation-test stimulus for the third stage in which N and NT trials were presented in extinction (N−/NT−). Finally, the last stage was a retardation test in which T was paired with the US (T+). Control groups were treated the same way except that pseudo-CI training (L+/LT−) was given in the second stage on training.

Experiments conducted by Marchant, Mis, and their numerous helpers addressed the parametric features of CI with an eye to testing implications of the Rescorla–Wagner model. One prediction of the model is that the strength of CI (negative V) should be directly related to strength of the conditioned excitator on which it is based. In our experiments, this meant that factors designed to promote the rate of growth and asymptotic level of V_L should promote the rate of growth and asymptotic level of V_T, the latter of course having negative value. The Rescorla–Wagner model and results from prior experiments in our laboratory (e.g., Ashton, Bitgood, & Moore, 1969) suggested that manipulation of US intensity would be a suitable way to test the prediction. The US in these, as in all our experiments, was paraorbital electric shock. With no further suspense, suffice to say that we found no differences in CI strength, as assessed by summation and retardation tests, over a range of US-intensity values.

Accepting of the null hypothesis is an unsatisfying means of rejecting so appealing a theory as the Rescorla–Wagner model. We therefore pressed on to another parameter, CS intensity. We assumed that the salience of T, α_T, would be positively correlated with this variable. Here we hit pay dirt—as predicted by the Rescorla–Wagner model, CI strength following L+/LT− training was an increasing function of T's intensity.

Mahoney, Kwaterski, and Moore (1975) investigated CI as a function of interstimulus onset interval (ISI). Training and testing consisted of four stages: In the first stage, all animals were conditioned to N and L (N+/L+) with an ISI of .5 sec. In the second stage, half the rabbits received CI training (L+/LT−) and half received simple single-component differential conditioning (L+/T−). Each of these groups was divided into three ISI conditions: .25, .75, and 1.5 sec. The values refer to reinforced (+) trials. Nonreinforced (−) trials consisted of CSs of duration equal to the ISI plus .05 sec, the duration of the US. In the third stage, all animals received a summation text (N−/NT−) with CS durations of .55 sec, and the fourth stage was a retardation test (T+) with an ISI of .5 sec.

The two longer ISIs produced significant inhibitory summation for both types of training, L+/LT− and L+/T−. The .25 ISI did not. However, T+ training was significantly retarded following both types of training with this ISI. When considered together with the results of the summation test, this retardation of T+ training suggests that T tended to lose salience during stage two rather than gain inhibition.

The finding that both types of training produced equivalent inhibition of T, at least with the longer ISI, was not anticipated. The Rescorla–Wagner model predicts that L+/LT− training should produce more inhibition than L+/T− training. This is because T is decreased according to the equation $\Delta V = \alpha\beta\,(O - \bar{V})$, where $\bar{V}$ is the sum of all the associative values of stimuli present on a nonreinforced trial. Clearly, then, more inhibition is to be expected when nonreinforcement occurs on the presence of L, the reinforced component, than when the target stimulus is presented alone.

This evident failure of the Rescorla–Wagner model worried us for some time, but we ultimately traced the difficulty to the use of N as the summation test stimulus. Ron Tintner working in our laboratory substituted a tactile stimulus for N in the design employed by Mahoney, Kwaterski, and Moore (1975). This tactile stimulus was a mild electric shock (8–10 v ac) delivered across safety pins implanted across the scapulae. Tintner had previously satisfied himself that this stimulus would not produce pseudo-conditioning. Tinther found that L+/LT− training endowed T with inhibitory properties but L+/T− training did not. This finding supports the Rescorla–Wagner model. Evidently the initial conditioning to N in the Mahoney et al. (1975) study produced enough generalization to T as to mitigate the expected advantage for inhibition of L+/LT− training over L+/T− training.

Disinhibition of CI

The Rescorla–Wagner model is virtually unique in making explicit the idea that a single-component CS may have either positive or negative (or zero) associative value, but not both simultaneously. This point stands in contrast to some Hullian theorizing. Hull (1943), for example, allowed for a stimulus to develop habit strength ($_{S}H_{R}$) and conditioned inhibition ($_{S}I_{R}$) simultaneously. Then there are the theories derived from Pavlov that describe inhibition as a process that overlaps excitation. Disruption of inhibition (disinhibition) reveals the underlying excitation.

Marchant reasoned that a conditioned inhibitor in the Rescorla–Wagner sense of the term should not be subject to disinhibition because an inhibitor cannot simultaneously be an excitor. Accordingly, as his doctoral dissertation Marchant (1975) set about to determine whether the model was correct. First, he found that it was possible to obtain "disinhibition" of an extinguished response; that is, a distracting stimulus (Tintner's back shock, BS) given just before a CS in extinction would produce a temporary recovery of the CR. This recovery, we felt, could best be interpreted as a reinstatement of salience rather than as a release from inhibition. Marchant next used the distractor to see whether it could provoke CRs in a CS undergoing CI training in the L+/LT− paradigm. In a typical experiment, rabbits were initially conditioned to both L and T (L+/T+) before going into L+/LT− training. The back shock stimulus was applied on probe trials at various points in training. No evidence of recovery of CR strength following the back shock was evident. Although not altogether satisfying, this negative evidence suggested that true conditioned inhibitors are not susceptible to disinhibition, thereby supporting the Rescorla–Wagner model.

Evidently CI in the A+/AB− paradigm is not particularly sensitive to variations of the distribution of trials or training sessions. The acquisition and strength of CI appears impervious to a range of protocols ranging from 200 trials per session with an intertrial interval (ITI) of 15 sec to 20 trials per session with an ITI of 60 sec (see Allan, Desmond, Stockman, Romano, Moore, Yeo, & Steele-Russell, 1981). These results stand in marked contrast with several studies showing that the rate of conditioned excitation in this preparation is inversely related to distribution of trials over a similar range. The invariance of CI training as a function of ITI represents a back-door success for the Rescorla–Wagner model, because it is uncluttered by concepts such as reactive inhibition or similar time-dependent processes.

Also consistent with the Rescorla–Wagner model is our observation of retention of CI over periods up to 1 month and beyond. These observations have been made in connection with lesions studies in which animals are given CI training, operated upon, and then retrained following a recovery interval. If the lesion does not affect neural elements essential for CI, rapid savings is the rule. The spontaneous dissipation of inhibition in the fashion of Pavlov and Hull is not a feature of the model.

Blocking of CI

The most striking failure of the Rescorla–Wagner model in our hands has been our inability to demonstrate blocking of CI (see Suiter & Lolordo, 1971). Nancy Goodell carried out a methodical series of experiments in which CI training in the usual A+/AB− paradigm was followed by A+/ABC− training. The Rescorla–Wagner model predicts that the prior CI training targeted to B should block the development of conditioned inhibition to C, as assessed by summation and retardation tests, against a variety of control procedures. Different experiments employed L, T, and back shock (BS), in various roles; i.e., as A, B, or C. For example, in one experiment the blocking group received 20 sessions of BS+/BST− training followed by 20 sessions of BS+/BSTL−. Control animals received BS+/BS− (i.e., 50% random partial reinforcement) for 20 sessions followed by 20 sessions of BS+/BSL− (i.e., typical CI training of L.) The to-be-blocked inhibitor, L, was equally inhibitory in both cases.

Excitatory Conditioning through Reinforcement

We have encountered difficulties with yet another prediction of the Rescorla–Wagner model. Independent experiments by Andrea Allan and Ellen Stockman attempted to demonstrate the acquisition of positive associative value attendant upon nonreinforcement in the presence of a conditioned inhibitor (Baker, 1974). The idea of conditioning without reinforcement may be the most bizarre prediction of the model. Allan gave rabbits the usual A+/AB− training and then introduced a novel stimulus, C, presented without reinforcement with the inhibitor B. At the same time the inhibitory properties of B was maintained. This phase of training can be coded as follows: A+/AB−/BC−. Control animals, matched with the experimental group on the basis of initial conditioning, received A+/AB−/C− in this phase of training. The excitatory potential of C in the two groups was assessed by a summation test in which C was paired with A in extinction. This test included A− trials, as well; thus, A−/AC−. There was no difference in either group between conditioned responding to A− or AC−, and both groups were equivalent to each other.

Stockman performed a similar experiment except that any conditioning to C in the experimental group was assessed by a savings (C+) test. The rate of conditioning to C was the same as in the matched control group. Although these results are inconsistent with the Rescorla–Wagner model, the rate of conditioning C+ trials was so rapid that a ceiling effect could have obscured any excitatory effect of nonreinforcement of C in the presence of B.

Conclusions Regarding Rescorla–Wagner Model

The Rescorla–Wagner model has difficulty accounting for many aspects of CI investigated in our laboratory. Despite its suitability for the more salient features of CI (see Wagner, Mazur, Donegan, & Pfautz, 1980), its failures in our laboratory and those of other researchers (e.g., failure of conditioned inhibitors to extinguish through nonreinforcement; Zimmer–Hart & Rescorla, 1974) prompted us to try another approach.

ANTIASSOCIATIONS

Antiassociations are predictive relationships between an event E1 and the nonoccurrence of another event E2 (E1–notE2). Our thesis is that antiassociations can have strength or value just as can the predictive associative relationship between E1 and the occurrence of E2. In this regard it resembles Hull's conceptualization of CI. As in Hullian theory and in most extant learning theories that have addressed the problem of CI, an antiassociation can grow in strength only when the stage has been set for the expectation of E2 by prior learning. An antiassociation begins with a disconfirmed expectation and gains strength as that expectation is repeatedly disconfirmed. On the response side, antiassociations directly oppose whatever behavior is controlled by its antitheses: If an existing association E1–E2 pulls the organism in one direction, a simultaneously active antiassociation, be it E1–notE2 or E3–notE2, pulls in the other direction. Such conflicts are resolved by algebraic summation as in Hullian theory.

Because these ideas have a long history in learning theory, and the basic ideas are shared by at least a few models of CI, the only thing novel in our approach is to exchange the one set of terminology for another: What presently passes under the term *excitatory conditioning,* dating from Pavlov's time if not earlier, here refers to the acquisition of a "predictive associative relationship," association, for short. What has been referred to as "inhibitory conditioning," here refers to the acquisition of an antiassociation. We feel that this change is desirable because it stresses the separation between the "knowledge domain" or "data base" on which an animal bases its actions and a description of the action itself.

In more traditional language, the change simply reemphasizes the distinction between learning (acquisition of knowledge or beliefs) and performance (what is done with this knowledge).

The principal departure from other theories is that we have incorporated the foregoing ideas into a real-time computational theory that, although by no means wholly adequate, overcomes the difficulties of the Rescorla–Wagner model while at the same time retaining most of its virtues. The next section describes this theory formally and illustrates its workings. Specifically, the theory represents an adaptation of our model of attentional-associative networks based on Mackintosh's (1975) attention theory. We have previously applied the model to the problem of understanding how removal of the hippocampal formation produces deficits of latent inhibition (CS preexposure effect) and Kamin blocking (Moore & Stickney, 1980). A subsequent paper considered how the model might be applied to spatial learning tasks such as sign and goal tracking, particularly for describing the peculiarities in such tasks of hippocampal animals (Moore & Stickney, 1982).

Moore-Stickney Model

To review, we propose that the predictive associative value of a target CS (A) increases on reinforced trials to an asymptote of λ according to the following equation:

$$\Delta V_A = \alpha_A \Theta \tau \, (\lambda - V_A), \tag{1}$$

where α_A is the salience (associability) of A ($0 \leq \alpha_A < 1$), Θ is the rate parameter contributed by the reinforcer (the target of the association), τ ($0 < \tau \leq 1$) is the rate parameter contributed by degree of predictive contiguity, i.e., a function of the interstimulus interval. The salience parameter α_A may vary from one occasion to the next according to rules described later.

On nonreinforced trials, the associative value of A for predicting the reinforcer decreases to an asymptote of 0 by the following amount

$$\Delta V_A = \alpha_A \, \Theta' \tau \, (0 - V_A). \tag{2}$$

The parameter Θ' is the rate parameter associated with nonreinforcement and α_A and τ are the same as Equation 1. In keeping with other models that predict the growth of V_A under partial reinforcement schedules of .5 or less, Θ' is assumed to be less than Θ. Therefore, $0 < \Theta' < \Theta < 1$.

The parameter τ is governed by the degree of contiguity between the CS and the reinforcer. Letting Δt be the interstimulus interval, the value of τ is governed by the equation

$$\tau = \tau(\Delta t) = e^{k(q - \Delta t)} \tag{3}$$

where $\Delta t \geq 0$. The constant q is positive and equal to the optimal interstimulus interval (ISI) for conditioning; k ($0 < k < 1$) is the rate constant, arbitrarily set equal to .67 in our computer simulations. For convenience in the computer simulations, the parameter τ was assigned the arbitrarily small value of .067 whenever $0 \leq \Delta t < q$. Whenever $\Delta t < 0$ (backward conditioning), τ was assigned a value of zero. By Equations 1 and 2, therefore, predictive associative relationships do not change when the reinforcer precedes the target CS.

The salience of the target CS, α_A, may increase, decrease, or remain unchanged as a consequence of its reinforcement history *in relation to* other potential predictors of reinforcement. If stimulus X is the best alternative predictor of reinforcement among all stimuli present at the time A occurs, i.e., the one with the highest associative value besides A, then

$$\Delta\alpha_A = c(1 - \alpha_A)\,\{D_X - D_A\}/\lambda \tag{4}$$

whenever $D_A < D_X$ and

$$\Delta\alpha_A = -\alpha_A c\{D_A - D_X\}/\lambda \tag{5}$$

whenever $D_A \geq D_X$. The term $D_A = |\lambda - V_A|$, and the term $D_X = |\lambda - V_X|$; c is a rate constant arbitrarily set equal to .2 in the computer simulations described later.

Equations 4 and 5 capture the essence of Mackintosh's (1975) attention theory by permitting α_A to increase when it is the best predictor of reinforcement. All other stimuli presented simultaneously with A lose salience. Conversely, if A is second best (or worse) in predicting the reinforcer among all available stimuli, it loses salience, whereas α_X increases in value by an equation similar to Equation 4. Computations concerning α_A occur instantaneously and are not changed until such time as A is presented again. Notice that the intervening period of time contains opportunities for V_A and other predictive relationships to change and therefore affect the direction of α_A on subsequent occasions when A occurs.

In order to develop an internally consistent rationale for the phenomenon of latent inhibition, Moore and Stickney (1980) developed the idea that predictive associative relationships besides those concerning the nominal reinforcer contribute to the growth or decline of α_A. Thus, for example, α_A is affected by how well it predicts other stimuli, e.g., X, that are encountered in the course of training. The contribution of such predictive relationships to $\Delta\alpha_A$ is weighted in relation to the nominal reinforcer, the latter contributing the greatest share. In addition, α_A is affected by A's predictive relationship to itself! These ideas may be formalized as follows.

A given training epoch contains the onset and/or offset through time of any integral number (> 1) of formally designated stimulus events. If A is but one of a set of r such stimuli, then α_A is potentially affected by r^2 predictive associative relationships. It is convenient to designate these relationships using subscripts to denote the predictor and superscripts to denote the target of the association, the predict*ee*.

To see how this works, suppose we consider a simple acquisition experiment consisting of a context (X, not necessarily the X of Equations 4–5), a CS (A), and a US. There are nine associative relationships in the network that evolve through time in such an experiment: V_A^{US} (same as V_A in Equation 1–2), V_A^X, V_A^A, V_X^{US}, V_X^X, V_X^A, V_{US}^{US}, V_{US}^X, and V_{US}^A. Computations of α_A are based on the first grouping of three, computations of α_X are based on the second group of three, and α_{US} (the salience of the US as a predictor of the other stimuli) depends on the last grouping of three. For the purposes of this chapter and our computer simulations, these nine associative relationships varied according to Equations 1–2. We choose not to introduce stimulus-specific parameters for τ, θ, θ', and λ. However, the appropriate saliences parameters for each grouping of three are stimulus specific—α_A, α_X, and α_{US}, respectively.

Focusing only on $\Delta\alpha_A$, Equations 4–5 are adequate provided the predictive associative relationship with respect to the US (V_A^{US}) dominates the situation. In the complete model, $\Delta\alpha_A$ is actually a weighted average of stimulus-specific components. For $\Delta\alpha_A$ these components are $\Delta\alpha_A^{US}$, $\Delta\alpha_A^X$, and $\Delta\alpha_A^A$. Each component is computed using the appropriate terms in Equations 4–5 as follows:

$$\Delta\alpha_A^{US} = c(1 - \alpha_A)\{D_X^{US} - D_A^{US}\}/\lambda \tag{6}$$

whenever $D_X^{US} > D_A^{US}$, and

$$\Delta\alpha_A^{US} = -\alpha_A c\{D_A^{US} - D_X^{US}\}/\lambda \tag{7}$$

whenever $D_X^{US} \leq D_A^{US}$. The terms D_X^{US} and D_A^{US} are defined as before: $D_X^{US} = |\lambda - V_X^{US}|$; $D_A^{US} = |\lambda - V_A^{US}|$.

By introducing the appropriate superscripts, $\Delta\alpha_A^X$ and $\Delta\alpha_A^A$ are as follows:

$$\Delta\alpha_A^X = c(1 - \alpha_A)\{D_X^X - D_A^X\}/\lambda \tag{8}$$

whenever $D_X^X > D_A^X$, and

$$\Delta\alpha_A^A = -\alpha_A c\{D_A^X - D_X^X\}/\lambda \tag{9}$$

whenever $D_X^X \leq D_A^X$.

Similarly,

$$\Delta\alpha_A^A = c(1 - \alpha_A)\{D_X^A - D_A^A\}/\lambda \tag{10}$$

whenever $D_X^A > D_A^A$, and

$$\Delta\alpha_A^A = -\alpha_A c\{D_A^X - D_X^A\}/\lambda \tag{11}$$

whenever $D_A^X \leq D_X^A$.

Equations 6–11 have been written in a form consistent with our previously published description of the model (Moore & Stickney, 1980). By assuming that all values of λ are equal to one, as we have in our computer simulations, it becomes possible to rewrite Equations 6–11 as follows:

$$\Delta\alpha_A^{US} = c(1 - \alpha_A)\{V_A^{US} - V_X^{US}\} \tag{6a}$$

whenever $V_X^{US} < V_A^{US}$, and

$$\Delta\alpha_A^{US} = -\alpha_A c\,\{V_X^{US} - V_A^{US}\} \tag{7a}$$

whenever $V_X^{US} \geq V_A^{US}$;

$$\Delta\alpha_A^{X} = c(1 - \alpha_A)\{V_A^{X} - V_X^{X}\} \tag{8a}$$

whenever $V_X^{X} < V_A^{X}$, and

$$\Delta\alpha_A^{X} = -\alpha_A c\,\{V_X^{X} - V_A^{X}\} \tag{9a}$$

whenever $V_X^{X} \geq V_A^{X}$,

$$\Delta\alpha_A^{A} = c(1 - \alpha_A)\{V_A^{A} - V_X^{A}\} \tag{10a}$$

whenever $V_X^{A} < V_A^{A}$, and

$$\Delta\alpha_A^{A} = -\alpha_A c\,\{V_X^{A} - V_A^{A}\} \tag{11a}$$

whenever $V_X^{A} \geq V_A^{A}$.

Equations 6a–11a tacitly assume that X, the context, is always the best competitor to A in computing components of salience change. This is not necessarily always the case: If X, US, and A all occupy the same moment of time, it is conceivable that V_{US}^{US}, V_{US}^{X}, V_{US}^{A} would be substituted for V_X^{US}, V_X^{X}, and V_X^{A} in Equations 6–11 should the US have the higher momentary value for predicting the superscripted stimulus.

Once the various stimulus-specific components of $\Delta\alpha_A$ have been computed, they are combined as follows:

$$\Delta\alpha_A = \frac{\phi_{US}\Delta\alpha_A^{US} + \phi_X\Delta\alpha_A^{X} + \phi_A\Delta\alpha_A^{A}}{\phi_{US} + \phi_X + \phi_A} \tag{12}$$

The ϕs in Equation 12 may be thought of as the weight in long-term memory of context-dependent representations of each stimulus. If ϕ_{US} is very much larger than $\phi_X + \phi_A$, then $\Delta\alpha_A \cong \Delta\alpha_A^{US}$ and, for practical purposes the entire model collapses to Equations 1–5.

Equations of the form of Equations 4–5 and 6–11 are slightly different from those presented by Moore and Stickney (1980). Aside from the simplifying assumption setting all λs equal to one, the earlier version allowed for the possibility that the value of predictive associative relationships might be *negative in sign,* as is the case for conditioned inhibitors in the Rescorla–Wagner model. This allowance for negative value was reflected in our earlier equations for $\Delta\alpha_A$ that *normalized* expressions of the form $\{D_X - D_A\}$ to 2λ rather than λ. We pointed out at the time that 2λ was the maximum value of $\{D_X - D_A\}$. This would arise when $D_X \cong |\lambda - V_X| = 2\lambda$, i.e., when $V_X \cong -\lambda$ and $V_A \cong \lambda$. In words, X could be a full-fledged conditioned inhibitor and A could be an asymptotic conditioned excitator. Thus, we felt obligated to allow for negative associative value without having at that time a mechanism for bringing this about.

Conditioned Inhibition: Antiassociations

For each predictive associative relationship in the attentional-associative network, there exists the potential for an antipredictive relationship (antiassociation). Thus, where V_A^{US} might be viewed as the strength of an organism's belief that the US will follow A, the target CS, there is the potential for an *equally strong* belief that the US will not follow A. We denote this variable $V_A^{\overline{US}}$. *The two associative values can coexist independently of each other.* It is therefore possible for the organism to both anticipate the US in the presence of A and to be equally prepared for the likelihood that the US might not be forthcoming. Clearly, then, there is room for conflict and doubt, and rules must be made as to how the coexistence of V_A^{US} and $V_A^{\overline{US}}$ might be transcribed into resolution of conflict and prediction of behavior. Luckily, many gifted theorists have confronted this question in the past. Following their lead and our own intuition, a simple summation rule at the "output" stage will suffice for present purpose. Thus, if $V_A^{US} = a$ and $V_A^{\overline{US}} = b$ then behavior of stimulus control by A with respect to the US may be inferred by the quantity $a–b$. The quantity is the *net* associative strength of A for predicting the US. It is symbolized as follows: $\dot{V}_A^{US}$. Our approach to conditioned inhibition is precisely that suggested by Mackintosh (1975, p. 288).

Given the limitation on the range of parameters such as described previously, the quantity $\dot{V}_A^{US}$ (net associative value of A with respect to the US) ranges between 1 and minus 1. Just how this rule applies to summation and retardation tests of conditioned inhibition is illustrated later. In the simplest case, the behavior (B) or response to A with respect to the US in context X will be an increasing function of the sum of net associative values:

$$B_{A,X}^{US} = \dot{V}_A^{US} + \dot{V}_X^{US}. \tag{13}$$

The antiassociation $V_A^{\overline{US}}$ grows according to the following expression:

$$V_A^{\overline{US}} = \alpha_A \Theta' \tau (1 - V_A^{\overline{US}}) \tag{14}$$

whenever there exists the expectation of the US at the time A occurs (or shortly thereafter) and the US does not occur. As before α_A, Θ', and τ are unchanged from the excitatory case, and τ reflects the ISI from prior excitatory conditioning. Transcribing these familiar ideas into formal terms, by "expectation of the US" we mean that $V_A^{US} + V_X^{US}$ exceeds some threshold L^{US} $(0 < L^{US})$ before Equation 14 is implemented. For maximum theoretical flexibility, L^{US} could vary as a function of situation parameters or overtraining (progressively increasing under partial reinforcement schedules, for example). In our simulations, we held L^{US} constant at .5. An alternative approach to computation of the antiassociation would be to make the rate parameter Θ' a continuous increasing function of $V_A^{US} + V_X^{US}$, ranging from zero to some maximum of less than one. It is a question of taste—in taking the threshold route, we simply followed a suggestion of Zimmer–Hart and Rescorla (1974, p. 844).

Just as the associative value of Stimulus A for predicting the US decreases when the US does not occur, so too does the antiassociation $V_A^{\overline{US}}$ decrease when the US occurs. The rate of this decrease depends on Θ, the rate-of-change parameter associated with the occurrence of the US:

$$\Delta V_A^{\overline{US}} = \alpha_A \Theta \tau \, (0 - V_A^{\overline{US}}). \tag{15}$$

Summary of the Model

Whenever the US is accurately predicted by a CS (A), the associative value of A for predicting the US increases by the expression

$$\Delta V_A^{US} = \alpha_A \Theta \tau (1 - V_A^{US}).$$

At the same time, the corresponding antiassociation decreases by the expression

$$\Delta V_A^{\overline{US}} = \alpha_A \Theta \tau (0 - V_A^{\overline{US}}).$$

Whenever the US is *not* predicted by Stimulus A, the associative value of A for predicting the US decreases by the expression

$$\Delta V_A^{US} = \alpha_A \Theta' \tau (0 - V_A^{US}).$$

At the same time, and provided that the aggregate associative value of all stimuli present with A exceeds some threshold L ($0 < L^{US} < \Sigma V^{US}$, where the summation is over all stimuli "in the time bin"), the corresponding antiassociation increases by the expression

$$\Delta V_A^{\overline{US}} = \alpha_A \Theta' \tau (1 - V_A^{\overline{US}}).$$

We have assumed that computations of $\Delta\alpha_A$ depend on associations, e.g., V_A^{US}, V_A^X, etc.; not on antiassociations. In other words, Stimulus A does not gain or lose salience because it is a better or worse predictor of the *nonoccurrence* of a stimulus than its nearest competitor. This computation tactic is a legacy of our previous paper (Moore & Stickney, 1980). We adhere to it for reasons of parsimony and reluctance to be drawn into quibbles about what constitutes a stimulus. We have also found support in Eliot Hearst's (1978) analysis of "feature positive" and "feature negative" discrimination learning. Hearst (1978) argues that one reason that feature positive learning is easier than feature negative learning is that a positive feature acquires noticeability by virtue of its close predictive relationship to the reinforcer. A negative feature, by contrast, may lose noticeability because it does not predict the reinforcer, only its nonoccurrence; however, the present model does not incorporate this possibility.

If our strategy concerning computations of α_A needs amending, our preference at this time would be simply to substitute *net* associative values for associative values in Equations 6a–11a, i.e., $\dot{V}_A^{US}$ and $\dot{V}_X^{US}$ substituted for V_A^{US} and V_X^{US}, etc. These equations in their original form, however, were employed in all computer simulations described in the next section.

APPLICATIONS AND SIMULATIONS

As in our previously reported computer simulations of the model (Moore & Stickney, 1980), each CS occupied two successive time bins. The US, when presented, occupies the second bin. Thus, all trials were of the forward-delay type with a CS and US terminating together. Such details deserve emphasis because the evolution of attentional-associative networks depends on the precise timing and degree of overlap among stimuli. The salience of a particular CS, for example, changes only in relation to the associative value of other stimuli present at the same time.

As indicated previously, the asymptotic limits (λs) of values for both associations and antiassociations was 1, and these could each decrease to 0 when "nonreinforced." Net associative values, therefore, could range between -1 and 1.

The threshold for triggering computations of antiassociations was set equal to .5 for all stimuli. The rate parameters for all computations were arbitrarily set as follows: $c = .20$, $\Theta = .12$, and $\Theta' = .02$. The weighting factors of salience change were as follows: $\phi_{US} = 1.00$, $\phi_X = .01$, $\phi_A = \phi_B = \phi_C = .16$.

The salience parameters were initialized as follows: $\alpha_{US} = 1.00$, $\alpha_X = .15$, $\alpha_A = \alpha_B = \alpha_C = .60$. Finally, with regard to computation of τ by Equation 3, $\tau(\Delta t < 0) = 0$ and $\tau(0 < \Delta t < q) = .067$, as indicated earlier.

Preliminary Acquisition Training (A+)

Simulations began with all potential associative values and antiassociations equal to zero. The initial run consisted of a series of 32 acquisition trials in which Stimulus A was paired with the US in context X in the forward-delay arrangement described previously and with an intertrial interval of 32 bins of width q, measured from one onset of A to the next. Recall that q is the nominal optimal interval for conditioning, rendering $\tau = 1$ in Equation 3.

With the parameters initialized as described in the preceding paragraphs, this training raised V_A^{US} from 0 to .76. However, the antiassociation $V_A^{\overline{US}}$ also developed during this training because, whereas A reliably predicted the occurrence of the US, it also provided reliable information about its absence in bins following a US presentation. The antiassociation grew from 0 to .18 so that by the end of 32 trials the net associative value $\dot{V}_A^{US}$ equaled .58. In addition, α_A increased from .60 to .98 over this period of training. These results are summarized in Fig. 7.2A.

The preliminary bout of A+ training also increased the net associative value of X, the context, to $\dot{V}_X^{US} = .12$; α_X decreased from .15 to .007. Because $\dot{V}_X^{US}$ remained stable at .12 throughout subsequent simulated phases of training, it was disregarded in predictions of behavioral response strength (Equation 13).

It is worth noting that the form of the acquisition function for $\dot{V}_A^{US}$ is uniformly negatively accelerating. It therefore appears to be the sort of learning curve obtained by linear models such as the Rescorla–Wagner model. However, S-shaped functions are also possible, e.g., when α_A is initially low in relation to α_X. In this regard the model resembles those that are more appropriate for rabbit NMR and eyeblink conditioning (e.g., Frey & Sears, 1978; Prokasy, 1972; Theios & Brelsford, 1966) than are the simpler linear models.

Conditioned Inhibition Training (A+/AB−)

Figure 7.1 illustrates the development of CI in the A+/AB− paradigm. The simulation began by initializing the variables associated with Stimulus A and the context, X, to their values after the preliminary acquisition series. In CI training, the two types of trials A+ and AB− alternated. As before each trial consisted of two successive bins of size q. The US occupied the second bin on (+) trials but not on (−) trials. The intertrial interval, onset to onset, was 32 bins as before.

The prominent feature of the simulation, summarized in the left-hand portion of Fig. 7.1, is the progressive decrease in $\dot{V}_B^{US}$ from a starting value of 0 to −.95 by the end of 128 AB− trials. Notice too that $\dot{V}_A^{US}$ declines from its near-

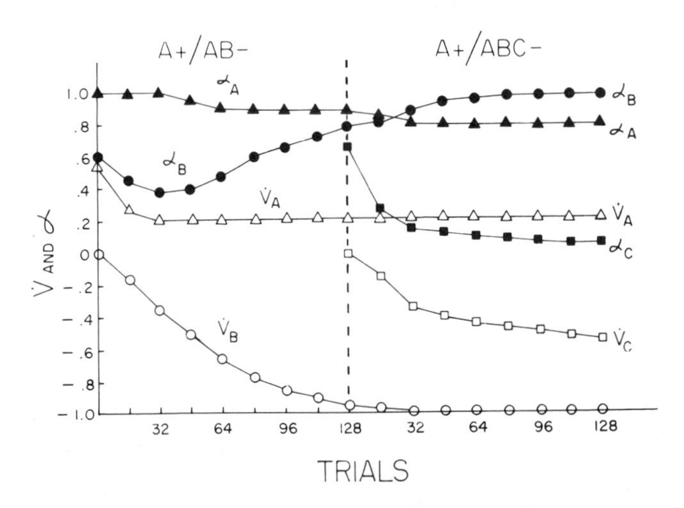

FIG. 7.1. Changes in salience (α) and net associative value ($\dot{V}$) in two phases of conditioned inhibition training. Phase 1: A+/AB−; Phase 2: A+/ABC−.

asymptotic value of .58 to a stable value of .20. This decrease in $\dot{V}_A^{US}$ is the result of extinction of V_A^{US} via Equation 2 and growth of $V_A^{\overline{US}}$ via Equation 14.

Another interesting feature of the simulation is the initial decline of α_B, which is followed by a slow increase that carries over into the right-hand portion of Fig. 7.1. The initial decrease of α_B is due to Stimulus B's inferior position as a predictor of the US in comparison with Stimulus A and the context, X. The component of $\Delta\alpha_B$ due to the US, $\Delta\alpha_B^{US}$, eventually becomes 0, arresting the decline of $\Delta\alpha_B$. This component is replaced by components of $\Delta\alpha_B$ that become positive in value, specifically $\Delta\alpha_B^A$ and $\Delta\alpha_B^B$. These components eventually stabilize to a point where overall increases in α_B become very small indeed.

These fluctuations of α_B can influence the course of a retardation test of Stimulus B's inhibitory potential. Figure 7.2B shows the change in $\dot{V}_B^{US}$ from negative to positive over 32 acquisition trials. Despite α_B's high value (> .90), retarded acquisition of Stimulus B is readily apparent when $\dot{V}_B^{US}$ is contrasted with $\dot{V}_A^{US}$ in Fig. 7.2A. Had retardation testing been initiated after, say, 48 trials of AB−, the growth of $\dot{V}_B^{US}$ would have been below the rate of growth of $\dot{V}_A^{US}$ for two reasons, a negative starting value and a diminished value of α_B.

Blocking of Conditioned Inhibition

The right-hand portion of Fig. 7.1 shows what happens with a third CS, Stimulus C, is added to the AB compound. Stimulus C loses salience, declining from .60

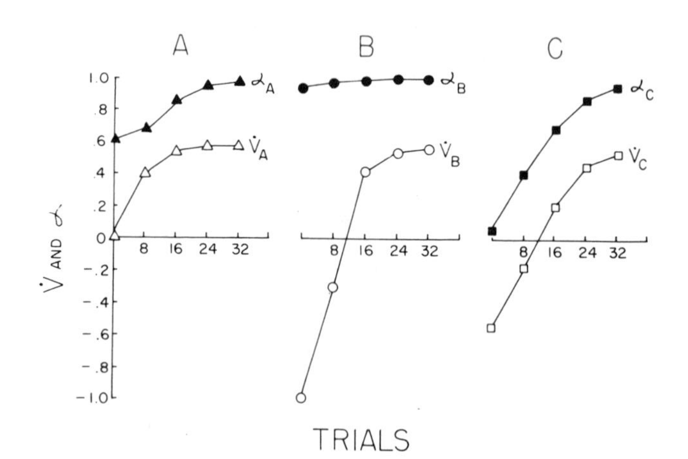

FIG. 7.2. A. Changes in salience (α_A) and net associative value $\dot{V}_A^{US}$ of Stimulus A for predicting the US. B. Retardation test: Changes in the salience (α_B) and net associative value $\dot{V}_B^{US}$ of Stimulus B following A+/AB− training as illustrated in Fig. 7.1. C. Retardation test: Changes in the salience (α_C) and net associative value $\dot{V}_C^{US}$ of Stimulus C following A+/ABC− training in Fig. 7.1.

to .10 in 64 presentations. It is this decline in α_C that is responsible for the arrested decline of $\dot{V}_C^{US}$ to approximately $-.50$ by the end of 128 trials. Thus, we see a *partial* blocking of CI. If Stimulus C had greater initial salience, its net inhibitory value would have been greater. This observation is germane to failure to detect blocking of CI in Goodell's experiment.

Retardation testing of Stimulus C is illustrated in Fig. 7.2C. Notice that the growth of $\dot{V}_C^{US}$ is retarded relative to $\dot{V}_A^{US}$, but the slow growth is the result of a negative starting value and a low initial value of α_C.

Summation Tests

Summation tests are straightforward. Suppose some CS, call it Stimulus D, had received acquisition training identical to that given Stimulus A during the preliminary training phase, i.e., until $\dot{V}_D^{US} = .58$. According to Equation 13, when combined with $\dot{V}_X^{US} = .12$, its behavioral effectiveness will be .70. When $\dot{V}_B^{US}$ is combined with Stimulus D, this expression is reduced by an amount that depends on the point where CI training of Stimulus B is terminated. However, in the simulation depicted in Fig. 7.1, $\dot{V}_B^{US}$ is $-.80$ after 96 AB− trials. Additional CI training would not yield a noticeable behavioral effect because behavioral response strength would be "below zero." The effect of further CI training, such as to move $\dot{V}_B^{US}$ closer to its limit of -1.00, would not be detected by a summation test unless $\dot{V}_D^{US}$ were higher.

A retardation test does not guarantee detection of any "excess" inhibition produced by CI training beyond 96 AB− trials. The reason for this is that the salience of the inhibitor can increase with extended training in the manner of α_B in Fig. 7.1. This increase in associability would tend to compensate for net inhibition by rapid growth toward net excitation.

Extinction of CI

Extinction of conditioned inhibition in our model comes about through extinction of antiassociations and concommitant growth of a parallel association. These processes require presentation of the US. In this regard our model is consistent with some other learning models, (e.g., Pearce & Hall, 1980), but unlike the Rescorla–Wagner model. The latter allows for extinction of conditioned inhibition by repeated nonreinforced presentation of the inhibitory CS. This particular prediction is a weak point of an unmodified Rescorla–Wagner model as several investigators have reported that inhibition is unaffected by nonreinforced presentation of the CS (Owren & Kaplan, 1981; Witcher, 1978; Zimmer–Hart & Rescorla, 1974).

We simulated a series of presentations of Stimulus B alone after a bout of CI training just to make certain that $\dot{V}_B^{US}$ retained its net inhibitory value. These tests yielded the desired result: $\dot{V}_B^{US}$ maintained its inhibitory potential with no change. The simulation therefore captures the empirical findings cited in the preceding paragraph.

Excitatory Conditioning by Nonreinforcement

The Rescorla–Wagner model's prediction that nonreinforcement of a novel stimulus in the presence of a conditioned inhibitor will make that stimulus a conditioned excitor has not been born out by most of the available evidence (e.g., Baker, 1974). Our model makes no such prediction: When a novel stimulus C ($V_C = \bar{V}_C = 0$), with an initial salience of .60, is compounded with a previously established inhibitor B ($\dot{V}_B^{US} = -.68$), its associative values V_C^{US} and $V_C^{\overline{US}}$ remain unchanged. This result is consistent with experiment by Stockman and Allan described in the main body of the chapter. In a simulation of 32 trials, α_C fell to a value of .38, whereas α_B rose from .50 to .70. These changes can be traced to subtle changes in the components of salience change not involving those weighted by the US, similar to those discussed in connection with Fig. 7.1.

Supernormal Conditioning

Unlike the Rescorla–Wagner model, our model does not predict supernormal conditioning, i.e., faster conditioning with reinforcement in the presence of a conditioned inhibitor compared with an appropriate control procedure (see Rescorla, 1971), and we have no evidence from our laboratory on this question. It is worth noting, however, that minor changes in our algorithm for computing salience changes could produce such an effect. The changes would entail, first, using $\dot{V}$ instead of V in salience-change equations (e.g., 6a) and, secondly, choosing as the reference stimulus the one with the greatest *absolute* net value. If stimulus A in Equation 6a were the target CS, the component of $\Delta\alpha_A$ contributed by $\Delta\alpha_A^{US}$ would be greater if $\dot{V}_B^{US}$ were used instead of V_A^{US} than is presently the case. Increases in α_A^{US} would be a function of ($\dot{V}_A^{US} - \dot{V}_B^{US}$) in the superconditioning case, a factor that would be greater than ($V_A^{US} - \dot{V}_X^{US}$) in the case where $\dot{V}_A^{US}$ is matched against the available stimulus with higher value according to algebraic sign. The greater differential in the former case would lead to faster increments of $\Delta\alpha_A^{US}$ and therefore faster increments of $\Delta\alpha_A$. This, in turn, would encourage faster growth of $\dot{V}_A^{US}$ (via increases in V_A^{US}) and thereby produce a supernormal conditioning effect. A similar salience-based mechanism to account for supernormal conditioning has been offered by Pearce and Hall (1980).

Correlations Between CS and US

Baker and Mackintosh (1977) reported that uncorrelated presentation of a CS and US retards subsequent CI training in a way analogous to the way such "learned irrelevance'' retards subsequent excitatory conditioning. In addition, Baker (1977) demonstrated that a negative correlation between CS and US occurrences produces some CI. Furthermore, this CI is dependent on excitatory conditioning of contextual stimuli in a manner consistent with the Rescorla–Wagner model. These two phenomena, retarded CI following uncorrelated presentation of a CS and US and CI as a result of a negative correlation between these two events, are consistent with our model.

The first result can be easily understood in terms of a loss of salience of the target CS (Stimulus A). Stimulus A would tend to lose salience in competition with X, the context, because of X's (presumed) greater associative value for predicting the US: On those occasions when A precedes the US closely in time, e.g., with an ISI equal to the optimal value q, the principal component of $\Delta\alpha_A$, $\Delta\alpha_A^{US}$, would decrease by Equation 7a. At the same time, these occasions would extinguish any antiassociative strength, $V_A^{\overline{US}}$, that might have developed on those occasions when Stimulus A was remote in time from the US. The decrease of α_A due to competition between V_A^{US} and V_X^{US} would be enough to predict retarded CI with A as the inhibitory CS.

The development of CI from a negative correlation between a CS and US is equally straightforward. If X, the context is a sufficient predictor of the US, such as to sustain an above-threshold expectation of the US, the CS will acquire antiassociative strength. This inhibition would show up as in summation tests. However, negative correlation between CS and US would likely lead to decreases in salience of the CS through competition with X. For this reason a retardation test would not be an appropriate vehicle for assessing inhibitory strength in these cases (see Baker, 1977).

OTHER CONTEMPORARY THEORIES OF CI

To summarize, the simulations provide evidence that our model of attentional-associative networks can be extended to encompass a number of features of our behavioral experiments on CI using the rabbit NM response. The key features of this model are these: (1) CI reflects the growth of antiassociations; (2) antiassociations compete directly with associations to determine (in part!) the strength of a behavioral CR at a given moment. Learned behavior, then, is some increasing function of the net associative value of aggregate stimuli; (3) antiassociations grow as a function of nonreinforcement, but only provided that the nonreinforcement occurs when the aggregate of associative values with respect to the reinforcer exceeds some threshold value; (4) antiassociations undergo extinction whenever, and to the extent that, the reinforcer follows the target CS in time.

The idea of competing associative tendencies, one excitatory and the other inhibitory, is a very old one (e.g., Konorski, 1948), and similar ideas can be found in more recent theories such as that of Pearce and Hall (1980). It is interesting that Rescorla has recommended a return to theories of this form (e.g., Rescorla, 1979), and Wagner has recently suggested that the V in the Rescorla–Wagner model might be thought of as a variable that "summarizes the *net* excitation and inhibition" engendered by a history of positive and negative Vs (Wagner, Mazur, Donegan, & Pfautz, 1980, p. 384, emphasis added). However, unlike the Rescorla–Wagner model, or closely allied variants such as those of Frey and Sears (1978) and Sutton and Barto (1981), ours is not a "zero-sum" model in which inhibition comes about whenever, and to the extent that, reinforcement is merely lesser in magnitude or effectiveness than that which the organism has come to anticipate. Pearce and Hall's (1980) model is similar to ours in implicitly allowing for antiassociations (see Pearce & Hall, 1980, p. 543), but it has a built-in zero-sum mechanism that serves as a negative feedback system for maintaining aggregate associative strength near to some level appropriate to the strength of the reinforcer.

Does conditioned inhibition arise when the reinforcer occurs on a trial but is of lesser magnitude than on other trials? Our approach to the problem of generating inhibition when outcomes are merely less than expectations would follow Logan's (1977) strategy in treating reinforcers of differing strength as different stimuli, each capable of independent associative relationships with other elements of the attentional-associative network. Before proceeding with our analysis, it is appropriate to call attention to the existence of arguments favoring the Rescorla–Wagner model over the "micromolar" approach of Logan's (1977) hybrid theory (see Wagner et al., 1980).

The relevant training paradigm for testing whether a decrease of US intensity, rather than its omission, is sufficient to generate conditioned inhibition is one similar to that employed by Wagner et al. (1980) with a rat CER task. With H and L designation high-and low-shock intensities, respectively, training in an A-H/AB-L procedure endowed Stimulus B with inhibitory properties according to a summation test. By treating H and L as distinct stimuli rather than as different levels of the same stimulus, our model would take into account both $\dot{V}_B^H$ and $\dot{V}_B^L$. To escape the dilemma posed by "zero-sum" models, we need only allow for the possibility that the negative value of $\dot{V}_B^H$ that evolves in the A-H/AB-L paradigm can more than offset the positive value of $\dot{V}_B^L$ that develops on AB-L trials. The trick, in other words, is to treat the A-H/AB-L paradigm as being a composite of two simpler procedures, an A-H/AB- (yielding negative $\dot{V}_B^H$) and AB-L (yielding positive $\dot{V}_B^L$). Summing the two variables and combining the result with other terms in Equation 13 can yield the desired summation results. Clearly, the problem of inhibition due to weak rather than absent events need not present insurmountable obstacles to further development of the model.

In closing, we wish to draw attention to some other features of our model and, where appropriate, to draw parallels with some of the thinking of others. Firstly, we have already indicated how the idea of antiassociations came to us from Mackintosh's (1975) attention theory. Similar ideas may be found in the formal model of Pearce and Hall (1980). Secondly, the idea that changes in salience of associability may be decomposed into stimulus-specific components resembles ideas presented by Dickinson and Macintosh (1979). Thirdly, our model is unusual in its formal deterministic treatment of the contribution of other associative relationships to change in salience and therefore to all associative relationships. By considering these tangential associations, our model conveys at least the flavor of a number of recent experimental reports that analyze the role of within-compound associations as they influence behavior appropriate to a given reinforcing event (e.g., Rescorla, 1980; Speers, Gillan, & Rescorla, 1980).

A last word concerns the role of memory for events in the computation of salience change by Equation 12. The ϕs in that equation are assumed to be stable in value and as entering into all computations after just one experience of a given subscripted stimulus event. How reasonable is the assumption? If the evidence of Leaton (1976) is typical, the assumption may not be too far off the mark. Leaton (1976) reports evidence of habituation following a single presentation of a loud tone up to a month after the event.

CONCLUSIONS

That attentional-associative networks lie behind much learned behavior is an article of faith. We have developed a model suitable for real-time applications that is based on some well-established principles of excitatory and inhibitory learning. Nevertheless, we have found it convenient from time to time to suggest modifications of the model that would bring it into closer alignment with empirical findings. Our suggestions regarding supernormal conditioning are a case in point. Learning theory is nevertheless confronted by a cornucopia of models that show reasonable promise for describing both excitatory and inhibitory learning. There is the Rescorla–Wagner model and its variants, Pearce and Hall's (1980) salience-loss model, both of which we have mentioned, and then there are promising variants of Corbit and Solomon's opponent-process theory that are coming to the fore (see Schull, 1979; Wagner, 1981). The fundamental differences among these models seems to revolve around two questions. The first is whether a "zero-sum" rule is necessary (we think not), and the second is whether stimulus salience increases or decreases through successful prediction of events. We believe with Mackintosh (1975) that the stimulus that has been the best available predictor of other stimuli will be the stimulus that most readily enters into an associative relationship with new, previously unexperienced stimuli.

REFERENCES

Allan, A. M., Desmond, J. E., Stockman, E. R., Romano, A. G., Moore, J. W., Yeo C. H., & Steele–Russell, I. (1980). Efficient conditioned inhibition of the rabbit's nictitating membrane response. *Bulletin of the Psychonomic Society, 16*, 321–324.

Ashton, A. B., Bitgood, S. C., & Moore, J. W. (1969). Auditory differential conditioning of the rabbit nictitating membrane response: III. Effects of US shock intensity and duration. *Psychonomic Science, 15,* 127–128.

Baker, A. G. (1974). Conditioned inhibition is not symmetrical opposite of conditioned excitation: A test of the Rescorla–Wagner model. *Learning and Motivation, 5,* 369–379.

Baker, A. G. (1977). Conditioned inhibition arising from a between-sessions negative correlation. *Journal of Experimental Psychology: Animal Behavior Processes, 3,* 144–155.

Baker, A. G., & Mackintosh, N. J. (1977). Excitatory and inhibitory conditioning following uncorrelated presentations of CS and UCS. *Animal Learning & Behavior, 5,* 315–319.

Brown, P. L., & Jenkins, H. M. (1967). Conditioned inhibition and excitation in operant discrimination learning. *Journal of Experimental Psychology, 75,* 255–266.

Dickinson, A., & Mackintosh, N. J. (1979). Reinforcer specificity in the enhancement of conditioning by posttrial surprise. *Journal of Experimental Psychology: Animal Behavior Processes, 5,* 162–177.

Frey, P. W., & Sears, R. J. (1978). Model of conditioning incorporating the Rescorla–Wagner associative axiom, a dynamic attention process, and a catastrophe rule. *Psychological Review, 85,* 321–340.

Hearst, E. (1969). Excitation, inhibition and discrimination learning. In N. J. Mackintosh & W. K. Honig, (Eds.), *Fundamental issues in associative learning* pp. 1–41). Halifax: Dalhousie University Press.

Hearst, E. (1978). Stimulus relationships and feature selection in learning and behavior. In S. D. Hulse, H. Fowler, & W. K. Honig, (Eds.), *Cognitive processes in animal behavior* (pp. 51–88). Hillsdale, NJ: Lawrence Erlbaum Associates.

Hull, C. L. (1943). *Principles of behavior.* New York: Appleton–Century–Crofts.

Jenkins, H. M. (1965). Generalization gradients and the concept of inhibition. In D. I. Mostofsky, (Ed.), *Stimulus generalization* (pp. 55–61). Stanford, CA: Stanford University Press.

Jenkins, H. M., & Sainsbury, R. S. (1969). The development of stimulus control through differential reinforcement. In N. J. Mackintosh & W. K. Honig (Eds.), *Fundamental issues in associative learning* (pp. 123–161). Halifax: Dalhousie University Press.

Keller, F. S., & Schoenfeld, W. N. (1950). *Principles of Psychology.* New York: Appleton-Century-Crofts.

Konorski, J. (1948). *Conditioned reflexes and neuron organization.* New York: Cambridge University Press.

Leaton, R. N. (1976). Long-term retention of the habituation of lick suppression and startle response produced by a single auditory stimulus. *Journal of Experimental Psychology: Animal Behavior Processes, 3,* 248–259.

Logan, F. A. (1977). Hybrid theory of classical conditioning. In G. H. Bower (Ed.), *The psychology of learning and motivation* (Vol. II). New York: Academic Press.

Mackintosh, N. J. (1975). A theory of attention: Variations in the associability of stimuli with reinforcement. *Psychological Review, 82,* 276–298.

Mahoney, W. J., Kwaterski, S. E., & Moore, J. W. (1975). Conditioned inhibition of the rabbit nictitating membrane response as a function of CS–UCS interval. *Bulletin of the Psychonomic Society, 5,* 177–179.

Marchant, H. G., III. (1975). *Disinhibition of the rabbit's conditioned nictitating membrane response.* Doctoral dissertation. University of Massachusetts, Amherst.

Marchant, H. G. III, Mis, F. W., & Moore, J. W. (1972). Conditioned inhibition of the rabbit's nictitating membrane response. *Journal of Experimental Psychology, 95,* 408–411.

Marchant, H. G., III, & Moore, J. W. (1974). Below-zero conditioned inhibition of the rabbit's nictitating membrane response. *Journal of Experimental Psychology, 102,* 350–352.

Mis, F. W., Lumia, A. R., & Moore, J. W. (1972). Inhibitory stimulus control of the classically conditioned nictitating membrane response of the rabbit. *Behavior Research Methods & Instrumentation. 4,* 297–299.

Mis, F. W., Norman, J. B., Hurley, J. W., Lohr, A. C., & Moore, J. W. (1974). Electrical brain stimulation as the reinforced CS in Pavlov's conditioned inhibition paradigm. *Physiology & Behavior, 12,* 689–692.

Moore, J. W. (1972). Stimulus control: Studies of auditory generalization in rabbits. In A. H. Black, & W. F. Prokasy, (Eds.), *Classical conditioning II: Current research and theory* (pp. 206–230). New York: Appleton–Century–Crofts.

Moore, J. W., & Stickney, K. J. (1980). Formation of attentional-associative networks in real time: Role of the hippocampus and implications for conditioning. *Physiological Psychology, 8,* 207–217.

Moore, J. W., & Stickney, K. J. (1982). Goal tracking in attentional-associative networks: Spatial learning and the hippocampus. *Physiological Psychology, 10,* 202–208.

Owren, M. J., & Kaplan, P. S. (1981). *On the failure to extinguish Pavolvian conditioned inhibition: A test of a reinstatement hypothesis.* Paper presented at the 53rd annual meeting of the Midwest Psychological Association, Detroit.

Pearce, J. M., & Hall, G. (1980). A model for Pavlovian learning: Variations in the effectiveness of conditioned but not of unconditioned stimuli. *Psychological Review, 87,* 532–552.

Prokasy, W. F. (1972). Developments with the two-phase model applied to human eyelid conditioning. In A. H. Black & W. F. Prokasy (Eds.), *Classical conditioning II: Current theory and research.* New York: Appleton–Century–Crofts.

Rescorla, R. A. (1971). Variations in the effectiveness of reinforcement and nonreinforcement following prior inhibitory conditioning. *Learning and Motivation, 2,* 113–123.

Rescorla, R. A. (1979). Conditioned inhibition and extinction. In A. Dickinson & R. A. Boakes (Eds.), *Mechanisms of learning and motivation: A memorial volume to Jerzy Konorski.* Hillsdale, NJ: Lawrence Erlbaum Associates.

Rescorla, R. A. (1980). Simultaneous and successive associations in sensory preconditioning. *Journal of Experiment Psychology: Animal Behavior Processes, 6,* 207–216.

Rescorla, R. A., & Lolordo, V. M. (1965). Inhibition of avoidance behavior. *Journal of Comparative and Physiological Psychology, 59,* 406–412.

Rescorla, R. A., & Wagner, A. R. (1972). A theory of Pavlovian conditioning: Variations in the effectiveness of reinforcement and nonreinforcement. In A. H. Black & W. F. Prokosy (Eds.) *Classical conditioning II: Current theory and research.* New York: Appleton–Century–Crofts.

Schull, J. (1979). A conditioned opponent theory of Pavlovian conditioning and habituation. In Bower, G. H. (Ed.), *The psychology of learning and motivation,* (Vol. 13, pp. 57–89). New York: Academic Press.

Speers, M. A., Gillan, D. J., & Rescorla, R. A. (1980). Within-compound associations in a variety of compound conditioning procedures. *Learning and Motivation, 11,* 135–149.

Stickney, K. J., & Donahoe, J. W. (1983). Attenuation of blocking by a change in US locus. *Animal Learning & Behavior, 11,* 60–66.

Suiter, R. D., & LoLordo, V. M. (1971). Blocking of inhibitory Pavlovian conditioning in the conditioned emotional response procedure. *Journal of Comparative and Physiological Psychology, 76,* 137–144.

Sutton, R. S., & Barto, A. G. (1981). Toward a modern theory of adaptive networks: Expectation and prediction. *Psychological Review, 88,* 135–170.

Theios, J., & Brelsford, J. W., Jr. (1966). A Markov model for classical conditioning. Application to eye-blink conditioning in rabbits. *Psychological Review, 73,* 393–408.

Wagner, A. R. (1981). SOP: A model of automatic memory processing in animal behavior. In N. E. Spear & R. R. Miller (Eds.), *Information processing in animals: Memory mechanisms.* Hillsdale, NJ: Lawrence Erlbaum Associates.

Wagner, A. R., Mazur, J. E., Donegan, N. H., & Pfautz, P. (1980). Evaluation of blocking and conditioned inhibition to a CS signalling a decrease in US intensity. *Journal of Experimental Psychology: Animal Behavior Processes. 6,* 376–385.

Witcher, E. S. (1978). *Extinction of Pavlovian conditioning inhibition*. Doctoral Dissertation, University of Massachusetts.

Zimmer–Hart, C. L., & Rescorla, R. A. (1974). Extinction of Pavlovian conditioned inhibition. *Journal of Comparative and Physiological Psychology, 86*, 837–845.

8 Opponent Processes and Pavlovian Inhibition

Allan R. Wagner and Mark B. Larew
Yale University

In *Hamlet* there is a dialogue between the King, Claudius, and Laertes that will provoke Laertes to avenge his father in the fatal dual with Prince Hamlet. Claudius encourages Laertes to act quickly upon the love of his father, for he says, "There lives within the very flame of love a kind of wick or snuff that will abate it."

In this chapter we have no sinister or provocative intent. But we advocate that what Claudius says about the abating of love is worth repeating in relationship to any conditioned response. Students of conditioning have been challenged since the time of Pavlov to characterize the process whereby conditioned responses are diminished during experimental extinction and conditioned inhibition training. We suppose that the "wick or snuff" is an integral part of the very response to the unconditioned stimulus as well as the conditioned stimulus.

There are several theories of inhibitory learning that make the same point, if somewhat less poetically. In the sections that follow we first briefly review several of these theories, with special emphasis on one (Wagner, 1981) that has been developed in our laboratory. Then we present some new evidence directly relevant to a central expectation of these theories, namely inhibitory backward conditioning via a single US–CS pairing. And, finally, we indicate, again with the support of new data, how inhibitory backward conditioning may be implicated in several well-known conditioning phenomena involving multiple trials, where there is the opportunity for effective US–CS juxtapositions.

THEORETICAL BACKGROUND

Konorski

As is often the case, it is appropriate to begin with Konorski. In his 1948 attempt to provide a Sherringtonian interpretation of Pavlovian phenomena, he supposed that the functional linkages developed between ''CS centers'' and ''US centers'' can be either ''excitatory'' or ''inhibitory,'' and that, in the general case, the behavioral result of any conditioning arrangement can be understood in terms of the degree and relative balance of the two kinds of linkages that are developed. If a CS center acquires excitatory linkages to a US center, then activation of the CS center through presentation of the CS would, presumably, also provoke activation of the US center and a consequent conditioned response. If a CS center acquires inhibitory linkages to a US center, then activation of the CS center would, presumably, act through these linkages to reduce the excitability of the US center when it might otherwise be activated, for example, by excitatory linkages from the same CS (as in experimental extinction) or from another CS (as in assessments of ''conditioned inhibition'').

Konorski assumed that the development of either excitatory or inhibitory linkages was dependent on concurrent activity dynamics of the CS and US centers. He presumed that when either a CS or US is presented for a duration of time there will be a rise and then eventual fall in activity of the respective center, and that for either type of linkage to become functional there must be potential connections whereby ''impulses'' from the CS center can reach the US center during its period of activity. Konorski (1948) specifically assumed that for the development of functional excitatory linkages ''impulses set up by the . . .[CS center]. . . must reach the unconditioned center in the stage when the latter's activity is growing'' (p. 106). And he assumed that functional inhibitory linkages ''arise on the contrary when excitation of the . . .[CS center]. . . coincides with the abrupt fall of active excitation of the . . .[US center]'' (p. 106).

With these propositions it was easy to rationalize why the development of conditioned responding was favored by CS-then-US asynchrony during training: The CS initiated impulses should be timed to overlap the rise of activation of the US center but not overlap the fall of such activation. And Konorski saw them as naturally accounting for inhibitory phenomena that had led Pavlov (1927) to propose mechanisms that were out of keeping with a Sherringtonian conception of the memory system. If a CS center, by virtue of prior excitatory conditioning, activated a US center but the US was not presented, there would be a subsequent fall of excitation of the US center. If another CS center were activated in appropriate time relationship to this fall of excitation, that CS center would be expected to develop singularly inhibitory connections to the US center. Thus we could understand the development of conditioned inhibition. If impulses from the initiating CS center continued to reach the US center during the fall of excitation, that center would also develop inhibitory linkages. Thus we could understand the phenomenon of experimental extinction.

Konorski (1948) rested his interpretation of the essential conditions for inhibitory learning primarily upon a phenomenon that he said had left Pavlov himself "extremely puzzled." This is the phenomenon said to have been shown by Kreps, Pavlova, Pietrova, and Vinogradov in Pavlov's laboratory, that "if an indifferent and an unconditioned stimulus are applied in reverse overlapping sequence (the unconditioned stimulus being precurrent) then not only does a positive conditioned reflex fail to be established, but the indifferent stimulus is transformed into a strong inhibitory stimulus" (pp. 135–136). Konorski (1948) proposed that the conditioning "obtained by the reverse sequence of . . .[the CS and US]. . .is the direct converse of the excitatory reflex formed by a normal sequence of stimuli" and that the training arrangement is the "primary and direct method of forming a stable and strong inhibitory reflex" (p. 137).

We return later to comment upon Konorski's view of inhibitory conditioning, but it should be clear that Shakespeare would have had it right. In the activity of the US center that provokes a UR or a CR, there lives the potential for fall that provides the wick or snuff of inhibitory learning.

Opponent Process Theory

In 1974 Solomon and Corbit published an influential paper indicating how many phenomena involving affective stimuli could be understood in terms of an opponent-process mechanism similar to that in certain theories of color vision (e.g., Hurvich & Jameson, 1974). The reasoning and the theoretical terminology are by now well known. Briefly, when a hedonic stimulus event is presented it is presumed to produce a direct process (called *a*) for as long as the stimulus is presented. The *a* process is presumed to initiate a "slave" reaction (called *b*) that acts to oppose it. Whereas the *a* process rapidly follows stimulus onset and termination, the *b* process is relatively sluggish, slower to recruit and slower to dissipate. The assumed general form of these processes in relationship to stimulation is depicted in Fig. 8.1, as is their presumed effective, "manifest" consequence, which is the sum of the *a* process minus its opponent, *b* process. As may be seen, the general expectation is that stimulus onset will be followed by a so-called A state ($a > b$), which peaks and then continues in diminished amount until the stimulus terminates. Stimulus termination will, in turn, be shortly followed by a "B state" ($b > a$), which peaks and eventually dissipates.

There have been several different proposals (Schull, 1979; Solomon, 1977; Solomon & Corbit, 1974) concerning the manner in which the presumed opponent processes could be influential in Pavlovian conditioning. Those of Schull (1979) are in the most continuous relationship between the theory of Konorski (1948) that has been mentioned and the tenets of SOP (Wagner, 1981) to which we turn shortly. Schull supposes that any US that is capable of supporting

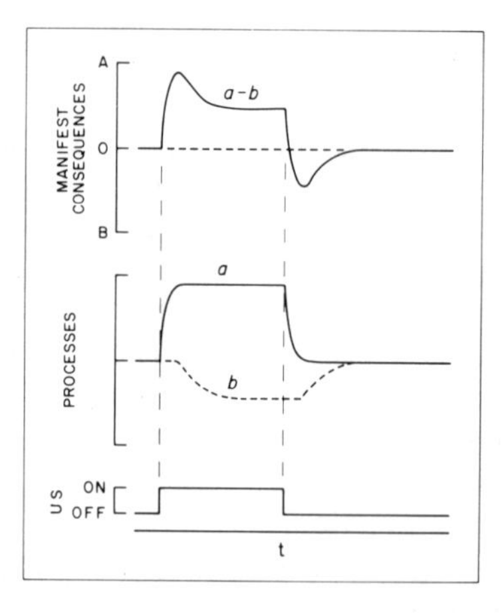

FIG. 8.1. The general forms of the temporal consequences assumed by opponent-process theory (e.g., Schull, 1979; Solomon & Corbit, 1974) to be occasioned by a hedonic US. The upper curve indicates the expected manifest consequences (A state or B state) of US presentation, obtained by summating the *a* and *b* processes depicted in the lower curves. (After Solomon & Corbit, 1974.)

Pavlovian conditioning produces a characteristic set of processes as depicted in Fig. 8.1. Then he assumes that when a conditioned stimulus is "paired with" an A state of a US, there will be excitatory learning between the CS and US such that the CS will come to elicit a customary (unspecified) CR. In contrast, when a conditioned stimulus is "paired with" a B state of a US, there will be inhibitory learning such that the CS will have diminished excitatory effects and may act like a conditioned inhibitor.

This much is obviously quite similar to Konorski's (1948) supposition about the conditions that are responsible for excitatory and inhibitory learning, with the substitution of an "A state" for the rise in excitation of the US center, and a "B state" for the fall in excitation of the US center. Schull (1979), however, carefully avoids any theoretical commitments to the notion of developing linkages between the likes of a CS center and a US center that would imply that activity induced in the US center via the CS is responsible for the CR observed. Instead he assumes that the *b* process, but not the *a* process, is a conditionable response, and, like all other CRs, reflects the occasions for excitatory and inhibitory learning: When a CS is paired with the A state of the US, excitatory learning occurs that causes the CS itself to be capable of eliciting the *b* process of the US. When a CS is paired with the B state of the US, inhibitory learning occurs that causes the CS to have tendencies that oppose the conditioned elicitation of the *b* process.

There are a large number of interesting consequences of Schull's (1979) formulation. In terms of our present emphasis, it is important that Schull, like Konorski (1948), treats "backward" pairings of US-then-CS as the fundamental way to produce a conditioned inhibitor. This is presumably the most direct experimental procedure for arranging the pairing of a CS with the B state of a US. Because an excitatory conditioned stimulus, however, will also elicit a conditioned *b* process, the pairing of a novel CS with a (nonreinforced) excitatory CS, in the manner of conditioned inhibition training, will also produce inhibitory learning to the novel CS. And when an excitatory CS produces a conditioned *b* process and is nonreinforced, that also is an occasion for pairing of the CS itself with a B state of the US, leading to inhibitory learning and experimental extinction. What lives within the very flame of the unconditioned response and the conditioned response to abate conditioned responding is the *b* process.

SOP

The "Sometimes-Opponent-Process" model that was developed in our laboratory (Donegan & Wagner, in press; Mazur & Wagner, in press; Wagner, 1981) was formulated as a quantitative rendering of certain notions about the effects of priming of short-term memory (e.g., Wagner, 1976, 1978) and the variation in "rehearsal" under conditions of expected versus surprising stimulation (e.g., Wagner, Rudy, & Whitlow, 1973). However, as Mazur and Wagner (in press) point out, it was also predicated largely on a need to deal with phenomena of inhibitory learning that had been ignored in these background theories. The relevant assumptions that were adopted are in many ways congruent with those of Konorski (1948) and Schull (1979).

We assume as did Konorski (1948), but not Schull (1979): (1) that the memory system may be usefully conceptualized as a graph structure, with nodes or "centers," representing stimuli, connected via directional associative links that may be either excitatory or inhibitory; and (2) that the behavioral consequences of Pavlovian conditioning are attributable to the acquired ability of activity in a CS node to influence, through the aforementioned linkages, the activity in a node normally responsive only to a US. We further assume that activity of any stimulus node has both a primary component, A1, and a secondary component, A2, that could be considered analogous to the "rise" and "fall" in Konorski's formulation, but have more likeness to the *a* and *b* processes, respectively, in Schull's treatment. In the spirit of both Konorski and Schull, it is assumed that it is the differential "pairing" of CS nodal activity with either the A1 activity or the A2 activity of the US node that is responsible, respectively, for excitatory or inhibitory learning. However, we assume like Schull (1979), but not Konorski (1948), that excitatory learning involves an acquired tendency for a CS to excite a US center to the A2 (*b*-like) state, without similar acquired activation to the A1 (*a*-like) state, whereas inhibitory learning involves a tendency for a CS to antagonize conditioned activation of a US center to the A2 state.

The model was articulated so as to have a degree of determinacy lacking in both the Konorski (1948) and Schull (1979) formulations. Whereas Konorski pointed to the importance of whether impulses from the CS center arrive during the rise or fall of excitation in the US center, the relevant activity dynamics were left almost completely unspecified. Thus, for example, one cannot anticipate whether presenting a neutral CS along with an excitatory CS, that would produce a (conditioned) rise and then fall in a US center, should produce excitatory learning (i.e., second-order conditioning) or inhibitory learning (i.e., conditioned inhibition). Likewise, Schull (1979) presents only a schematic suggestion of the forms of the *a* and *b* processes, following the drawings of Solomon and Corbit (1974), and does not specify the temporal relationships of a CS with either process that would constitute critical "pairings."

To approach these matters with relatively simple and well-understood mathematical tools, Donegan and Wagner (in press) suggested that any stimulus node be conceived as a large but finite collection of elements, and that the nodal activity dynamics be treated in terms of changes in the proportion of elements among the states of inactivity (I), A1, and A2. It was assumed that, in each moment a stimulus is present, the node representing that stimulus will have some proportion (p_1) of its elements in the I state activated to the A1 state, from which the elements will "decay" to the A2 state (with momentary probability, p_{d1}), and from there decay back to the I state (with momentary probability, p_{d2}). It was further assumed that p_1 increases with increasing stimulus intensity and, following arguments developed by Mazur and Wagner (in press), that p_{d1} and p_{d2} reflect capacity limitations of active memory with $p_{d1} > p_{d2}$. On this reasoning, given the initial activity values for a node such that $p_I + p_{A1} + p_{A2} = 1.0$, and the parameters p_1, p_{d1}, and p_{d2}, one can easily specify the nodal activity state following any episode of stimulation. For example, Fig. 8.2a depicts the expected course of p_{A1} and p_{A2} following a single momentary presentation of a US to an inactive node, where $p_1 = 1.5p_{d1}$ and $p_{d2} = .2p_{d1}$. For comparison, Fig. 8.2b depicts the quite differently shaped course of p_{A1} and p_{A2} that is expected when the US is more protracted. By these examples one can appreciate the expressiveness of the stochastic formalization. Indeed, it is interesting to observe that SOP allows the characteristic temporal manifestations emphasized by Schull (1979) as a special case. Notice that with the protracted US of Fig. 8.2b, p_{A1} initially increases to a peak value and then decreases to a lower, equilibrium value while p_{A2} monotonically increases to its equilibrium value, before p_{A1} more quickly decays than p_{A2} following US termination. As a consequence, the *relationship* between p_{A1} and p_{A2} over time behaves rather similarly to the relationship between the *a* and *b* processes depicted in Fig. 8.1. To illustrate this general correspondence, Fig. 8.2c indicates a family of functions, $w_1p_{A1} - w_2p_{A2}$, based on the course of p_{A1} and p_{A2} in Fig. 8.2b, that can be compared to the manifest *a–b* function in Fig. 8.1. In SOP, the parameters, w_1 and w_2, that weight the relative influences of p_{A1} and p_{A2}, must be specified for determining any selected UR (see Wagner, 1981) and for determining associative learning (see following).

Given this manner of conceptualizing the activity dynamics of any stimulus node, Mazur and Wagner (in press) offered the following rules for specifying the excitatory and inhibitory learning that may occur between a CS and US in any episode. It is assumed that an increment in excitatory linkages from one node to

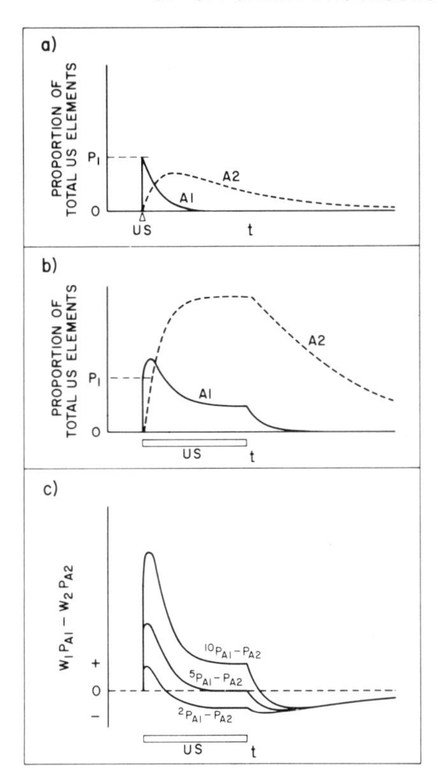

FIG. 8.2. Simulations of the processing assumed by SOP in response to a US: (a) expected proportions of US nodal elements in the A1 and A2 states of activity following presentation of a momentary US; (b) expected proportions of US nodal elements in the A1 and A2 states of activity during and following an extended US; (c) weighted sums of the proportions of US nodal elements in the A1 and A2 states of activity in panel b. (See text for explication.)

another occurs only in moments in which both nodes have elements in the A1 state, and that the size of the increment is proportional to a measure of the degree of joint, A1 activity. Stated more specifically, for the linkage from a CS node to a US node, it is assumed that the increment in excitatory strength (ΔV^{+}_{CS-US}) in any moment is the product of $p_{A1,CS}$, the proportion of CS elements in the A1 state, times $p_{A1,US}$, the proportion of US elements in the A1 state, times L^{+}, an excitatory learning rate parameter. Generalizing the consequences over any duration of *t* moments, this becomes

$$\Delta V^{+}_{CS-US} = L^{+} \sum^{t} (p_{A1,CS} \times p_{A1,US}). \qquad (1)$$

It is assumed that an increment in inhibitory linkages from one node to another occurs only in moments in which the former node has elements in the A1 state, while the latter node has elements in the A2 state, and that the size of the increment is proportional to a measure of the degree of such joint activity. Stated more specifically, for the linkage from a CS node to a US node, it is assumed that the increment in inhibitory strength (ΔV^{-}_{CS-US}) in any moment is the product of $p_{A1,CS}$, the proportion of CS elements in the A1 state, times $p_{A2,US}$, the proportion of US elements in the A2 state, times L^-, an inhibitory learning rate parameter. Generalizing over any duration of *t* moments, this becomes

$$\Delta V^-_{CS-US} = L^- \sum^{t} (p_{A1,CS} \times p_{A2,US}). \qquad (2)$$

These assumptions treat the ''pairings'' that are necessary for associative learning as occasions when the CS node has elements in the A1 state while the US node has elements in the A1 state (producing excitatory learning) and/or the A2 state (producing inhibitory learning). At any moment and over any episode of time both kinds of conjoint activity may occur so that the net change in associative strength will be

$$\Delta V_{CS-US} = \Delta V^+_{CS-US} - \Delta V^-_{CS-US}. \qquad (3)$$

To illustrate the associative consequences of such assumptions, Wagner (1981) and Mazur and Wagner (in press) reported a series of simulations involving a variety of basic Pavlovian conditioning arrangements. One set is especially relevant to the present context and is shown in Fig. 8.3. Figure 8.3a presents representative simulations of the theoretical events assumed to be involved in a

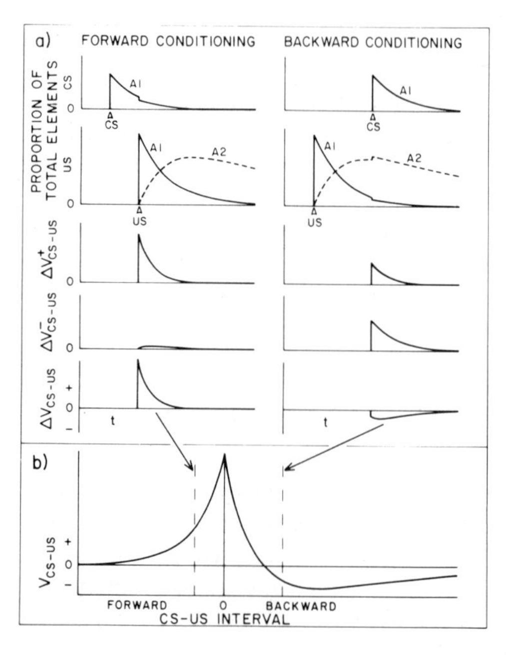

FIG. 8.3. Simulation of the processes assumed by SOP to be involved in a single (initial) conditioning trial with a punctate CS and US at different interstimulus intervals: (a) cases of a forward, CS–US, and a backward, US–CS, pairing where the top two graphs in each column indicate the expected proportions of CS and US nodal elements in the A1 and A2 states of activity and the bottom three graphs in each column indicate the expected changes in V^+_{CS-US}, V^-_{CS-US}, and V_{CS-US}; (b) expected outcomes of single conditioning trials, expressed as V_{CS-US}, across a range of CS–US intervals. The broken vertical lines show where the two examples from panel a fall in the function. (Reprinted from Wagner, 1981.)

forward, trace-conditioning trial (left column of graphs) in which a momentary CS precedes by some distance a momentary US, and in a similar backward conditioning trial (right column of graphs) in which the CS follows the US. The top two graphs in each case describe the relevant activity dynamics of the CS and US nodes. The parameters chosen in these simulations differed for the CS and US only in setting $p_{1,CS} = .5p_{1,US}$ (on the reasonable supposition that the CS is typically less intense than the US in actual experimental conditions), which is without consequence for the general conclusions to be demonstrated. The portrayal of the CS nodal dynamics in Fig. 8.3a has also been simplified to depict only $p_{A1,CS}$, which, along with $p_{A1,US}$ and $p_{A2,US}$, is all that is relevant to the development of V_{CS-US}.

The three lower graphs in each column of Fig. 8.3a trace, in order, the momentary ΔV^+_{CS-US}, ΔV^-_{CS-US}, and net ΔV_{CS-US} that result from the nodal activity in the two conditioning arrangements. In the case of the forward conditioning example, at the moment of application of the US there is a large increment in excitatory association, ΔV^+_{CS-US}, due to the substantial value of $p_{A1,CS}$ that persists from the prior application of the CS. In successive moments of time post-US, further increments accrue, diminishing in size with the gradual decay in both $p_{A1,CS}$ and $p_{A1,US}$. And there are small increments in the inhibitory association, ΔV^-_{CS-US}, that first increase and then decrease in size, due to the overlap of the diminishing $p_{A1,CS}$ with the increasing and then decreasing $p_{A2,US}$. That the values of ΔV^+_{CS-US} are greater than those of ΔV^-_{CS-US} is due in part to the use of differential learning-rate parameters in the simulations, $L^+ = 5L^-$ (see Wagner, 1981, for discussion of the reasoning behind this general assumption of SOP), but, more germane to the comparisons of Fig. 8.3, is also due to the preferential overlap that is afforded the A1 activity of the CS node with the initial A1 activity of the US node versus the later occurring A2 activity of the US node. This may be appreciated by inspection of the functions for ΔV^+_{CS-US} and ΔV^-_{CS-US} in the backward conditioning example. In this case, where $p_{A1,US}$ has largely decayed but $p_{A2,US}$ is substantial at the time of application of the CS, the succeeding increments in excitatory association, ΔV^+_{CS-US}, are consistently less than the corresponding increments in inhibitory association, ΔV^-_{CS-US}. The result is that, whereas the algebraic consequence, ΔV_{CS-US}, is uniformly positive in the forward conditioning case, it is uniformly negative in the backward conditioning case.

The overall associative outcomes of the two conditioning arrangements are obtained by summing the successive momentary values of ΔV_{CS-US} over the episodes depicted. To place the outcomes in context, Fig. 8.3b summarizes the results of equivalent simulations with a range of CS–US temporal relationships and indicates by the broken vertical lines where the examples from Fig. 8.3a would fall in relationship to other stimulus arrangements. As may be seen, all forward conditioning arrangements are expected to result in an excitatory

V_{CS-US}, as in the example case of Fig. 8.3a, with the magnitude of the excitatory learning decreasing exponentially as the CS–US interval increases. In contrast, all backward conditioning arrangements are *not* expected to have the same directional net effect.

SOP, like the formulations of Konorski (1948) and Schull (1979), supposes that the backward arrangement of US-then-CS is an effective procedure for bringing about the critical circumstances for inhibitory learning, that is, the concurrent activation of a CS node to the A1 state and US node to the A2 state. However, because backward conditioning can also produce excitatory learning due to the concurrent activation of a CS node to the A1 state and US node to the A1 state, only under certain conditions will the net consequence be inhibitory. In Fig. 8.3b it may be seen that net inhibitory effects are most prominent with an intermediate duration US–CS interval. Net excitatory conditioning is anticipated with sufficiently short backward intervals, to be gradually replaced by inhibitory conditioning, before a decrease in any associative effect, as the US–CS interval is increased.

The expectation of maximum net inhibitory conditioning with intermediate US–CS intervals is a relatively general prediction of the model. However, it should be apparent that a number of other variables besides the US–CS interval should also be of consequence. Inspection of Fig. 8.2b, for example, should suggest how variation in US duration would have a nonmonotonic effect on the relative sizes of $p_{A1,US}$ and $p_{A2,US}$ at the termination of the US, and hence upon the degree of inhibitory backward conditioning that would be produced with a constant US–CS interval.

The full treatment of inhibitory learning by SOP involves the assumption that the development of excitatory linkages between a CS node and a US node allows activity in a CS node to excite inactive elements of the US node directly to the A2 state (with momentary probability, p_2, as a function of $p_{A1,CS}$ and V_{CS-US}). This "conditioned" A2 activation of the US node in overlap with the A1 activation of a CS node is presumed to produce inhibitory learning (e.g., in experimental extinction or conditioned inhibition training) just as in the case of backward conditioning that has been described. And the effects of a backward arrangement of US–CS should be influenced by cues that may precede the US to promote A2 activity of the US node prior to application of the US (see, e.g., Wagner & Terry, 1975). The reader must be referred to Mazur and Wagner (in press) and Wagner (1981) for a full specification of the assumptions involved and for the simulation of a variety of cases of inhibitory learning.

For the moment it should suffice to appreciate that SOP, like the formulations of Konorski (1948) and Schull (1979), supposes that backward conditioning should (under specifiable circumstances) produce inhibitory learning. The wick or snuff that is responsible for inhibitory learning is presumed to live within the A2 activity of the US node that is assumed to be an essential component of US processing as well as the mediator of conditioned responding.

INHIBITORY BACKWARD CONDITIONING

The several theoretical accounts outlined in the preceding section predict that a CS will acquire inhibitory associative tendencies if appropriate CS processing coincides with a secondary consequence of US presentation, that is, a fall in excitation of a US center (Konorski, 1948), a B state (Schull, 1979), or US elements in an A2 state of activity (Wagner, 1981). Because, in general, these secondary consequences of US presentation are expected to persist beyond the termination of the US, a CS that is presented at an appropriate time following US termination should acquire inhibitory tendencies. Our main objectives in this section are briefly to review existing experimental evidence directly relevant to this possibility and to present some unpublished observations from our laboratory that address an important theoretical issue with regard to inhibitory backward conditioning.

There is little doubt that a training schedule of backward, US-then-CS, pairings that would provide the opportunity for CS processing coincident with a secondary consequence of US presentation *can* produce an inhibitory CS (e.g., Moscovitch & LoLordo, 1968; Siegel & Domjan, 1971). What has been in doubt is whether such findings, when obtained, are due to the backward temporal arrangement of the CS and US per se, or whether they are due to some other characteristic of the backward conditioning procedure. For example, a typical backward conditioning schedule, involving multiple US–CS trials separated by relatively long intertrial intervals, can be viewed as a special instance of "explicitly unpaired" training, if "pairing" is taken to imply a forward CS–US relationship: The CS occurs in a context in which USs are regularly experienced but is never followed closely by a US. A backward conditioning schedule can then be proposed to be effective in producing inhibitory learning for the reason that any explicitly unpaired schedule is conventionally presumed to be effective (see, e.g., Wagner & Rescorla, 1972). In fact, when inhibitory backward conditioning has been observed empirically, it has been after extensive training, as might be necessary to train a discrimination between the context alone "signaling" the US and the CS, added to the context, "signaling" a period free from the US.

In the remainder of this section, we emphasize experimental studies that have attempted to separate the effects of backward pairings from the effects of the general class of explicitly unpaired training procedures. We begin with investigations of backward conditioning employing a relatively large number of US–CS trials and end with investigations of one-trial backward conditioning.

The US–CS Relationship in Inhibitory Backward Conditioning

Moscovitch and LoLordo (1968) were the first to address experimentally the question of whether inhibitory conditioning results from backward pairings of the

CS and US, per se, or from some other feature of backward conditioning procedures. They proposed that the temporal relationship of the CS to the preceding US in a backward conditioning procedure might be unimportant, insofar as the CS might acquire inhibitory properties only if it signals a relatively long period of time during which the US is not presented and thereby serves as a "safety signal" (or, more generally, as a signal for nonreinforcement under conditions when reinforcement is otherwise expected).

In an initial experiment, Moscovitch and LoLordo (1968) demonstrated that a 5-sec tone CS acquired inhibitory properties for a group of dogs (Group B1) when the CS was presented 1 sec following termination of a 4- to 6-sec shock US, and the intertrial interval, timed from termination of the CS in one trial to onset of the US in the next trial, ranged from 2 to 3 min. Training involved five Pavlovian conditioning sessions during each of which 24 US–CS trials were presented. Conditioned inhibition was indexed by a reduction in the rate of responding on a Sidman avoidance baseline when the CS was presented in subsequent test sessions.

In their second experiment, which was designed to determine whether the US–CS interval or the intertrial interval was important for the development of inhibitory backward conditioning, Moscovitch and LoLordo (1968) assessed the conditioned properties of a CS for two additional groups of dogs. Group B15 differed from Group B1 in the previous experiment only in the duration of the US–CS interval, which was 15 rather than 1 sec. Group B1R differed from Group B1 only in the range of intertrial interval durations, which was from 0 sec to 15 min rather than from 2 to 3 min (the mean duration of the intertrial interval was 2.5 min for all groups). When the CS was superimposed on the Sidman avoidance baseline in subsequent test sessions, the rate of avoidance responding was reduced significantly for Group B15, as it had been for Group B1, but there was no significant change in response rate for Group B1R. The level of response rate reduction for Group B15 was comparable to that observed for Group B1, and both of these groups differed reliably from Group B1R.

The outcome of these experiments is consistent with a safety signal account. Comparison of the results for Groups B1, B15, and B1R would suggest that inhibitory backward conditioning will develop if the CS consistently signals a relatively long period of time during which the US is not presented, regardless of the US–CS interval duration, but not if there are occasions on which the US occurs immediately or shortly following the CS. This conclusion is based both on the failure to detect evidence of conditioned inhibition for Group B1R and on the failure to detect a difference in the level of inhibition for Groups B1 and B15.

It remains possible, however, to interpret the results in a manner that is consistent with the postulation of an important role for the temporal relationship of the CS to the preceding US in the development of inhibitory backward conditioning. Although it is clear from the obtained difference between Groups B1 and

B1R that the duration of the intertrial, CS–US interval can affect the outcome of a backward conditioning procedure, it is not necessary to conclude that the failure to obtain evidence of conditioned inhibition for Group B1R reflected the inadequacy of the backward relationship of the CS and US to produce conditioned inhibition. Instead, it can be proposed that any inhibitory properties acquired by the CS in Group B1R as a result of the backward relationship of the CS and US were masked by excitatory properties acquired by the CS as the result of occasional, effective, forward CS–US pairings when the intertrial interval was very short. The inhibitory and excitatory properties could have summated to produce the appearance of a neutral CS in the test sessions. According to this reasoning, a group of subjects that received the same schedule of CS–US intervals as Group B1R but a substantially longer US–CS interval would be expected to display evidence that the CS had acquired excitatory properties.

The failure to detect a difference between Groups B1 and B15 suggested that variations in the US–CS interval do not produce differences in inhibitory conditioning. Nevertheless, subsequent investigators, using different experimental preparations, have reported differences in inhibitory conditioning across a wider range of US–CS intervals. Using a rabbit nictitating membrane response measure and a constant intertrial interval of 55 sec, Plotkin and Oakley (1975) found no difference in the level of conditioned inhibition, assessed in a retardation of acquisition test, between groups of subjects for which the US–CS interval was either 200 or 500 msec during backward conditioning training, but both of these groups were retarded in acquisition of excitatory conditioned responding relative to a group of subjects for which the US–CS interval was 55 sec. The group for which the US–CS interval was 55 sec did not differ reliably in the retardation test from control groups that did not receive backward conditioning training. Maier, Rapaport, and Wheatley (1976), likewise, reported evidence to indicate that a CS became more inhibitory for a group of rats for which the US–CS interval was 3 sec during backward conditioning training than for a group for which the US–CS interval was 30 sec. (The intertrial interval for the two groups commonly ranged from 2 to 4 min.) The 30-sec US–CS interval group did not differ reliably from a control group that received the same schedule of US presentations but random presentations of the CS. In this experiment, conditioned properties of the CS were assessed by superimposing the CS on an ongoing Sidman avoidance baseline.

The results of the Plotkin and Oakley (1975) and Maier et al. (1976) experiments indicate that the duration of the US–CS interval can affect the outcome of a backward conditioning procedure in the absence of differences in the duration of the intertrial interval. Evidence of conditioned inhibition was obtained when the US–CS interval was relatively short but not when the US–CS interval was relatively long. On the basis of these results, it is reasonable to propose that the 15-sec US–CS interval in the Moscovitch and LoLordo (1968) study was simply not long enough to produce a difference in inhibitory learning relative to a 1-sec US–CS interval.

Before concluding that the duration of the US–CS interval has a direct effect on the outcome of backward conditioning procedures, perhaps by modifying the occasions of CS processing relative to the time of US processing, it must be noted that procedures in which the US–CS interval is varied while the intertrial interval is held constant result in differences in the rate of presentation of the US that could, in turn, be responsible for differences in the conditioned inhibition produced. The difference in rate of presentation of the US between groups receiving short or long US–CS intervals was relatively small in the Maier et al. (1976) study but was relatively large in the Plotkin and Oakley (1975) study. Such differences in rate of US presentation could conceivably produce differences in the level of excitatory strength of the experimental context; that is, higher rates of US presentation would be expected to result in higher levels of excitatory strength of the context. And it is a familiar expectation of the Rescorla–Wagner model (e.g., Wagner & Rescorla, 1972) that, as the level of excitatory strength of the contextual stimuli increases, the amount of conditioned inhibition resulting from nonreinforced (i.e., unpaired) presentations of the CS in that context should increase as well. This account can therefore explain differences in the level of conditioned inhibition between groups receiving different US–CS intervals but the same intertrial, CS–US interval without appealing to a direct effect of the duration of the US–CS interval: Longer US–CS intervals are expected to produce lower levels of excitatory strength of contextual stimuli and, consequently, lower levels of conditioned inhibition from nonreinforced presentations of the CS.

Although there is experimental evidence to indicate that the temporal relationship of a CS to the preceding US importantly determines the outcome of a backward conditioning procedure (Maier et al., 1976; Plotkin & Oakley, 1975), any conclusions regarding the basis for this effect must remain tentative. In order to provide conclusive evidence for a direct effect of the US–CS interval in backward conditioning procedures, it is necessary to demonstrate differences in conditioning as a function of the duration of the US–CS interval in the absence of differences in the duration of the intertrial interval and in the rate of presentation of the US. One way to do this is to employ a within-subject comparison, in which two CSs are presented in different backward temporal relationships to the US but in the same forward relationship to the US. The within-subject procedure avoids the possibility of different levels of excitatory strength of the experimental context. We present the results of within-subject manipulations of the US–CS interval using indirect measures of inhibitory learning in a later section. A second way to investigate the effects of the US–CS interval in backward conditioning without differences in the intertrial interval or the rate of US presentation is to use only a single conditioning trial. We turn now to a consideration of one-trial backward conditioning.

One-Trial Backward Conditioning

The striking fact apparent in studies that have observed the associative product of a small number of backward pairings of US-then-CS is that the product has been *excitatory* rather than inhibitory. Early studies of backward conditioning (e.g., Champion & Jones, 1961; Spooner & Kellogg, 1947) often reported a transient phase of apparent conditioned responding to the backward CS, although for lack of appropriate comparisons to rule out pseudoconditioning and sensitization reviewers (e.g., Mackintosh, 1974) have generally been skeptical of the associative underpinnings of the responding. More recently, a number of experiments have provided convincing demonstrations that excitatory conditioning can result from a relatively small number (i.e., less than 25) of US–CS pairings, including experiments involving only a single trial (e.g., Heth & Rescorla, 1973; Mahoney & Ayres, 1976; see Spetch, Wilkie, & Pinel, 1981, for a review of the older and more recent literatures).

The critical comparisons in the recent literature involve treatments with different temporal relationships between the CS and US. For example, Keith–Lucas and Guttman (1975), using a one-trial backward conditioning procedure with exposure to a toy hedgehog as the CS, reported that groups of rats for which the CS was presented for 60 sec 1, 5, or 10 sec after presentation of a .75-sec shock US avoided the CS in a subsequent test session significantly more than a group for which the CS was presented 40 sec after presentation of the US. This latter group did not differ in the avoidance measure from CS-alone and US-alone control groups. Using a more conventional CS in a one-trial backward conditioning procedure, Mahoney and Ayres (1976) reported that a licking baseline was disrupted more by presentation of a tone CS for groups of rats that had received a 4-sec presentation of the CS either simultaneous with or 4 sec after termination of a 4-sec shock US, compared to a group that received the CS 150 sec prior to the US presentation.

The finding of excitatory backward conditioning is certainly not in itself embarrassing to the class of theories exemplified by those of Konorski (1948), Schull (1979), and Wagner (1981). SOP, for example, explicitly allows for excitatory backward conditioning under specifiable conditions when it may be assumed that CS processing coincides principally with US nodal elements in the A1 state rather than the A2 state (see Fig. 8.3 and related text). What must be of concern, however, is that excitatory conditioning appears to be the rule with a small number of backward conditioning trials, whereas inhibitory conditioning appears to be demonstrable only with a large number of trials (Heth, 1976).

We have already acknowledged a conventional argument for why inhibitory conditioning might become more likely with extended as compared to minimal backward training. With successive US–CS trials there is increasing opportunity for background, contextual cues to acquire associative strength. Because the CS is presented in this context and not reinforced by any immediately following

USs, there is increasing possibility of inhibitory conditioning resulting (e.g., Wagner & Rescorla, 1972) in spite of, or in addition to, the associative consequence of the backward US–CS arrangement. With sufficient training the subject could, in effect, discriminate the CS as a "safety signal" a la Moscovitch and LoLordo (1968; see also Jones, 1962). This reasoning is supported by Hinson's (1982) demonstration that inhibitory backward conditioning is augmented by pretreatment in which US-alone presentations are given in the experimental context prior to backward training, and by Fowler, Kleiman, and Lysle's data (this volume) that indicate that such pretraining is less effective if the USs involved are signaled so as to "block" conditioning of contextual cues.

The reasoning of SOP with regard to the consequences of increasing the associative strength of the contextual cues is somewhat more complicated, as was anticipated by Wagner and Terry (1975). What must be assumed is that, as the contextual cues become more excitatory, a greater proportion of US nodal elements will be caused to be in the A2 state rather than the I state at the time of US application. There will thus be less of an ability of the US of a US–CS occasion to produce A1 activity that can lead to excitatory backward conditioning. In such case, the overlap in processing of the CS and US will predominantly involve A2 activity of the US, thereby leading to inhibitory backward conditioning.

There are few data that are critical to this interpretation but those of Wagner and Terry (1975) are quite consistent with it. Using a rabbit, conditioned eyeblink preparation, they conducted a within-subject comparison of the outcome of backward conditioning trials in which the US was either signaled or not by a well-trained discriminative CS; that is, backward conditioning training was preceded by a discrimination training phase in which one CS (CS+) was established as a reliable signal for the US to be used in backward conditioning trials and another CS (CS-) was explicitly unpaired with the US. During backward conditioning training, two other target CSs were arranged to occur in the same backward temporal relationship to the US on different trials interspersed among continued discrimination training trials. For trials with one target CS, the US was always preceded by CS+. For trials with the other target CS, the US was always preceded by CS-. The reasoning was that CS+ should have functioned in the manner otherwise attributed to contextual cues following extensive backward conditioning training so as to decrease the likelihood of excitatory backward conditioning. In comparison, CS- should have functioned like the less excitatory context that might obtain without extensive training so as to allow excitatory backward conditioning. Over the course of 24 backward conditioning trials with each target CS, reliable conditioned responding, assessed during nonreinforced test trials, did not develop to the target CS that was trained with the signaled US but did develop to the target CS that was trained with the contrasignaled US. Presumably, if assessment had been made of potential inhibitory conditioning to the two target CSs, it would have been more likely to have been observed to the target CS that was trained in backward pairings with a signaled US (and showed no evidence of net excitatory conditioning) than to the target CS that was trained in backward pairings with a contrasignaled US (and showed excitatory effects). Such assessment, however, was not made.

It is important to recognize that we have described two potentially separable ways in which the associative strength of the context in which a backward US–CS pairing occurs can influence the excitatory versus inhibitory outcome of the CS exposure. The more excitatory the context is at the time of CS presentation, the more it should (1) favor inhibitory learning, independent of the preceding US and (2) reduce the A1 processing of the US necessary for excitatory backward conditioning to the consequent CS. In considering the effects of simple variation in the number of US–CS trials it can be assumed that with increasing training the overall context will become more excitatory and, consequently, will act to produce more inhibitory conditioning and less excitatory conditioning. In other instances, however, a manipulation can have opposite effects on the net associative strength of a CS via these two avenues. A case in point is the report of Fowler, Kleiman, and Lysle (this volume, Experiment 4) that net inhibitory learning to a backward CS was less when the USs were signaled rather than unsignaled, a finding that appears directly opposed to that of Wagner and Terry (1975). However, it can be noted that whereas Wagner and Terry used a within-group comparison so as to equate the associative strength of the background context in which the target CSs would be experienced, Fowler et al. used a between-group comparison in which the signaled-US condition would likely entail a reduced associative strength of the overall context, in addition to the differential signaling at the time of US presentation.

One-Trial Inhibitory Backward Conditioning

Whereas SOP clearly allows for excitatory backward conditioning under appropriate conditions and includes strong reasons to anticipate that the major evidence for inhibitory backward conditioning would come from studies with extended training, the fact remains that the model likewise predicts that inhibitory conditioning should be demonstrable following a single US–CS pairing.

If it is the case that studies employing a small number of US–CS pairings have produced excitatory conditioning because of parametric conditions that favored CS processing coincident with persisting US nodal elements in the A1 state of activity, then it should be possible to produce inhibitory conditioning by parametric changes, guided by SOP, that would favor CS processing coincident with US elements in the A2 state of activity. According to SOP, although the majority of active US elements will be in the A1 state for some period of time immediately following an initial presentation of a US, the majority of active elements will be in the A2 state during some subsequent period of time, before all elements eventually revert to the I state (see Fig. 8.3). The predicted consequence, as previously noted, is that inhibitory backward conditioning will be most likely with an *intermediate* US–CS interval.

We sought to evaluate this prediction in an unpublished one-trial, backward conditioning, CER study with rat subjects. Because we had to presume that a single conditioning trial would yield only modest levels of conditioned inhibition, two special procedures were employed in an attempt to provide a sensitive measure. First, rather than using a summation test, which would necessitate training a second CS as a conditioned excitor either before or after the single backward conditioning trial, or a conventional retardation test, which would require at least one forward CS–US pairing following the single backward conditioning trial before any test of inhibition could be made, a "concurrent retardation" procedure was employed. What this meant was that all subjects received a single forward pairing of CS and US, which was then followed or not by a backward presentation of the same CS at different intervals after the US. Conditioned properties of the CS could then be measured on immediately following CS-alone test trials. The assumption was that the forward CS–US pairing presented to all subjects would establish a common excitatory tendency, which would be added to or offset by any excitatory or inhibitory learning as a result of a subsequent, backward CS presentation (cf. Burkhardt & Ayres, 1978).[1]

Secondly, an attempt was made to reduce heterogeneity in the unconditioned responsivity to the CS by eliminating subjects that, after an initial phase of CS habituation given to all animals, displayed extreme percentage suppression scores (see later) to the forward CS on the single conditioning trial prior to differential treatment. An arbitrary cutoff of ±50% was adopted. Fifteen of 88 total subjects were eliminated on this basis, leaving 19, 20, 18, and 16 subjects in the four treatment groups from which data are reported.

All subjects were initially trained to bar press for food reward on a variable-interval 60-sec schedule and then received three 1-hr experimental sessions with this schedule in effect. In the first, habituation, session all subjects received four presentations of a 30-sec, 2000-Hz tone that was subsequently employed as the CS. In the following, conditioning, session all subjects received a single 32-sec presentation of the tone CS overlapping and terminating with a 2-sec, 1-mA footshock US, 20 min after the start of the session. The four groups were treated differently only in that one group (Group NO) received no further stimulation, whereas the remaining three groups received a 30-sec, post-US presentation of the CS, beginning either 1 sec (Group T1), 31 sec (Group T31), or 600 sec (Group T600) after the termination of the US. In the final, test, session all subjects received four 30-sec CS presentations during which conditioned sup-

[1]This reasoning ignores possible complications of competitive or synergistic stimulus processing, e. g., that the forward CS and/or the US might be differentially processed as a result of backward CS occurrence, to yield other than a common excitatory tendency. Donegan, Whitlow, and Wagner (1977) may be consulted for a discussion of relevant theoretical possibilities.

pression of bar pressing was assessed according to the formula $[(A - B)/A] \times 100$, where A was the rate of bar pressing in the 2-min periods immediately preceding each CS and B was the rate of bar pressing during the CS. (According to this measure, a percentage suppression score greater than 0 indicates a decrease in bar pressing during the CS relative to the pre-CS baseline, whereas a score less than 0 indicates a corresponding increase.)

The four selected groups were well matched in their mean levels of unconditioned suppression to the CS during the initial habituation session and on the forward pairing of the conditioning session. They differed, however, in mean conditioned suppression during the test session, following differential treatment, as may be seen in Fig. 8.4. The important trend was for suppression to be lowest in Group T31, in which the CS was presented at an intermediate post-US interval. Indeed, only in the case of this group was there statistical support for an inhibitory effect of the backward CS exposure. A decrease in percentage suppression in Group T1, T31, or T600 relative to Group NO would be indicative of some inhibitory consequence of the post-US presentation of the CS. However, Group T600 was anticipated, by virtue of the very long US–CS interval involved, not to acquire any inhibitory backward conditioning and a planned comparison showed no reliable difference between the mean suppression of Group T600 and Group NO. In comparison, Group T31 evidenced reliably less conditioned suppression than these two groups, $F(1, 50) = 4.23$, $p < .05$. The suppression of Group T1 was somewhat lower than, but not reliably different from, that of Group T600 and Group NO.

Although this outcome pattern is encouraging to the tenets of SOP in demonstrating an apparent instance of one-trial, inhibitory backward conditioning in Group T31, with an intermediate US–CS interval, the results from Group T1 are somewhat disconcerting. Certainly, a stronger theoretical case would have been made if this group had shown *more* suppression than Groups NO and T600, that is, evidenced excitatory backward conditioning as has been reported in CER

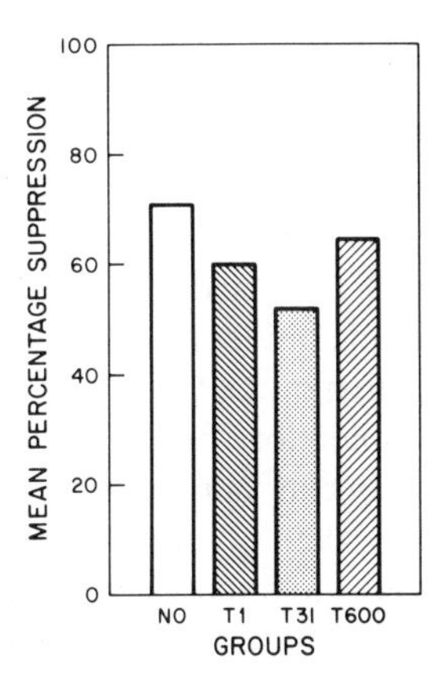

FIG. 8.4. Mean percentage suppression scores in the postconditioning test session for four groups receiving either no post-US presentation of the CS (Group NO) or a 30-sec presentation of the CS beginning 1 sec (Group T1), 31 sec (Group T31), or 600 sec (Group T600) following termination of the US in a concurrent retardation procedure.

following a single trial with a short US–CS interval (e.g., Mahoney & Ayres, 1976; Shurtleff & Ayres, 1981). On the other hand, the aforementioned reports of excitatory backward conditioning have typically employed shorter CSs (e.g., 4 sec) than the 30-sec tone employed in the present study, which makes the differential outcomes theoretically reasonable.

It is reasonable to assume that the initial seconds of processing of the backward CS in Group T1 were primarily in conjunction with US elements in the A1 state as compared to the A2 state, and if the CS had been terminated at this point the net consequence would have been excitatory. It can be supposed that as the 30-sec backward CS instead continued, while the US elements decayed into the A2 state, inhibitory conditioning accrued until the net effect was weakly inhibitory (see Fig. 8.4). Such an interpretation is obviously testable by probing the associative consequences of different US–CS intervals with shorter CSs than presently employed.

We have not as yet done this but have proceeded the other way in varying the CS duration, so as to attempt to produce more robust evidence of inhibitory backward conditioning with a single trial; that is, if one assumes that presentation of a CS from 1 to 31 sec post-US produced a weakly inhibitory CS (Group T1), while presenting the CS from 31 to 61 sec post-US produced a stronger inhibitory cue (Group T31), each because of the net overlap of the A2 versus A1 processing of the US that the CS processing encompassed, then it might be expected that a longer post-US CS that covered the entire 60-sec duration would become more clearly inhibitory. A small-scale study, which simply compared the effects of a 60-sec, post-US CS, initiated 1 or 600 sec post-US, to the effects of no post-US CS, explored this possibility.

The study was identical in design to that which has been described, with treatments comparable to those of the aforementioned Group NO, Group T1 and Group T600, except that the backward CS in Group T1 was 60 sec in duration (bridging the entire interval covered by the backward CSs in Group T1 and T31 of the preceding study) and in Group T600 was made equivalently 60 sec. The pre-US CS and the test CS were 30 sec as in the preceding study. From a total of 36 subjects, 10, 8, and 9 remained in the three groups after eliminating subjects that had extreme unconditioned responses as in the preceding study.

Figure 8.5 summarizes the test suppression in the three groups. The outcome is quite clear and consistent with the data of Fig. 8.4. There was no reliable difference between Group T600 and Group NO. In comparison, Group T1 showed substantially and reliably less suppression, $F(1, 24) = 6.33$, $p < .05$, suggesting that the application of the CS over the interval from 1 to 61 sec post-US produced appreciable inhibitory backward conditioning.

The results of our experiments using a concurrent retardation procedure indicate that a CS can acquire inhibitory associative strength in a single backward conditioning trial. The results join those of Plotkin and Oakley (1975), Maier et al. (1976), and others (e.g., Denny, 1971), to indicate that the US–CS interval is

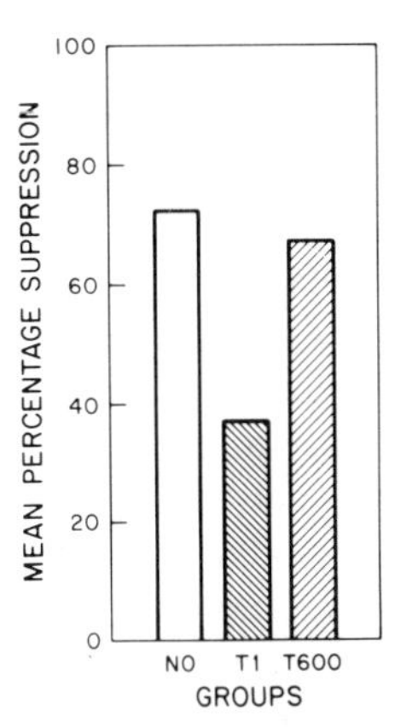

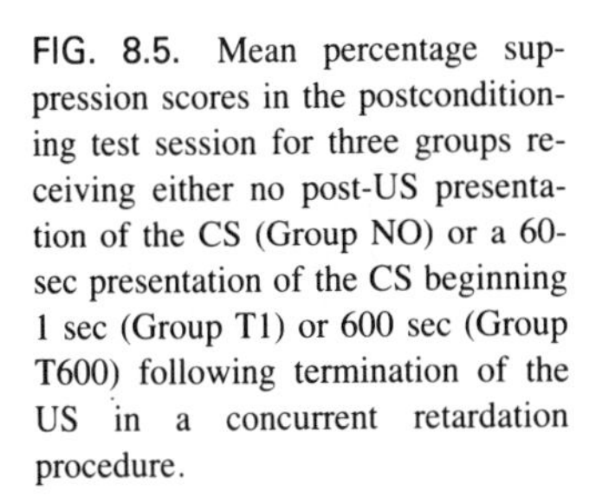
FIG. 8.5. Mean percentage suppression scores in the postconditioning test session for three groups receiving either no post-US presentation of the CS (Group NO) or a 60-sec presentation of the CS beginning 1 sec (Group T1) or 600 sec (Group T600) following termination of the US in a concurrent retardation procedure.

an important variable in the development of conditioned inhibition in one-trial as well as multiple-trial backward conditioning procedures. Although further investigations are clearly necessary, the existing data provide encouragement for the theories of Konorski (1948), Schull (1979), and Wagner (1981), which postulate a secondary consequence of US presentation that contributes to inhibitory backward conditioning. An appreciation of the inhibitory learning that results from the close temporal relationship of a CS to a preceding US, distinct from that which may otherwise result from explicitly unpaired presentations of a CS and US (e.g., Baker, 1977), may contribute importantly to our full understanding of various empirical phenomena. We turn now to some examples.

EXTRAPOLATION

According to the theoretical accounts of Konorski (1948), Schull (1979), and Wagner (1981), whenever a CS is presented relatively close to the termination of a US, there is the possibility that CS processing will coincide with a secondary consequence of US presentation and, therefore, that the CS will acquire inhibitory associative strength. This may occur not only in circumstances in which a backward "pairing" is explicitly arranged but in various circumstances in which, for other dictates of scheduling, a CS is presented within the time frame of the secondary consequence of a preceding US. In this section we consider two Pavlovian scheduling parameters the effects of which may be, in part, theoretically attributable to the differential opportunity for inhibitory backward conditioning that is afforded. The first is the effects of different intertrial intervals (ITIs) in forward conditioning procedures. With sufficiently short intervals between successive CS–US pairings, it is theoretically possible that the CS on trial n+1 will overlap with the secondary consequence of the US on trial n to produce inhibitory learning that would work to offset excitatory learning produced by the forward CS–US pairings. The second case is the effects of presentations of the US in the absence of the CS in "random" schedules. In this instance it is theoretically possible that some of the scheduled USs will so precede a CS that the processing of the CS will overlap with a persisting secondary consequence of US presentation to produce inhibitory learning that would serve to offset excitatory conditioning produced by other (forward) CS–US juxtapositions that occur. In order to evaluate these possibilities in relationship to existing accounts, it is necessary to provide experimental comparisons that are quite straightforward but have rarely been employed. We thus report three recent experiments conducted in our laboratory in collaboration with Michael Ewing. Although further assessments are necessary, our results are consistent with the suggestion that the aforementioned manipulations may be effective, in part, according to the differential inhibitory backward conditioning they produce.

Conditioning as a Function of the Duration of the ITI

It has been well documented that, given a constant forward CS–US pairing relationship capable of supporting excitatory conditioning, acquisition of conditioned responding proceeds more slowly or to a lower asymptote with relatively short as compared to relatively long ITIs (e.g., Gormezano & Moore, 1969). There are a number of common theoretical accounts of this phenomenon. One is in terms of the differential conditioning of contextual cues (e.g., Gibbon & Balsam, 1981; Miller & Schachtman, this volume). Another, which is closely related to the reasoning of SOP but does not appeal to differential inhibitory conditioning, was proposed by Wagner (1976, 1978) in his priming theory.

Gibbon and Balsam (1981) have proposed that the performance of conditioned responding to a CS depends on the level of excitatory conditioning to the CS relative to the level of excitatory conditioning to the experimental context. In their model, excitatory conditioning to a CS depends on the rate of US presentation during that CS and, similarly, excitatory conditioning to the context depends on the overall rate of US presentations in that context. Consequently, if the rate of occurrence of the US in the presence of the CS is held constant but the overall rate of occurrence of the US in the context is increased by shortening the ITI, conditioned responding should decrease. A review of experiments on the rate of acquisition of key pecking in autoshaping situations has been taken to support this reasoning (Gibbon & Balsam, 1981).

The basic assumption of Wagner's (e.g., 1976, 1978) priming theory is that when a stimulus representation is active in transient, short-term memory (STM), presentation of the stimulus itself will be less likely than it otherwise would be to provoke the full memory processing (rehearsal) necessary for excitatory associative learning. Manipulation of the ITI could therefore affect excitatory conditioning from a CS–US pairing by changing the probability that the representation of any of the trial stimuli will have remained in STM from the occasion of the preceding trial. Terry (1976) has demonstrated that presentation of a US a short time before CS–US trials retards the acquisition of a conditioned eyeblink response by rabbits in a manner consistent with a priming analysis. And what is supposed is that the US on one trial may similarly decrement the excitatory conditioning resulting from a subsequent CS–US pairing when the ITI is sufficiently short.

That neither a contextual conditioning nor a priming interpretation of the effects of differential ITIs may be totally adequate is suggested by the results of an experiment by Kaplan (in press). In this experiment, pigeons commonly received trace conditioning trials involving a 12-sec key-light CS followed, after a 12-sec trace interval, by 3 sec of access to food (the US). For different groups of subjects, the average ITI duration, timed from termination of the US on one trial to the onset of the CS on the next trial, was 15, 30, 60, 120, or 240 sec. Following extensive training with the trace conditioning procedure, the conditioned properties of the CS were assessed in a test session, during which no USs were presented, by recording the pigeons' tendencies to approach or to withdraw from the key-light CS. Approach to the CS was taken to indicate excitatory conditioning, whereas withdrawal from the CS was taken to indicate inhibitory learning (see Wasserman, Franklin, & Hearst, 1974). The data were very orderly: The 240-sec and 120-sec ITI groups tended to approach the CS, the 60-sec ITI group was indifferent to the CS, whereas the 30-sec and 15-sec ITI groups tended to withdraw from the CS. These data suggest that, with very short ITIs, not only does the CS of an otherwise effective forward "pairing" become less excitatory, but, instead, it becomes inhibitory. Neither the Gibbon and Balsam (1981) nor the Wagner (1976, 1978) accounts of variation in ITI anticipate the latter, inhibitory outcome.

An alternative explanation of the effects of ITI duration, including inhibitory conditioning with very short ITIs, is provided by SOP. Like priming theory, SOP predicts that a US may be less effective in producing excitatory conditioning if the US is closely, as compared to more distantly, preceded by another presentation of the same US. In terms of SOP, this "priming" effect occurs when US elements activated by one US presentation are still in the A2 state at the time of a second US presentation. In this case, because the effect of a US is to promote only a given proportion of the US elements in the I state to the A1 state, the number of elements promoted to the A1 state will be less than in the case when all US elements are in the I state at the time of US presentation. Consequently, there is presumably less opportunity for joint occurrence of CS elements and US elements in the A1 state, and less excitatory learning. SOP takes the further step, beyond priming theory, of supposing that if US elements thus persist in the A2 state until the next CS–US occasion, they will, in overlap with CS elements in the A1 state, produce inhibitory learning. With very short ITIs, the coincidence of CS processing with US elements persisting in the A2 state may be sufficient to produce a net inhibitory CS, as suggested by the results for the 30-sec and 15-sec

ITI groups in the Kaplan (in press) experiment. With somewhat longer ITIs, the effects of the smaller number of US elements persisting in the A2 state may be seen only in decrements in the net excitatory conditioning produced. Of course, with sufficiently long ITIs, all US elements in the A2 state would be expected to decay to the I state before the CS is presented, and the conditioning resulting from one trial would not be expected to be influenced by the occurrence of the US in the preceding trial.

Kaplan (in press) accepted such an interpretation as the "most complete" account of his findings but also noted that the data would be amenable to some extension of the Gibbon and Balsam (1981) theory that included appropriate suppositions about when a CS would act "inhibitory." It is certainly in the spirit of the Gibbon and Balsam theory that a CS with a constant relationship to following USs will become systematically less like an excitatory cue as the density of USs in the overall context is increased with decreasing ITIs. But it should be possible easily to distinguish between ITI effects, per se, as treated by SOP, and context conditioning effects, that might normally vary with ITI, as emphasized by Gibbon and Balsam (1981) and others (e.g., Rescorla & Wagner, 1972). It is simply necessary to establish conditioning to different CSs within the same overall context but with different ITIs regularly separating each from the immediately preceding trials involving the US. This was done in two companion experiments that we report here.

Both experiments involved CER conditioning of rats on a bar-pressing baseline supported by a VI 60-sec schedule of food reward. The CSs were 30-sec presentations of either a 2000-Hz tone, a 3-Hz interrupted light (from a shielded 3-W bulb) or, in the second experiment only, a 60-Hz vibratory stimulus (from an electromagnetic coil attached to the subject's chamber). The US was a .5-sec, 1-mA footshock delivered through the grid floor. Training and testing sessions were uniformly 90 min long and were conducted in cycles of 4 training days in which the ITIs preceding the experimental CSs were manipulated according to the experimental plan, followed by a nonreinforced test day in which each CS was presented twice (in counterbalanced order) at a common, long ITI (12–18 min) to evaluate conditioned suppression.

In the first experiment, each training session included six trace conditioning trials with the tone CS, intermixed with six similar trials with the light CS. The trace procedure, which was adopted to make the forward CS–US relationships similar to those in the Kaplan (in press) experiment, involved a 30-sec gap between the termination of the CS and presentation of the US. Successive trials were consistently separated by one of two ITIs (timed from US offset to CS onset), a relatively long interval of 600 sec or a relatively short interval of 60 sec. For all subjects, one of the two CSs was consistently preceded by a 600-sec ITI and designated as CS-Long (CS_L) and the other CS was consistently preceded by a 60-sec ITI and designated CS-Short (CS_S). The sequence of trials within conditioning sessions was otherwise arranged so that CS_L and CS_S equally often occurred as the first trial and were each equally often followed by CS_S trials (after a 60-sec interval) and CS_L trials (after a 600-sec interval). For four subjects CS_L was the light and CS_S was the tone, whereas for four subjects this cue assignment was reversed. (Two subjects were lost, one due to apparatus malfunction, the other due to failure to respond during baseline periods in the test sessions, leaving six subjects from which data are reported.)

Conditioning, as described, was conducted over three blocks of four sessions. Figure 8.6 presents the resulting mean percentage suppression scores to CS_L and CS_S, plotted separately for the two cue-identification subgroups, during the nonreinforced test session that followed each block. For both subgroups it may be seen that the level of conditioned suppression to CS_S was less than that to CS_L, with this difference especially notable in the final test session. Statistical analyses confirmed that there was significantly less overall suppression to the tone and light CSs when regularly trained with a 60-sec interval separating them from the preceding trial (CS_S) than when similarly trained with a 600-sec interval (CS_L), $F(1, 4) = 9.13$, $p < .05$.

These results indicate an effect of ITI duration on conditioning that is not attributable to differences in the overall rate of US presentation. There was less evidence of excitatory conditioning to a CS that consistently followed a short ITI than to a CS that consistently followed a long ITI despite the fact that both CSs were reinforced in the same experimental context with the same overall rate of US presentation. Accounts of the effect of ITI duration that propose that differences in the overall context in which CS–US pairings are presented may be responsible for the obtained differences in the conditioned responding (e.g., Gibbon & Balsam, 1981; Kaplan, in press) do not anticipate the present results. On the other hand, the lower level of conditioned suppression to CS_S than to CS_L is consistent with the possibility that CS_S was more likely to acquire inhibitory

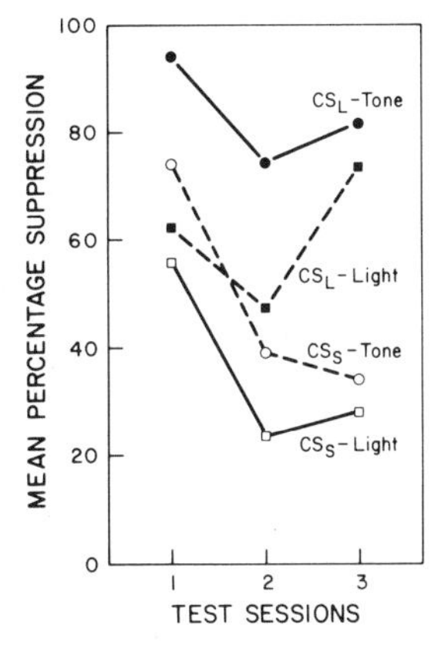

FIG. 8.6. Mean percentage suppression scores over three test sessions interspersed during acquisition involving within-subject manipulation of the ITI preceding each of two CSs. Scores are plotted separately for cue-assignment subgroups for which either the tone was preceded by a long ITI (CS_L) and the light by a short ITI (CS_S), or in which this designation was reversed.

associative strength as a result of the relatively short interval separating it from the previous occurrence of the US.[2]

The results of this experiment are less than fully convincing, however, of the involvement of inhibitory backward conditioning with the US–CS juxtapositions produced by short ITIs. It is just as possible that short ITIs served to reduce excitatory learning from the following CS–US "pairings" as opposed to reducing the net excitatory loading via an inhibitory contribution. This, indeed, would be a natural interpretation from Wagner's (1976, 1978) priming theory: It is possible that as the US representation was more likely to persist in STM from the preceding trial until the occasions of CS_S–US pairings as compared to CS_L–US pairings, the US would be less effective in producing excitatory conditioning in the former case. What is called for is a within-subject experiment similar to that which has been reported but in which such an interference-with-excitatory-learning account (a general-interference, or "overshadowing" account would be other examples) could be eliminated.

We attempted this in the second experiment by asking whether the *extinction* of a previously trained CS would proceed faster if the *nonreinforced* CS trials were always preceded by a CS–US trial with another CS at a shorter as compared to a longer ITI. A priming account would not predict differences insomuch as there was no US to be altered in effectiveness on the nonreinforced extinction trials. And this, or any other "interference" account, would not in any case suppose *greater* learning with shorter ITIs. On the other hand, an inhibitory backward conditioning interpretation, such as that of SOP, would predict that the shorter ITIs would facilitate extinction by causing CS processing to overlap the persisting A2 processing provoked by the preceding US.

Subjects were initially trained over a series of sessions, in each of which there were two reinforced presentations of the tone, light, and vibratory CS in a pseudorandom order and with 12-min ITIs. A difference between this experiment and the preceding one was that on all reinforced trials the 30-sec CS overlapped and terminated with the US. For eight animals this initial training phase involved one block of four sessions, whereas for another eight animals it involved two such blocks (with this difference being without consequence).

In the following extinction phase the tone and light CSs were consistently nonreinforced, whereas the vibratory CS continued to be reinforced. Each session included three tone presentations and three light presentations in irregular order but each immediately preceded in the sequence by a pairing of the vibratory CS and the shock US, at different ITIs. For all subjects, one of the two nonreinforced CSs was consistently separated from the preceding reinforced trial by a

---

[2]One could entertain the possibility that CS_S was experienced in a *local context* that had acquired greater associative strength than that in which CS_L was experienced (see related discussion by Miller & Schachtman, this volume) and thus acquired more inhibitory tendency. But there is no obvious basis for such possibility, given the balanced schedule of 60- and 600-sec ITIs employed.

600-sec ITI and designated CS_L, whereas the other CS was consistently separated from the preceding reinforced trial by a 60-sec ITI and designated CS_S. The sequence of trials within extinction sessions was further arranged so that CS_L and CS_S equally often occurred as the last trial and otherwise were consistently followed by a reinforced vibrator trial at a 300-sec ITI. For eight subjects CS_L was the light and CS_S was the tone, whereas for another eight subjects this cue assignment was reversed.

Extinction, as described, was conducted over four blocks of four sessions. Figure 8.7 presents the resulting mean percentage suppression scores to CS_L and CS_S, plotted separately for the two cue-identification subgroups, during the nonreinforced test sessions that followed each block. No differences in suppression were observed in the first two sessions. However, it may be seen that by the last two sessions there were substantial differences, such that there was less suppression to either the tone or the light if it was treated as CS_S as compared to CS_L. Statistical analyses confirmed that there was reliably less overall suppression to the CS that was extinguished via nonreinforced trials that regularly occurred 60 sec, as compared to 600 sec, following the US of a preceding trial, $F(1, 14) = 5.75, p < .05$.

This outcome would appear to provide substantial support for the supposition that inhibitory backward conditioning may make an important contribution to the diminished conditioned responding that is commonly observed with short as compared to long ITIs in a training sequence. The effect cannot, in light of the within-subject comparison involving a common overall context, be attributed to differential associative strength of contextual cues in the presence of which the CS occurs. And, as has been indicated, it does not appear amenable to other candidates for interpretation, such as reduced excitatory conditioning due to differential priming. This is not to say that these other influences cannot contribute to ITI effects. They most likely do in cases in which they are not prohibited or equated as they were in the present study.

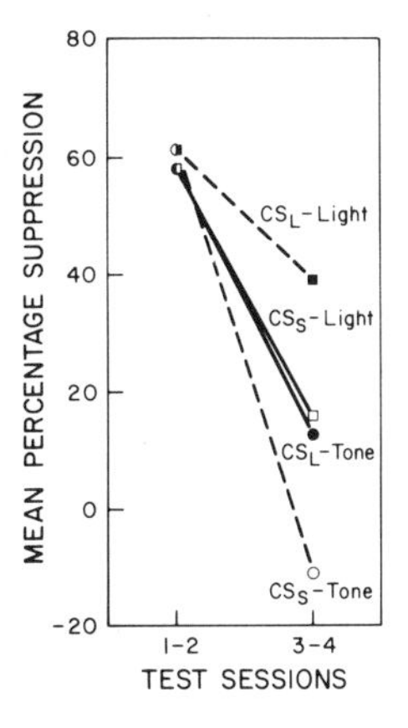

FIG. 8.7. Mean percentage suppression scores across blocks of two sessions interspersed during extinction involving the within-subject manipulation of the ITI preceding each of two CSs. Scores are plotted separately for cue-assignment subgroups for which either the tone was preceded by a long ITI (CS_L) and the light by a short ITI (CS_S), or in which this designation was reversed.

Before leaving these studies of ITI effects, it should be noted that the within-subject manipulations employed exemplify the type of comparison proposed in an earlier section to be necessary for evaluating the role of the US–CS interval in simple, multiple-trial backward conditioning, unconfounded by differences in contextual conditioning. At the moment we have better evidence of the reality of inhibitory backward conditioning due to the proximity of US and CS on different trials than due to the proximity of US and CS on the same trial.

The Effects of US–CS Intervals in a Random Schedule

There is a more general awareness to which we might be led through the ITI studies that have been described. When a CS is trained in an experimental context, the associative consequence may depend not just on the reinforcements that occur during the CS, but also on the density of reinforcement in periods that precede and follow the CS. Now, of course, this point was effectively made by Rescorla (e.g., 1969) via studies of CER that demonstrated that with a constant probability of reinforcement during (and shortly following) CS presentation the CS could become either excitatory, associatively neutral, or inhibitory, depending on whether the probability of reinforcement in the absence of the CS was, respectively, lower, the same, or higher. And there is a conventional interpretation (e.g., Rescorla & Wagner, 1972): USs that are programmed to occur in the absence of the CS act to condition the contextual cues so that the reinforcements and nonreinforcements of the CS in this context will be modulated to have effects that range from excitatory to inhibitory. But, as well known is the empirical contingency function (e.g., Rescorla, 1969), and as secure is the associated reasoning based on context conditioning (see, e.g., Baker, 1977; Durlach, 1983; Dweck & Wagner, 1970), we may have overlooked more local empirical effects and ignored the theoretical possibility that is salient in the present discussion. That is, it is possible that when a CS is made less excitatory or more inhibitory as a result of USs programmed in the "absence" of the CS, it is in part attributable to those USs that occur in the period prior to the CS so as to produce inhibitory backward conditioning, which may or may not be offset by excitatory conditioning produced by other USs that occur in the period following the CS. It is possible, for example, that a "random" schedule of CSs and USs that produces, asymptotically, no evidence of net excitatory or inhibitory association is one in which the occasions for backward, simultaneous, and forward pairings lead to substantial but equal inhibitory and excitatory conditioning.

A final study from our laboratory demonstrates the potential value of this view. It asked whether the conditioned responding that might be observed to a CS following a synthetic "random" schedule, in which the probability of the US was generally equivalent in the presence and absence of the CS, would be substantially influenced by prohibiting US occurrence in a period that immediately preceded each CS, as compared to prohibiting US occurrence in another period more distantly preceding each CS. Such an influence was observed, even though care was taken to ensure that the two treatments did not differ in the simultaneous or forward CS–US relationships.

The investigation used a CER preparation with general procedures similar to those of the preceding studies that have been reported, except that the conditioning was "off baseline," that is, with access to the bar prevented. The CS was tested in subsequent sessions when food-rewarded responding was again allowed but no USs were delivered.

The conditioning phase involved five daily sessions, each 130 min long and including six presentations of a 2-min tonal CS. The CSs were placed so that the first was always initiated 12 min after the beginning of the session, the last 30 min before the end of the session, and the intermediate ones 8, 12, 16, 20, and 24 min (in irregular orders) after the termination of the preceding CS. Twenty-six USs (.5-sec, 1-mA shocks) were scheduled to occur in each session, with placements such that the overall probability of US presentations per 1-min period was 0.2 in the presence and the absence of the CS.

Two groups of subjects (*ns* = 8) were treated differently only in the manner in which the daily scheduling of USs was constrained in the 4-min periods that preceded each of the six CSs (pre-CS periods) and in six other 4-min periods with which the pre-CS periods were matched (yoked periods). For Group Pre-.2, five USs were scheduled to occur in the pre-CS periods in each session and no USs were scheduled to occur in the yoked periods. For Group Pre-0, no USs were scheduled to occur during the pre-CS periods, but five USs were scheduled to occur during the yoked periods. The yoked periods were selected so that they matched the pre-CS periods in location *following* CS presentations: In each session one pre-CS period (the first) did not follow a CS, whereas the remaining pre-CS periods followed CSs by 4, 8, 12, 16, and 20 min; thus one yoked period was also selected to precede the first CS and the remainder to follow CSs by 4, 8, 12, 16, and 20 min. What this treatment thus provided was a comparison of two groups that experienced identical overall contingency between CS and US, identical possibilities for association between the CS and remote forward USs, but differences in the opportunities for backward US–CS pairings. For Group Pre-.2, USs occurred with the same relative frequency in the 4-min period preceding the CS as they did during the CS; for Group Pre-0, USs never occurred in the 4-min pre-CS period and, consequently, the minimum US–CS interval was 4 min.

Following the completion of training, subjects were allowed two sessions of bar pressing to recover baseline behavior and were then administered two 90-min test sessions on successive days. Each session involved three nonreinforced CS presentations at 18-min ITIs. (Scores for one subject were excluded due to equipment failure.) Figure 8.8 presents the mean percentage suppression scores for the two groups over the six test trials. As may be seen, the groups did not differ on the first test trial but Group Pre-0 consistently suppressed more than

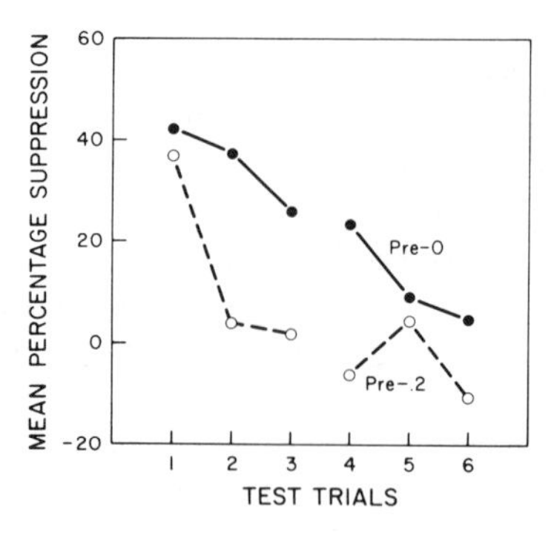

FIG. 8.8. Mean percentage suppression scores across six nonreinforced test trials following synthetic "random" schedules of CS and US presentations. During training, Group Pre-0 experienced no USs in the 4-min period preceding each CS and Group Pre-.2 experienced USs in the pre-CS period with a probability equivalent to the overall probability of 0.2.

Group Pre-.2 over the last five trials. Statistical analysis confirmed that the latter difference was reliable, $t(13) = 2.08$, $p < .05$, one-tailed test.

If one is willing to assume that the data from the first test trial may have been influenced by nonassociative factors, the overall data for Group Pre-.2 show what is commonly expected when the probability of the US is the same in the absence as in the presence of the CS, that is, no measurable conditioned responding. It should be instructive then that Group Pre-0 showed substantial and persistent conditioned suppression, even though it also experienced the same probability of the US in the presence and absence of the CS. Further research will obviously be required to determine whether or not this is attributable to the removal of opportunities for inhibitory backward conditioning, priming effects, or yet other possibilities. But there are local "backward" effects.

CONCLUDING COMMENTS

If the wick or snuff of inhibitory conditioning is a secondary process that lives within the overall response to the US, as suggested by Konorski (1948), Schull (1979), and Wagner (1981), then inhibitory backward conditioning should be a demonstrable effect that may have influence in many circumstances of training involving US–CS juxtapositions. There is an abundant literature, as we have noted, demonstrating inhibitory consequences of explicit backward conditioning training and suggesting that comparable consequences may be involved in other schedules that dictate the close following of US by CS. What has not been clear in this literature has been (1) whether or not the data patterns could be accounted for in terms of other mechanisms, such as those that would allow for the development of "safety signals" (e.g., Moscovitch & LoLordo, 1968; Wagner & Rescorla, 1972) or only diminished excitatory conditioning (e.g., Wagner, 1976); and (2) how, if at all, the supposition about inhibitory backward conditioning could be reconciled with the recent facts (e.g., Spetch et al., 1981) attesting to excitatory backward conditioning.

The issues are, in fact, sufficiently complicated and subtle that it is doubtful that they can be resolved outside of some relatively specific theoretical context and without relatively complex experimental designs. For example, our appraisal of the available data on one-trial backward conditioning is predicated on a detailed model (SOP), which suggests that excitatory conditioning is a likely outcome with short US–CS intervals and that inhibitory conditioning should more likely predominate with intermediate intervals. The model supposes that other parameters (such as US duration and predictedness) will likewise be important, rather than simply anticipating conditioned inhibition. Most of these suppositions remain to be fully evaluated, but they can be.

We have repeatedly called attention to the special research designs that are necessary to evaluate inhibitory backward conditioning per se, apart from other, potential influences, such as differential contextual conditioning. Perhaps the most that can be concluded at this time is that the initial data we have collected and report here are encouraging to the reality of inhibitory backward conditioning, as supposed by an opponent-process account of Pavlovian inhibition.

ACKNOWLEDGMENTS

The research reported and preparation of the chapter were supported in part by NSF Grant BNS 80–23399 to Allan R. Wagner. The authors are indebted to Michael Ewing (who collaborated on several of the studies reported) and John Dolan for stimulating discussions of the empirical and theoretical issues addressed.

REFERENCES

Baker, A. G. (1977). Conditioned inhibition arising from a between-sessions negative correlation. *Journal of Experimental Psychology: Animal Behavior Processes, 3,* 144–155.

Burkhardt, P. E., & Ayres, J. J. B. (1978). CS and US duration effects in one-trial simultaneous fear conditioning as assessed by conditioned suppression of licking in rats. *Animal Learning and Behavior, 6,* 225–230.

Champion, R. A., & Jones, J. E. (1961). Forward, backward, and pseudo-conditioning of the GSR. *Journal of Experimental Psychology, 62,* 58–61.

Denny, M. R. (1971). Relaxation theory and experiments. In F. R. Brush (Ed.), *Aversive conditioning and learning.* New York: Academic Press.

Donegan, N. H., & Wagner, A. R. (in press). Conditioned diminution and facilitation of the UCR: A sometimes-opponent-process interpretation. In I. Gormezano, W. F. Prokasy, & R. F. Thompson (Eds.), *Classical conditioning: Behavioral, neurophysiological, and neurochemical studies in the rabbit* (Vol. 3). Hillsdale, NJ: Lawrence Erlbaum Associates.

Donegan, N. H., Whitlow, J. W., & Wagner, A. R. (1977). Posttrial reinstatement of the CS in Pavlovian conditioning: Facilitation or impairment of acquisition as a function of individual differences in responsiveness to the CS. *Journal of Experimental Psychology: Animal Behavior Processes, 3,* 357–376.

Durlach, P. J. (1983). Effect of signaling intertrial unconditioned stimuli in autoshaping. *Journal of Experimental Psychology: Animal Behavior Processes, 9,* 374–389.

Dweck, C. S., & Wagner, A. R. (1970). Situational cues and correlation between CS and US as determinants of the conditioned emotional response. *Psychonomic Science, 18,* 145–147.

Gibbon, J., & Balsam, P. D. (1981). Spreading association in time. In C. M. Locurto, H. S. Terrace, & J. Gibbon (Eds.), *Autoshaping and conditioning theory.* New York: Academic Press.

Gormezano, I., & Moore, J. W. (1969). Classical conditioning. In M. H. Marx (Ed.), *Learning: Processes.* London: Macmillan.

Heth, C. D. (1976). Simultaneous and backward fear conditioning as a function of number of CS–UCS pairings. *Journal of Experimental Psychology: Animal Behavior Processes, 2,* 117–129.

Heth, C. D., & Rescorla, R. A. (1973). Simultaneous and backward fear conditioning in the rat. *Journal of Comparative and Physiological Psychology, 82,* 434–443.

Hinson, R. E. (1982). Effects of UCS preexposure on excitatory and inhibitory rabbit eyelid conditioning: An associative effect of conditioned contextual stimuli. *Journal of Experimental Psychology: Animal Behavior Processes, 8,* 49–61.

Hurvich, L. M., & Jameson, D. (1974). Opponent processes as a model of neural organization. *American Psychologist, 29,* 88–102.

Jones, J. E. (1962). Contiguity and reinforcement in relation to CS–UCS intervals in classical aversive conditioning. *Psychological Review, 69,* 176–186.

Kaplan, P. S. (in press). The importance of relative temporal parameters in trace autoshaping: From excitation to inhibition. *Journal of Experimental Psychology: Animal Behavior Processes.*

Keith–Lucas, T., & Guttman, N. (1975). Robust-single-trial delayed backward conditioning. *Journal of Comparative and Physiological Psychology, 88,* 468–476.

Konorski, J. (1948). *Conditioned reflexes and neuron organization.* Cambridge: Cambridge University Press.

Mackintosh, N. J. (1974). *The psychology of animal learning.* New York: Academic Press.

Mahoney, W. J., & Ayres, J. J. B. (1976). One-trial simultaneous and backward fear conditioning as reflected in conditioned suppression of licking in rats. *Animal Learning and Behavior, 4,* 357–362.

Maier, S. F., Rapaport, P., & Wheatley, K. L. (1976). Conditioned inhibition and the UCS–CS interval. *Animal Learning and Behavior, 4,* 217–220.

Mazur, J. E., & Wagner, A. R. (in press). An episodic model of associative learning. In M. L. Commons, R. J. Herrnstein, & A. R. Wagner (Eds.), *Quantitative analyses of behavior: Acquisition* (Vol. 3). Cambridge, MA: Ballinger.

Moscovitch, A., & LoLordo, V. M. (1968). Role of safety in the Pavlovian backward fear conditioning procedure. *Journal of Comparative and Physiological Psychology, 66,* 673–678.

Pavlov, I. P. (1927). *Conditioned reflexes* (G. V. Anrep, Trans.). London: Oxford University Press.

Plotkin, H. C., & Oakley, D. A. (1975). Backward conditioning in the rabbit (*Oryctolagus cuniculus*). *Journal of Comparative and Physiological Psychology, 88,* 586–590.

Rescorla, R. A. (1969). Conditioned inhibition of fear. In N. J. Mackintosh & W. K. Honig (Eds.), *Fundamental issues in associative learning.* Halifax: Dalhousie University Press.

Rescorla, R. A., & Wagner, A. R. (1972). A theory of Pavlovian conditioning: Variations in the effectiveness of reinforcement and nonreinforcement. In A. H. Black & W. F. Prokasy (Eds.), *Classical conditioning II.* New York: Appleton–Century–Crofts.

Schull, J. (1979). A conditioned opponent theory of Pavlovian conditioning and habituation. In G. H. Bower (Ed.), *The psychology of learning and motivation* (Vol. 13). New York: Academic Press.

Shurtleff, D., & Ayres, J. J. B. (1981). One-trial backward excitatory fear conditioning in rats: Acquisition, retention, extinction, and spontaneous recovery. *Animal Learning and Behavior, 9,* 65–74.

Siegel, S., & Domjan, M. (1971). Backward conditioning as an inhibitory procedure. *Learning and Motivation, 2,* 1–11.

Solomon, R. L. (1977). An opponent-process theory of acquired motivation: IV. The affective dynamics of addiction. In J. D. Maser & M. E. P. Seligman (Eds.), *Psychopathology: Experimental models*. San Francisco: Freeman.

Solomon, R. L., & Corbit, J. D. (1974). An opponent-process theory of motivation. *Psychological Review, 81,* 119–145.

Spetch, M. L., Wilkie, D. M., & Pinel, J. P. J. (1981). Backward conditioning: A reevaluation of the empirical evidence. *Psychological Bulletin, 89,* 163–175.

Spooner, A., & Kellogg, W. N. (1947). The backward-conditioning curve. *American Journal of Psychology, 60,* 321–334.

Terry, W. S. (1976). The effects of priming US representation in short-term memory on Pavlovian conditioning. *Journal of Experimental Psychology: Animal Behavior Processes, 2,* 354–370.

Wagner, A. R. (1976). Priming in STM: An information-processing mechanism for self-generated or retrieval-generated depression in performance. In T. J. Tighe & R. N. Leaton (Eds.), *Habituation: Perspectives from child development, animal behavior, and neurophysiology*. Hillsdale, NJ: Lawrence Erlbaum Associates.

Wagner, A. R. (1978). Expectancies and the priming of STM. In S. H. Hulse, H. Fowler, & W. K. Honig (Eds.), *Cognitive processes in animal behavior.* Hillsdale, NJ: Lawrence Erlbaum Associates.

Wagner, A. R. (1981). SOP: A model of automatic memory processing in animal behavior. In N. E. Spear & R. R. Miller (Eds.), *Information processing in animals: Memory mechanisms.* Hillsdale, NJ: Lawrence Erlbaum Associates.

Wagner, A. R., & Rescorla, R. A. (1972). Inhibition in Pavlovian conditioning: Application of a theory. In R. A. Boakes & M. S. Halliday (Eds.), *Inhibition and learning*. New York: Academic Press.

Wagner, A. R., Rudy, J. W., & Whitlow, J. W. (1973). Rehearsal in animal conditioning [Monograph]. *Journal of Experimental Psychology, 97,* 407–426.

Wagner, A. R., & Terry, W. S. (1975). Backward conditioning to a CS following an expected vs. a surprising UCS. *Animal Learning and Behavior, 3,* 370–374.

Wasserman, E. A., Franklin, S. R., & Hearst, E. (1974). Pavlovian appetitive contingencies and approach versus withdrawal to conditioned stimuli in pigeons. *Journal of Comparative and Physiological Psychology, 86,* 616–627.

9 The Nature of Conditioned Inhibition in Serial and Simultaneous Feature Negative Discriminations

Peter C. Holland
University of Pittsburgh

Conditioned inhibition is often described as the learning that two stimuli, usually an initially neutral conditioned inhibitory stimulus (CI) and an unconditioned stimulus (US), occur apart in time or space. Typically we assume this information to be represented as an inhibitory association between representations of the CI and the US. Thus, if excitatory conditioned stimuli (CSs) are thought to evoke conditioned behavior by activating a representation of the US (e.g., Konorski, 1948; Rescorla, 1974), CIs then modulate conditioned behavior by altering the activity of that US representation (Fig. 9.1, top panel.) In fact, the very techniques that we use to measure conditioned inhibition seem to demand acceptance of such assumptions (cf. Hearst, 1972). Consider for example the two most commonly used assessment procedures, the retardation and summation tests (Rescorla, 1969). Acquisition of excitation to a suspected CI is retarded presumably because the CI's previously established ability to depress the activity of the US representation must be overcome before net excitatory associations may be established between the CI and the US. Similarly, a summation test examines the ability of a putative CI to modulate behavior to an excitor established by pairings with the same US that was used to generate the CI.

However, there has been relatively little direct investigation of the nature or content of inhibitory learning. This lack of research is surprising given the recent interest in the content of excitatory learning and the extensive investigation of the impressive variety of procedures that generate inhibitory phenomema. This chapter considers several ideas about the nature of conditioned inhibition. The results of the experiments reported here suggest that the content of inhibitory learning may be substantially influenced by relatively minor procedural variations.

All these experiments used one of the more popular procedures for generating conditioned inhibition, the Pavlovian conditioned inhibition (Pavlov, 1927) or

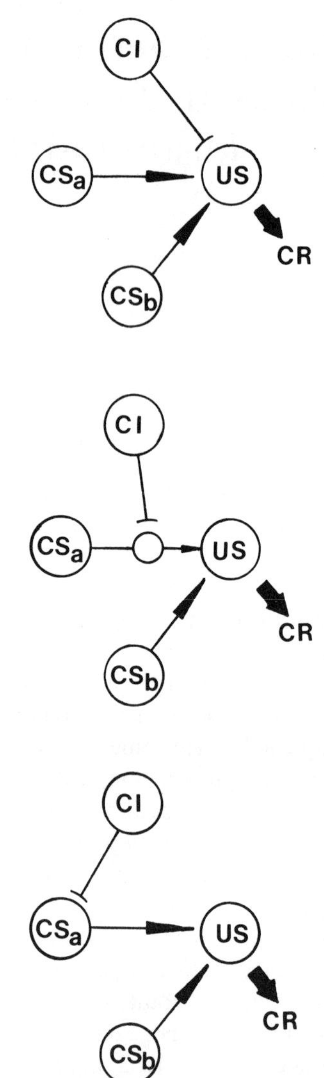

FIG. 9.1. Top panel: Conditioned inhibition acts by modulating the activity of the US representation. Center panel: Conditioned inhibition acts by modulating the action of the CS–US association. Bottom panel: Conditioned inhibition acts by modulating the activity of the CS representation. Arrows refer to excitatory associations, bars refer to inhibitory associations. CI = conditioned inhibitor, CS_a = conditioned stimulus originally compounded with the conditioned inhibitor on nonreinforced trials, CS_b = conditioned stimulus not previously paired with the conditioned inhibitor, US = unconditioned stimulus, CR = conditioned response.

feature negative (Jenkins & Sainsbury, 1969) procedure. In that (A+, AX−) procedure, one stimulus, A, is reinforced when presented alone, but not reinforced when presented in compound with another stimulus, X, the negative feature. Thus, X uniquely predicts nonreinforcement of the otherwise reinforced A stimulus. Using a conditioned suppression preparation with rats, we examined X's inhibitory power when feature negative training involved the simultaneous presentation of A and X on nonreinforced compound trials and when it involved nonreinforced serial compounds in which X preceded A. The data reported here suggest that although inhibitors established using the simultaneous procedure may modulate conditioned behavior by acting on a representation of the US, inhibitors established with the serial procedure may act on particular CS–US associations (Fig. 9.1, center panel.) Casually speaking, a simultaneously trained negative feature may signal a period of safety or nonoccurrence of the US, whereas a serially trained negative feature may indicate that the A–US relation will not hold.

Our study of this hierarchical signaling notion, in which a stimulus acts by signaling the validity of an upcoming relation between two events rather than by signaling the occurrence or nonoccurrence of a single event, stems from a series of experiments conducted in collaboration with Dr. Robert Ross (Holland, 1983; Ross, 1983; Ross & Holland, 1981, 1982; Ross, Orr, Holland, & Berger, 1984.) Those experiments examined rats' solution of feature *positive* (AX+, A−) discriminations in which one stimulus, A, was reinforced when presented in compound with a feature stimulus, X, but was not reinforced when presented alone. Briefly, our findings suggested that when A and X were presented simultaneously on reinforced compound trials, rats simply formed associations between the more predictive positive feature, X, and the US, and not between A and the US. But when X preceded A on reinforced compound trials, X additionally acquired the ability to signal when the A–US relation would be in effect. This ability of a serially arranged positive feature to "set the occasion" for a subsequent CS–US relation proved in many cases to be affected quite differently by a variety of experimental manipulations than was simple response evocation powers of such a stimulus, presumably due to its direct association with the US. Those findings, together with those reported here from parallel investigations using the feature negative procedure, encourage us to suggest that many Pavlovian conditioning procedures may establish quite sophisticated conditional control functions which are substantially different from simple response evocation powers, both in their manner of influencing learned behavior and in their representation in memory.

The first part of this chapter describes the results of a number of experiments that investigated the content of inhibitory learning established by serial and simultaneous feature negative procedures. The second section relates those data to analogous data obtained from the feature positive paradigm and the third section considers in a more general fashion the solution of feature positive and other conditional discriminations.

Comparison of the Nature of Inhibitory Learning in Serial and Simultaneous Feature Negative Discriminations

We used two general strategies for studying the content of inhibitory learning. The first involved various transfer tasks (e.g., Rescorla, 1975; Rescorla & Holland, 1977) that examined the ability of a CI established in feature negative procedures using a particular CS and US to inhibit conditioned responding to other CSs and to inhibit conditioned responding based on other USs. The second strategy involved posttraining counterconditioning of the CI, which tested the independence of a CI's inhibitory powers and its excitatory association with the US.

All the experiments reported here used similar subjects and general procedures, and the same apparatus. The subjects were albino rats of both sexes, 100–120 days old. They were maintained at 80% of their ad lib body weights by limiting their access to food. The experiments were conducted in eight identical, standard operant chambers enclosed in light- and sound-attenuating closets. Prior to the experimental training sessions, the rats were taught to press a lever on a VI 2-min schedule for a .3 ml delivery of 8% sucrose solution. All experimental sessions were conducted on this baseline. Unless otherwise noted, all CSs were 1 min in duration and all sessions were 88 min long. The measure of conditioning used was a standard suppression ratio (Annau & Kamin, 1961), computed by dividing the response rate during CS presentations by the sum of response rates during and for 2 min prior to CS presentations. With this measure, a score of 0 indicates complete suppression (conditioning), a score of .50 indicates no effect of stimulus presentation, and a score of more than .50 indicates a facilitary effect of CS presentation.

Transfer of Inhibition Across CSs: Experiment 1. If a conditioned inhibitor modulates behavior by somehow depressing the activity of a US representation, that inhibitor should be capable of inhibiting any conditioned responding mediated by that representation. Thus we would expect an inhibitory feature (CI in Fig. 9.1), established by nonreinforced presentations in compound with one excitor, CSa, to be capable of inhibiting responding to other excitors, never before presented in compound with the inhibitor (e.g., CSb in the top panel of Fig. 9.1.) On the other hand, if an inhibitory feature acts to depress the activity or operation of a particular excitor–reinforcer (CSa–US) association, that feature might not be expected to inhibit conditioned responding evoked by any stimulus other than CSa (e.g., CSb in the center panel of Fig. 9.1.)

In a series of experiments (Holland & Lamarre, in press), Jennifer Lamarre and I examined the ability of serially and simultaneously established negative features to inhibit conditioned responding evoked by an excitor that was not involved in the feature negative training. Experiment 1a considered the question straightforwardly (the top portion of Table 9.1 gives an outline of the procedures of Experiment 1a). Rats first received noise–shock and tone–shock pairings until both auditory stimuli controlled substantial suppression. Then the rats were divided into four groups (ns = 8) and discrimination training was begun. Group Ser received two reinforced tone presentations and six nonreinforced *Ser*ial light-

TABLE 9.1
Procedures of Experiments 1a and 1b

Experiment 1a

Group	Conditioning	Discrimination	Summation Tests 1	Summation Tests 2
Ser	N+, T+	T+, L → T− (48 sessions)	N−, L → N− T−, L → T−	N−, LN− T−, LT−
Sim–L	N+, T+	T+, LT− (48 sessions)	N−, LN− T−, LT−	N−, L → N− T−, L → T−
Sim–E	N+, T+	T+, LT− (9 sessions)	N−, LN− T−, LT−	N−, L → N− T−, LT−
Con	N+, T+	T+, L− (48 sessions)	N−, L → N− T−, L → T−	N−, LN− T−, LT−

Experiment 1b

Group	Conditioning	Discrimination	Extinction	Summation Test
Ser–E	N+, T+	N+, L → N− (24 sessions)	N−	T−, L → T−
Ser–N	N+, T+	N+, L → N− (24 sessions)	no trials	T−, L → T−
Sim–E	N+, T+	N+, LN− (12 sessions)	N−	T−, LT−
Sim–N	N+, T+	N+, LN− (12 sessions)	no trials	T−, LT−

Note: N = noise; T = tone; L = houselight; + = shock reinforcement; − = nonreinforcement.

then-tone trials in each of 48 sessions. Because previous data indicated that serial feature negative discrimination learning required considerably more conditioning trials, we were concerned about confounding either amount of training or performance level with the serial versus simultaneous manipulation. So in this experiment we ran two simultaneous discrimination groups. Both simultaneous groups reveived two reinforced tone and six nonreinforced tone + light compound presentations in each session. Group Sim–L (*Sim*ultaneous training, *L*ate testing) received the same number of training sessions as Group Ser. The rats in Group Sim–E (*Sim*ultaneous training, *E*arly testing) received nine sessions, that is, only enough training for them to reach the same level of discrimination performance as was reached by Group Ser in 48 sessions. Group Con (*Con*trol) served as a control for the occurrence of conditioned inhibition. The rats in that group received simple discrimination training between the reinforced tone and nonreinforced light-alone presentations, a procedure that in our laboratory generates little conditioned inhibition (but see LoLordo, this volume). Group Con received the same number of training sessions as Groups Ser and Sim–L. Finally, all rats received two summation test sessions which assessed responding to the "original" tone excitor used to establish the discrimination, the "test" noise excitor not involved in discrimination training, and to light + noise and light + tone compounds. The stimuli within both compounds were delivered in the same temporal arrangement as was used during acquisition, i.e., serially in Group Ser and simultaneously in Groups Sim–E and Sim–L. Half the rats in Group Con received serial compounds in this test and half received simultaneous compounds.

The various discriminations were acquired abnormally slowly in this experiment relative to our previous and subsequent experiments, so they are not shown here. Nevertheless, a discrimination criterion of .15 or greater suppression ratio difference between S+ and S− responding over 3 consecutive sessions was reached in 42 sessions in Group Ser and in 9 sessions in Group Sim–E.

Figure 9.2 shows the results of the summation test. Clearly, the light feature readily inhibited responding to both the tone (original) and noise (test) excitors in the two simultaneous groups, although transfer was less than complete. This observation is of course not in the least novel (e.g., Marchant, Mis, & Moore, 1972; Rescorla & Holland, 1977). However, the light feature in Group Ser showed no evidence of inhibiting responding to the test excitor, even though its ability to inhibit responding to the *original* excitor was comparable to that of the feature in Group Sim–E. Further, the serially trained light feature in Group Ser

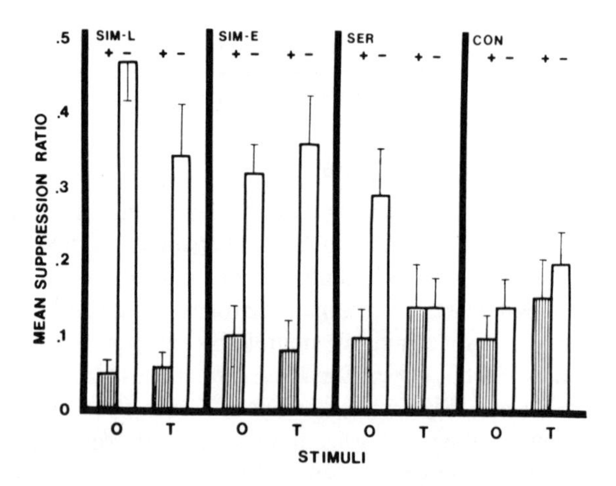

FIG. 9.2. Summation test responding in Experiment 1a. Serial and simultaneous compound responding of the rats in Group Con were identical and so are combined. O and T on the abscissa refer to the original (tone) and test (noise) excitors, respectively. + and − refer to presentations of the excitor alone and within the feature–excitor compound, respectively. Brackets represent standard errors of the means.

was no more able to inhibit responding to the test excitor than the light in Group Con, which had been presented alone on nonreinforced training trials. Thus, the results of the transfer test are consistent with the notion that simultaneous feature negative procedures establish inhibitors that act on a representation of the US, but serially established features act on specific CS–US pairs.

Although we presumed the differential outcomes of the transfer tests to be the consequence of different types of learning in the two training procedures, it should be noted that in Experiment 1a the temporal arrangement of stimuli in discrimination training was confounded with that in summation testing; that is, subjects trained with serial compounds were tested with serial compounds, and subjects trained with simultaneous compounds were tested with simultaneous compounds. Thus the possibility remained that the two test procedures were differentially likely to show transfer performance, regardless of the learning that occurred during training.

To examine the potential contribution of performance effects to Experiment 1a's data, we administered to the same rats another set of summation tests identical to the last ones except that Group Ser received tests with simultaneous compounds and Groups Sim–E and Sim–L received tests with serial compounds. Group Ser showed as much or more inhibition of responding to the original excitor when tested simultaneously (an S+ vs S− difference in suppression ratio of .21) as when tested serially, but no evidence of transfer to the test excitor (a difference score of −.01). On the other hand, when tested serially, the simultaneously trained features had no inhibitory power on either their original excitors (difference score of 0) or within novel compounds (difference score of .05). We have obtained this lack of transfer from simultaneous training to serial testing on other occasions; we suspect it to be the result of a postinhibitory rebound effect akin to Pavlov's (1927) positive induction. Nevertheless, the observation that serially trained features did not inhibit responding to a new excitor even in simultaneous testing suggests that the differential transfer of inhibitory powers across CSs found here was apparently the result of a difference in learning rather than in performance.

Although we favored the possibility that the nature of inhibitory learning differed as a function of the temporal arrangement of stimuli, a somewhat different account for the data of Experiment 1a is also feasible. This alternate account explains the lack of transfer with serial procedures while assuming identical inhibitory processes in both serial and simultaneous procedures. Rescorla (1982b) noted that transfer of inhibition across excitors may often be masked by responding due to within-compound associations. The pairing of the light feature and the noise excitor on compound trials may endow the light with the ability to arouse a representation of the noise excitor. Consequently, in the transfer test the excitation to the tone might sum with that evoked by the surrogate noise to produce a level of excitation exceeding the level of inhibition controlled by the light. Presumably, such excitation borrowed from the noise would not sum with that of the noise itself, so the inhibitory effect of the light would appear considerably larger when tested in compound with the original noise excitor than when tested in compound with the tone, which had never been presented in compound with the light. If we assume that light–noise associations are stronger after training with serial light–noise compounds than after simultaneous ones, we would expect less transfer after serial than after simultaneous discrimination training, as observed in the preceding experiments. That assumption seems warranted for several reasons. First, in Experiment 1a we observed substantial second-order conditioning to the serially presented light feature (not shown here) in serial groups but not in the simultaneous groups. Second, Rescorla (1973) found considerably greater second-order conditioning using serial pairings than using simultaneous pairings. Third, Rescorla (1982a) found that although simultaneous procedures may generate second-order conditioned suppression comparable to that observed with serial procedures involving only two presentations of the first-order excitor alone, simultaneous second-order conditioning was minimal when the first-order excitor was frequently presented alone (as in the experiments reported here).

Experiment 1b evaluated this within-compound association account for the lack of transfer in the serial groups. An outline of the procedures of Experiment 1b is found in the bottom portion of Table 9.1. All rats first received tone–shock and noise–shock pairings. Two groups of rats received simultaneous and two groups received serial feature negative training, with noise as the excitor and a light as the inhibitory feature. Then one *Ser*ial (Ser–E) and one *Sim*ultaneous (Sim–E) group received repeated nonreinforced (*E*xtinction) trials with the original noise excitor, whereas the other two groups (Ser–N and Sim–N) received *N*o explicit events but simply continued to lever press on the VI schedule. Finally, responding to the tone and tone + light compounds was assessed in a transfer test. If the lack of transfer of the serially trained light's inhibitory power was due to its borrowing excitatory power from the noise in the transfer test, extinction of the noise prior to that test should enhance transfer (Cunningham, 1981; Rescorla, 1981, 1982b).

Feature negative discrimination learning occurred rapidly in all groups (Fig. 9.3.) This rapid learning is more typical of the performance we have observed in our research; we have no explanation for the much slower acquisition observed in Experiment 1a. Figure 9.4 shows responding in the transfer test of Experiment 1b. There was substantial transfer in the two simultaneously trained groups, Sim–E and Sim–N, reliably more than in Group Ser–N, which received serial training. Unlike in the previous experiments there was numerical, but not statistically significant, evidence of transfer in Group Ser–N. However, there was no evidence of transfer in Group Ser–E, which received intervening extinction of the original noise excitor. Thus because noise extinction *reduced* rather than facilitated transfer, it seems unlikely that a within-compound mechansim such as

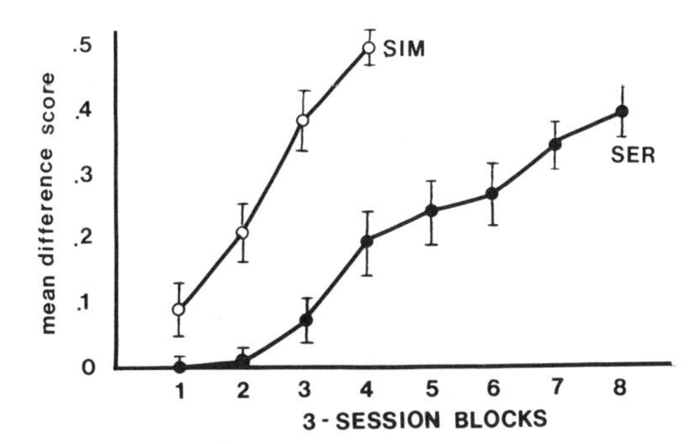

FIG. 9.3. Acquisition of the feature negative discriminations in Experiment 1b. Brackets represent standard errors of the means.

that described previously is an important contributor to the lack of transfer found after serial training procedures.

The elimination of transfer by noise extinction in Group Ser–E suggests that the small transfer observed in Group Ser–N was the result of generalization between the noise and tone; that is, in the transfer test the serially trained light would be capable of inhibiting responding evoked by stimulus elements shared by tone and noise. But extinction of the noise would abolish those shared elements' ability to evoke CRs and hence abolish the light's ability to inhibit CRs evoked by the tone in testing (see Rescorla & Holland, 1977).

It could be argued that the within-compound associations formed between the light and noise were "S–R" in nature, encoding the response evoked by the noise, but not its stimulus properties. In that case, extinction of the noise would not be expected to reduce excitation to the light nor to enhance transfer. However, if within-compound excitation to the light was S–R in nature and not mediated by stimulus aspects of the subsequent excitor, that excitation should be manifest regardless of the identity of the excitor with which the light was compounded. Thus, within-compound S–R associations would not be expected to affect the magnitude of inhibition in a transfer test any differently than it affected the magnitude of inhibition in original training. Further, another manipulation

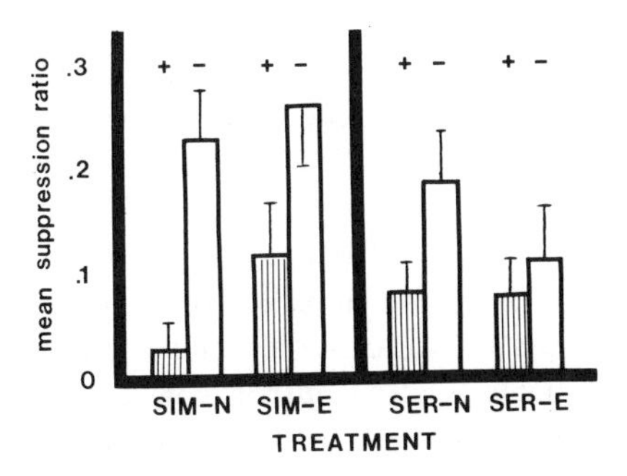

FIG. 9.4. Summation test responding in Experiment 1b. + and − refer to presentations of the excitor alone and within the feature–excitor compound, respectively. Brackets show the standard errors of the means.

that would be expected to reduce feature-excitor associations of either S–R or S–S nature, nonreinforced feature-alone presentations, was also found not to enhance transfer (see Experiment 4, later.) Thus, differences in transfer seem more appropriately attributed to differences in inhibitory learning than to variations in within-compound associations.

Our final experiment in this series investigated the generality of our observations of no across-CS transfer of inhibition after serial feature negative training. Both Experiments 1a and 1b used the same three stimuli in the same roles. Rescorla (1980) noted that the choice of stimuli used in second-order conditioning experiments can substantially affect both the rate of acquisition and the content of second-order conditioning. For example, the use of similar stimuli both speeds the acquisition of second-order conditioning and encourages the formation of S–S rather than S–R associations. In Experiment 1c we gave eight groups of rats serial feature negative discrimination training using various combinations of steady or intermittent houselights and steady or intermittent noises or tones as the elements of the nonreinforce compounds. Subjects learned the serial discriminations with each stimulus combination, but there were substantial differences in the levels of discrimination obtained after the 24 sessions given in this experiment, depending on the stimulus combination used. Despite the variety of stimulus conditions and the levels of responding to the training S− (ranging from .20 to .55), no transfer of inhibitory effects to a new excitor was observed in any instance. Thus, stimulus variables that had substantial effects on the rate of serial feature negative learning had no observable effect on the content of that learning (cf. Rescorla, 1980).

In summary, although simultaneous feature negative discrimination procedures produced inhibitors that inhibited responding to CSs not involved in the feature negative training, serial feature negative procedures generated inhibitors whose action was specific to the CS with which they were trained. This differential specificity of serially and simultaneously established inhibition was not an artifact of amount of training (Experiment 1a), level of discrimination performance (Experiments 1a and 1c), within-compound associations (Experiment 1b), or the choice of particular stimuli as excitors and inhibitors (Experiment 1c). Instead, we believe that, whereas simultaneous procedures establish inhibitory associations between the negative feature and the US, serial procedures establish an occasion setting or gating function which permits the negative feature to signal when a particular CS–US relation will not obtain.

Transfer of Inhibition Across USs: Experiment 2. The lack of transfer of inhibition across CSs after serial feature negative training observed in the previous experiments is also consistent with another view of inhibition, that the feature acquires inhibitory associations with the excitor itself. This notion is more parsimonious than the occasion setting just described, because no new higher order or hierarchical function is demanded of the negative feature. Within this viewpoint (expressed by Pavlov, 1927), in a summation test the feature acts directly on the original excitor (or more properly, on its internal representation) to depress its responding (bottom panel of Fig. 9.1). Because there is no opportunity for the negative feature to develop inhibitory associations with a "new" excitor, no transfer would be anticipated. However, within this view, transfer across USs would be expected, because activity of the CS representation would be depressed regardless of what responding that CS might engender.

In Experiment 2 a noise was first paired with shock and a serial light-then-noise compound was nonreinforced (Table 9.2 provides an outline of the procedures of Experiment 2). After good discrimination was established, the noise was paired with food delivery until it no longer evoked shock-related responses but instead evoked food-related behavior. Finally, the ability of the light feature to inhibit food-related responding to the noise was examined. If the light inhibited responding to the noise by acting directly on the noise representation, it should inhibit any responding established to that noise.

This experiment used rats that had previously been subjects in Group Ser/Ser of Experiment 3b (later). After a brief retraining period designed to reestablish two serial discriminations (noise vs. lightA-then-noise and tone vs. lightB-then-tone), the subjects were placed in a different apparatus used to observe the behavior of subjects via television equipment. This apparatus was similar to their original conditioning chambers except there were no response levers. All subjects were first given a summation pretest consisting of a single 1-min noise and a single lightA-then-noise compound so we could be sure that the conditioning acquired in one context transferred to another context and could be revealed in the animals' behavior directly, without use of a lever-pressing baseline. Next, all subjects received 8 presentations of the 1-min noise paired with the delivery of two 45-mg food pellets in each of 8 sessions. The ability of lightA to inhibit food-based responding to the noise, as well as lightB's continued ability to inhibit shock-based responding to the tone, was then examined in a single summation test. That test included two presentations each of the noise, the tone, the lightA-then-noise compound, and the lightB-then-tone compound, all nonrein-

TABLE 9.2
Procedure of Experiment 2

Discrimination	*Pretest*	*Conditioning*	*Test*	*Retraining*	*Retest*
N → shock, A → N− T → shock, B → T−	N−, A → N−	N → food	N−, A → N− T−, B → T−	N → shock	N−, A → N− T−, B → T−

Note: N = noise; T = tone; A = light A; B = light B; − = no reinforcement. The discrimination phase was conducted on a lever-pressing baseline in one set of experimental chambers. All other phases were conducted in different chambers without response levers.

forced. If inhibitory associations were established between the negative features and their excitatory partner CSs, then lightA should inhibit food-based responding to the noise. Finally, all rats received two sessions with two noise–shock pairings and were given a final summation retest session identical to the test session, in order to determine if lightA retained its previously acquired ability to inhibit shock-based responding to the noise.

The measures of conditioning used in this experiment were freezing, defined as a motionless crouching posture (see Holland, 1979, for a further description), as the measure of shock-based conditioning, and magazine behavior, defined as insertion of the head into the food magazine, as the measure of food-based conditioning. Rats' behavior was monitored at 1.25-sec intervals throughout the 60-sec CSs. Figure 9.5 shows the outcomes of the summation pretest, test, and retest sessions. In the pretest session (left panel) the noise CS alone evoked moderate amounts of freezing, but the noise presented within a compound evoked very little freezing. Freezing during the noise was gradually supplanted

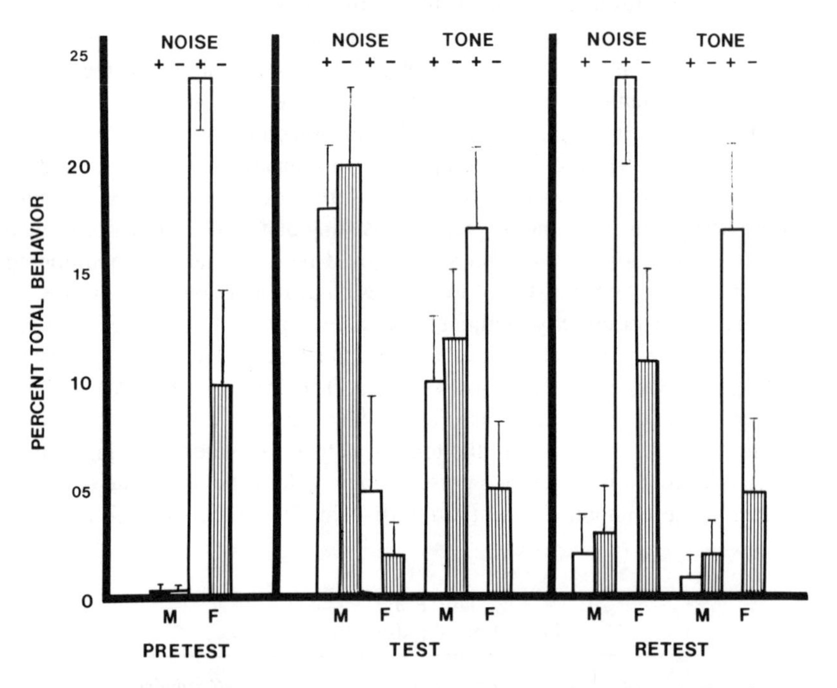

FIG. 9.5. Results of the summation pretest, test, and retest sessions in Experiment 2. The ordinate measure was calculated by dividing the number of observations of a particular behavior by the total number of observations made. M and F on the abscissa refer to magazine and freezing behaviors, respectively. + and − refer to responding to the excitor alone and to the excitor within feature–excitor compounds, respectively. Brackets show the standard errors of the means.

by magazine behavior over the course of noise–food pairings. The summation test (center panel) showed that magazine behavior controlled by the noise was not inhibited by presentation of lightA. Thus lightA's power to inhibit freezing to the noise did not transfer to the food-based magazine behavior. Also of interest in the first summation test was the subjects' response to the tone and lightB-then-tone compounds which had not previously been presented in this experimental context. The tone alone evoked low levels of both freezing and magazine responding, indicating generalization of both extinction and conditioning acquired by the noise. More interestingly, lightB inhibited freezing evoked by the tone, but not magazine behavior. Finally, the retest after noise–shock retraining showed that both lightA and lightB maintained their abilities to inhibit shock-based responding to their respective excitors.

Thus, in Experiment 2 there was no evidence of the features' inhibition acting directly on its excitor. Given that serially established inhibition was both CS- and US- specific, it seems reasonable to claim that it is specific to particular CS–US pairs, i.e., that it operates by modulating the activity of another relation, rather than of a representation of a single event, as may be the case in simultaneously established inhibition.

Counterconditioning of Inhibitory Features: Experiment 3. Experiments 3a and 3b (Holland, in press) examined the effects of pairing simultaneously and serially established features with the US after feature negative training. Presumably, if a feature X modulates behavior by acquiring inhibitory associations with the US or otherwise depressing the activity of the US representation, subsequent X–US pairings should result in relatively slow acquisition of excitatory associations between X and the US. Indeed, this procedure is the frequently used retardation test of conditioned inhibition (e.g., Rescorla, 1969). Further, once X's ability to depress activity of the US representation is counteracted by the acquisition of those excitatory associations, X should no longer be capable of inhibiting responding to A; that is, X should not be able to simultaneously predict both the occurrence and nonoccurrence of the US (Fig. 9.6, top panel). Conversely, if X modulates behavior by affecting a specific A–US association, as we proposed was the case in serial feature negative discrimination learning, somewhat different outcomes might be anticipated. It is conceivable that a serially established X's ability to provide information about the A–US relation may be independent of its ability to signal the US. Thus, serial feature negative training may not endow X with the ability to predict the nonoccurrence of the US outside the context of A presentation. If so, then it is possible that X's acquisition of excitation when paired with the US would not be retarded; that is, that a retardation test might give no evidence of inhibition after serial feature negative training. Similarly, the establishment of X–US associations might have little

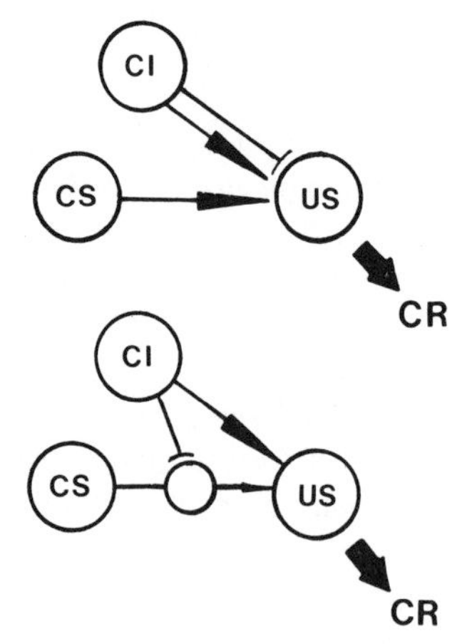

FIG. 9.6. Upper panel: Associative structure after a simultaneously trained conditioned inhibitor is paired with the unconditioned stimulus. Lower panel: Associative structure after a serially trained conditioned inhibitor is paired with the unconditioned stimulus. Arrows symbolize excitatory associations, bars symbolize inhibitory associations. CI = conditioned inhibitor, CS = conditioned stimulus, US = unconditioned stimulus, CR = conditioned response.

effect on X's ability to signal the invalidity of the A–US relation: X might well inform the subject that the X will be followed by shock, but that should an A occur, that A will not be followed by shock (Fig. 9.6, lower panel).

In Experiment 3a rats received either simultaneous or serial feature negative discrimination training until comparable discrimination criteria were reached in both conditions (.35 difference in suppression ratio between S+ and S− over three consecutive sessions). This training took six sessions for simultaneously trained subjects and 24 sessions for serially trained subjects. Then the light feature (the conditioned inhibitor) was paired with shock until responding to it was completely suppressed in both conditions. Finally, the light's ability to inhibit responding to the original excitor was assessed in a summation test akin to the original training conditions. The effects of these manipulations were assessed against the performance of control subjects that received equivalent experience with a simple discrimination procedure involving reinforced noise and nonreinforced light trials, known to generate little inhibition to the light in our laboratory. The top portion of Table 9.3 provides a summary of the procedures of Experiment 3a.

Figure 9.7 shows the results of pairing the light with shock in all four groups, a retardation test of inhibition. Not surprisingly, acquisition of excitatory conditioning to the light was significantly slowed in Group Sim, in which the light was trained as an inhibitor using *Sim*ultaneous procedures, relative to the *c*ontrol group (Sim–C) that received the same number of light pairings in a simple discriminative procedure. However, acquisition of excitatory conditioning to the light was not slowed in Group Ser, which had received *Ser*ial feature negative training, relative to either Group Ser–C or Group Sim. Thus, even though by the end of discrimination training the light had equal inhibitory powers in Groups Sim and Ser as measured by the summation testing inherent in the training procedure, a retardation test gave evidence of inhibition only in Group Sim.

TABLE 9.3
Procedures in Experiments 3a and 3b

Experiment 3a				
Group	*Phase 1*	*Phase 2*	*Retardation Test*	*Summation Test*
Sim	N+	N+, LN− (6 sessions)	L+	N−, LN−
Sim/C	N+	N+, L− (6 sessions)	L+	N−, LN−
Ser	N+	N+, L → N− (24 sessions)	L+	N−, L → N−
Ser/C	N+	N+, L− (24 sessions)	L+	N−, L → N−

Experiement 3b				
Group	*Phase 1*	*Phase 2*	*Phase 3*	*Summation Test*
Sim/Sim	N+, T+	N+, HN−, T+, PT−	H+ or P+	N−, HN−, T−, PT−
Ser/Ser	N+, T+	N+, H → N−, T+, P → T−	H+ or P+	N−, H → N−, T−, P → T−
Sim/Ser	N+, T+	N+, HN−, T+, PT−	H+ or P+	N−, H → N−, T−, P → T−
Ser/Sim	N+, T+	N+, H → N−, T+, P → T−	H+ or P+	N−, HN−, T−, PT−

Note: N = noise; T = tone; H = houselight; P = panel light; + = shock reinforcement; − = no reinforcement.

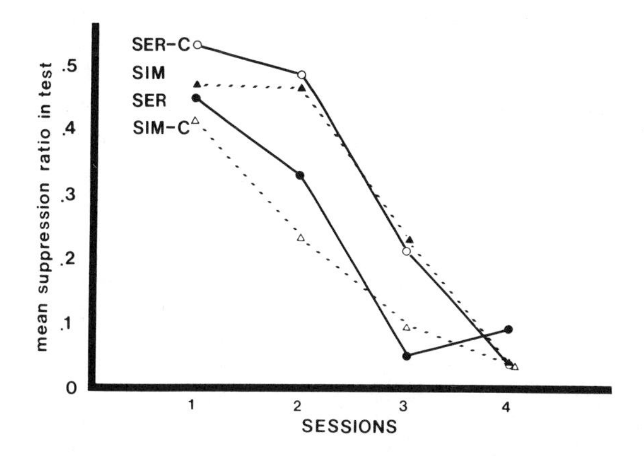

FIG. 9.7. Retardation test responding in Experiment 3a. (From Holland, in press. © 1984 by the American Psychological Association. Reprinted by permission.)

The other differences in Fig. 9.7 deserve comment. The reliably slower acquisition of excitation in Group Ser–C than in Group Sim–C is likely the consequence of the former group's receiving four times as much training, which could have enhanced either conditioned or latent inhibition. The significantly slower acquisition of excitation in Group Ser–C than in Group Ser may have been the result of residual excitation due to within-compound associations in Group Ser or of the disruption of latent inhibition to the light in Group Ser by the postlight presentations of the noise (see Lubow, Schnur, & Rifkin, 1976). None of those differences, however, weaken the basic observation of differential susceptibility to retardation test procedures of serially and simultaneously trained inhibitors.

Figure 9.8 shows responding in the summation test administered after the light came to control complete suppression in all groups. Clearly, light–shock pairings destroyed the light's ability to inhibit responding to the noise in Group Sim but left it fairly strong in Group Ser. Thus, the results of both the retardation test and the postretardation summation test indicate that a serially trained feature's inhibitory powers are relatively independent of excitatory associations established between that feature and the US, whereas the inhibitory and excitatory powers of a simultaneously trained feature are mutually incompatible.

Experiment 3b replicated the basic effects of Experiment 3a in a somewhat different design (bottom portion of Table 9.3.) All subjects received feature negative discrimination training involving two common element excitors and two feature inhibitors, i.e., A+, AX−, B+, BY−. Two groups received two serial discriminations and two groups received two simultaneous discriminations. Then one feature/inhibitor (X) was paired with shock and the subjects given a summation test that examined their responding to both excitors, both inhibitors, and both compounds. Experiment 3b added several features in addition to simply replicating the results of Experiment 3a. First, because the effects of feature–shock pairings on summation test behavior were examined within subjects, Experiment 3b provided an indication of the specificity of those effects. Second, it provided a contemporaneous comparison of the effects of feature–shock pairings on summation test behavior in each group rather than the before–after comparison provided by Experiment 3a. Third, some serially trained subjects were

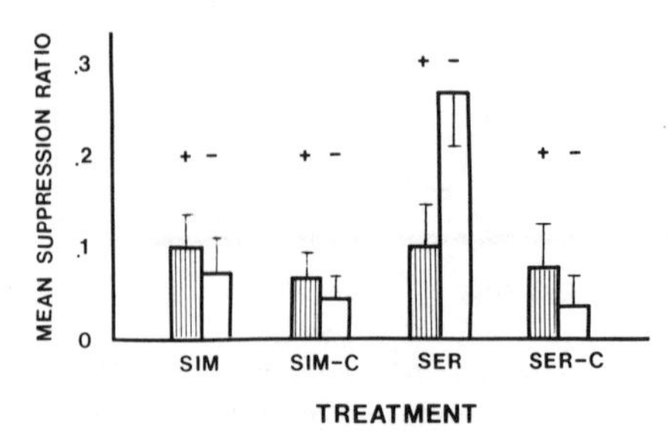

FIG. 9.8. Summation test responding in Experiment 3a. + and − refer to responding to the excitor presented alone and within the feature–excitor compound, respectively. Brackets represent standard errors of the means. (From Holland, in press. © 1984 by the American Psychological Association. Reprinted by permission.)

tested with simultaneous compounds and some simultaneously trained subjects were tested with serial compounds. Thus, Experiment 3b examined the possibility that the differential effects of inhibitor–shock pairings observed in Experiment 3a were artifacts of different test procedures rather than learning effects. Finally, it equated the amount of feature negative training across simultaneous and serial compound conditions. In Experiment 3a the serial and simultaneous groups were trained to equal performance criteria; it is possible that the critical variable was amount of training rather than the temporal arrangement of the elements within the compound; that is, extended simultaneous feature negative training might generate an inhibitory feature that shows the same independence of inhibitory and excitatory effects exhibited by the serially trained feature in Experiment 3a.

All subjects first received intermittent (2.5 Hz) tone–shock and steady noise–shock pairings for two sessions. Then Groups Sim/Sim and Sim/Ser received 42 sessions of discrimination training, each comprising one reinforced tone and one reinforced noise trial, three nonreinforced simultaneous noise + houselight compounds, and three nonreinforced simultaneous tone + synchronously intermittent (2.5 Hz) panel light trials. Groups Ser/Ser and Ser/Sim received identical discrimination training except that the two nonreinforced compounds were arranged serially, that is, houselight-then-noise and panel light-then-tone. Next, all subjects received two presentations of one of the visual feature stimuli (X) paired with shock in each of five sessions. The shocked feature was the houselight for half of the rats in each group and was the panel light for the other half. Finally, all rats received two test sessions, each of which included two nonreinforced presentations each of the two excitatory CSs, noise and tone, and the two compound stimuli, houselight + noise and panel light + tone. The compounds were presented serially in Group Ser/Ser and Group Sim/Ser and simultaneously in Group Sim/Sim and Ser/Sim.

As in previous experiments, the simultaneous discriminations were learned more rapidly than the serial discriminations. By the end of training, mean S− responding was .40 in the serially trained groups and .59 in the simultaneously trained groups. Although the discriminations involving the tone were learned less rapidly than those involving the noise, performance was equivalent by the end of training.

As in Experiment 3a, feature–shock conditioning was more rapid in the serially trained groups than in the simultaneously trained groups. In the former groups, suppression to the light feature changed from .33 on day 1 to .04 on day 4; in the latter, from .47 to .29. Of course, given the absence of explicit controls and the different initial levels of the simultaneous and serial groups, it is difficult to make inferences about the degrees of retardation observed.

The primary data of Experiment 3b were those of the summation tests. Figure 9.9 shows the results of the summation test. Serially trained subjects showed no differential effect of feature–shock trials on that feature's ability to inhibit sup-

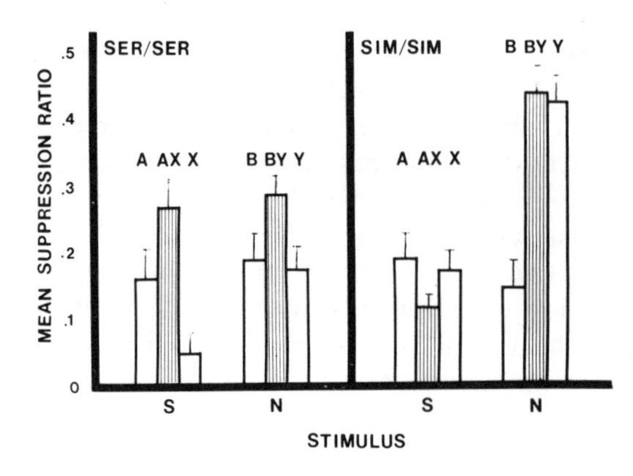

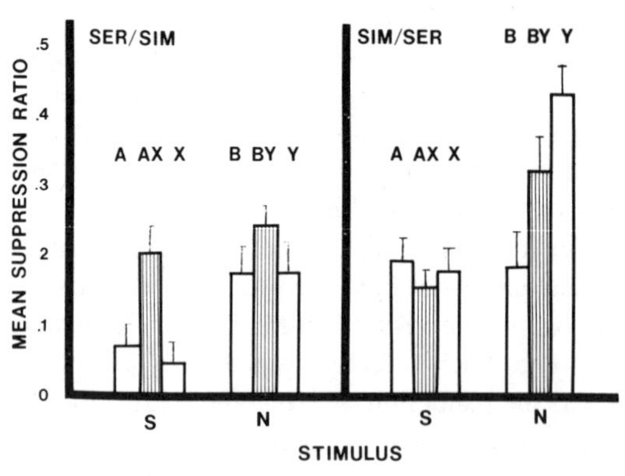

FIG. 9.9. Responding during the summation test in Experiment 3b. S on the abscissa refers to tests involving the feature that had been paired with Shock prior to testing, N refers to tests involving the feature Not paired with shock. X refers to responding to the shocked feature alone, A refers to responding to the excitor that had been originally paired with X, and AX refers to responding to the compound itself. Y, B, and BY refer to responding to the unshocked feature, its excitatory partner, and their compound presentation, respectively. Brackets show the standard errors of the means. (From Holland, in press. © 1984 by the American Psychological Association. Reprinted by permission.)

pression to a subsequent excitor: Responding to the excitor (A) presented after the feature (X) that had been paired with shock was equivalent to that to the excitor (B) presented after the feature (Y) that had not been paired with shock. Nevertheless, suppression to X was substantially greater than that to Y. Thus, as in Experiment 3a, the inhibitory and excitatory powers of the feature were relatively independent in serially trained and tested subjects. Further, those effects held whether subjects were tested serially (Group Ser/Ser, upper left) or simultaneously (Group Ser/Sim, lower left).

However, feature–shock pairings destroyed the feature's inhibitory powers in simultaneously trained subjects. Suppression to the AX compound was greater than that to A alone; further, suppression to the AX compound was considerably greater than that to the BY compound. Again, this pattern of data was evident in both simultaneous testing (Group Sim/Sim, upper right) and serial testing (Group Sim/Ser, lower right), although the inhibitory effects of Y were somewhat larger when tested simultaneously. Thus, also as in Experiment 3a, the inhibitory and excitatory powers of the feature were mutually incompatible in simultaneously trained subjects.

Summary of Feature Negative Data. All the data presented so far suggest that the content of inhibitory learning established in serial and simultaneous feature negative procedures is different. The successful transfer of simultaneously trained inhibition across CSs (Experiment 1), the retardation of excitatory conditioning (Experiment 3), and the destruction of a simultaneously trained feature's inhibitory powers after excitatory training (Experiment 3) are all consistent with the view that the locus of action of inhibition established using simultaneous procedures is the US representation, that is, simple inhibitory associations are formed between representations of the CI and US.

Conversely, the locus of action of inhibition established using the serial procedure is apparently the association between the common element excitor and the US. This view is suggested by the observation that serially established inhibition did not transfer across either CSs (Experiment 1) or USs (Experiment 2), and Experiment 3's findings that serially established inhibitors were not retarded in their acquisition of excitation, nor were their inhibitory powers appreciably modified by excitatory training. Experiment 3's findings of independence of excitatory and inhibitory powers of a serial feature are perhaps the most convincing evidence favoring this view, because they indicate clearly that excitation and serially established inhibition operate at different loci. Although the feature could not possess both net excitatory and net inhibitory associations with the US, it could be associated with the US in an excitatory manner but inhibit the action of the common element-US association (Fig. 9.6, bottom).

Are Serially Established Negative Features True Conditioned Inhibitors? One may question whether the inhibition observed in the serial feature negative discriminations described here is "true" inhibition. After all, not only did serial features fail to inhibit responding to other excitatory CSs in transfer summation tests (Experiments 1a–c, above), but also they failed to show slowed acquisition of excitation in retardation tests (Experiments 3a–b, above). Given that conditioned inhibition has been operationally defined almost exclusively in terms of the outcomes of summation and retardation tests (Rescorla, 1969), and theoretically as the symmetrical opposite of excitation, the answer would have to be no. However, it is likely that the more hierarchical type of inhibition described here and elsewhere (Jenkins, this volume; Rescorla, this volume) is at least as pervasive and as basic to the conditioning process as the "true" inhibition defined earlier. Moreover, it too may possess a symmetrical opposite in *positive occasion setting* (below). Thus, although the serially established inhibitors described here fail to meet conventional criteria of conditioned inhibition, they clearly reflect an important inhibitory process. Our concepts of conditioned inhibition must be broadened to encompass such findings, just as our views of conditioned excitation must be broadened to encompass recent findings that suggest occasion setting (following) or facilitation (Rescorla, this volume).

Relation to Occasion Setting in Feature Positive Discriminations

Occasion Setting in Feature-Positive Discriminations. Robert Ross and I recently concluded that the content of learning in serial and simultaneous feature *positive* discriminations also differs (Holland, 1983; Ross, 1983; Ross & Holland, 1981, 1982). In a feature positive (AX+, A−) discrimination the feature signals when the common element-US relation is in effect. Ross's and my studies used an appetitive conditioning preparation in which 5-sec CSs were followed by the delivery of two 45-mg food pellets. Behavior was scored using the observation technique described earlier in Experiment 2. This preparation has the unusual and analytically useful feature that the form of conditioned responding is determined by the nature of the CS as well as that of the US (e.g., Holland, 1977). For instance, discrete visual CSs evoke *rear* behavior (standing on the hindlegs, but not grooming), but auditory CSs evoke head-jerk behavior (rapid horizontal and/or vertical head movements).

Ross and I observed rats' performance in a variety of serial and simultaneous feature positive discriminations. When the elements of the AX compound were presented simultaneously, the rats acquired responding to the compound of the form characteristic to the predictive feature; that is, if a tone + light compound was paired with food and the tone alone was nonreinforced, the compound (as well as the light alone if it was presented alone in test) came to evoke rear behavior, and the tone evoked no conditioned responding. Similarly, if a tone + light compound was reinforced and the light alone was nonreinforced, the compound (and/or tone alone) came to evoke head-jerk behavior, and the light evoked no CR. These outcomes are consistent with the expectation of a variety of modern conditioning theories that the more predictive feature would acquire associations with the US, but the less predictive common element stimulus would not.

However, we observed a very different pattern of conditioned behavior when serial compounds were used in feature positive training. When a light-then-tone compound was followed by food but the tone alone nonreinforced, the rats acquired three kinds of responding. First, the tone common element acquired head-jerk behavior, but only when presented within the serial compound. Second, the light feature also acquired head-jerk behavior normally characteristic of tone responding. Third, as in the simultaneous discriminations, the light feature also acquired rear behavior. Considerable experimentation led us to suggest that the light feature's evocation of rear behavior reflected simple feature–US associations, its evocation of head-jerk behavior indicated the acquisition of within-compound light–tone associations, and the occurrence of differential head-jerk behavior to the tone within and outside of the compound was due to the light feature's acquisition of an ability to conditionally signal that the tone will be followed by food. We termed this last ability *occasion setting,* after earlier discussions by Moore, Newman, and Glasgow (1969) and Skinner (1938).

Ross and I conducted a number of experiments using the feature positive procedure that parallel the experiments reported here using feature negative procedures. For example, the ability of a serially trained feature to "set the occasion" for responding to a subsequent excitor was apparently specific to the excitor used in training: A light feature trained initially with a tone common element facilitated head-jerk behavior to that tone, but not to a noise that had been paired with food for the same proportion of trials as the tone, but not as part of a feature positive discrimination (Holland, 1983).

Similarly, a number of experiments indicated that the feature's simple associative (feature–US) and occasion-setting [feature—(common element–US)] relations are quite separate, just as Experiments 3a and 3b preceding suggested the independence of a negative feature's associative and "negative occasion-setting" relations. First, extending the feature–US interval was detrimental to feature–US associations but, within the range of intervals examined, enhanced occasion setting (Ross & Holland, 1981). Second, enhancing feature–US associations by administering feature–US training prior to introducing feature-positive training slowed the acquisition of occasion setting (Ross, 1983). Third, the administration of common element–US pairings prior to feature positive training blocked the acquisition of feature–US associations but had little effect on the acquisition of occasion setting (Ross, 1983).

One of Ross's (1983) findings is especially interesting in the present context. He noted that repeated nonreinforced presentations of the feature after serial feature positive training resulted in a disruption of its occasion-setting ability, although that ability was quickly regained when the feature positive training regime was reinstated. Ross argued that the loss occurred because the feature was no longer followed by the common element–US relation; that is, maintenance of the occasion-setting function of a feature stimulus may demand the *occurrence* of the events specified in the relation signaled by that feature. John Gory and I speculated that *negative* occasion setting, i.e., the inhibition acquired using serial feature negative procedures, might be similarly extinguished. Following Ross's logic, that function might be decremented by feature-alone presentations, because the feature would no longer be followed by the common element's countermanding of the common element–US relation.

Because the question of extinction of inhibition is of interest for a variety of theoretical reasons (see Baker & Baker, this volume) and most published experimental studies using simultaneous feature positive procedures have found no evidence of such extinction of inhibition (DeVito, 1980; Pearce, Nicholas, & Dickinson, 1982; Rescorla, 1982; Zimmer–Hart & Rescorla, 1974), Gory and I compared the effects of nonreinforced feature presentation after simultaneous and serial feature negative training.

Extinction of Inhibition: Experiment 4. In Experiment 4 all groups first received feature negative discrimination training. In each of fifteen 60-min sessions, the rats in Group Sim–E and Group Sim–N received 1 noise–shock pairing and 3 nonreinforced presentations of a *Sim*ultaneous light + noise compound. The rats in Group Ser–E and Group Ser–N received 1 reinforced noise presentation and 3 nonreinforced presentations of a *Ser*ial light-then-noise compound in each of 36 sessions. The simultaneous and serial compound groups were given different numbers of sessions in an effort to equate the level of inhibition established with the two different temporal arrangements.

Next, the rats in Groups Sim–*E* and Ser–*E* received training designed to *E*xtinguish the inhibitory powers of the light feature stimuli. Four nonreinforced light presentations were delivered in each of 10 sessions. The rats in Groups Sim–*N* and Ser–*N* were placed in the experimental chambers during those sessions, but *N*o stimuli were delivered.

All rats then received a single summation test session to evaluate the effects of the extinction procedure. The rats in Groups Sim–E and Sim–N received two nonreinforced noise presentations and two nonreinforced simultaneous light + noise compound presentations and the rats in Groups Ser–E and Ser–N received two nonreinforced noise presentations and two nonreinforced serial compound presentations.

As in previous experiments the serial discrimination was acquired more slowly than the simultaneous discrimination. However, by the final three discrimination sessions performance on the two discriminations was the same, both groups showing a .22 difference in suppression ratio between S+ and S−. Responding to the light feature did not change over the course of the 10 extinction sessions.

TABLE 9.4
Procedures in Experiment 4

Group	*Conditioning*	*Discrimination*	*Extinction*	*Summation Test*
Sim–E	N+	N+, LN− (15 sessions)	L−	N−, LN−
Sim–N	N+	N+, LN− (15 sessions)	no trials	N−, LN−
Ser–E	N+	N+, L → N− (36 sessions)	L−	N−, L → N−
Ser–N	N+	N+, L → N− (36 sessions)	no trials	N−, L → N−

Note: N = noise; L = houselight; + = shock reinforcement; − = nonreinforcement.

The mean suppression ratios were .49 and .53 on the first and last days, respectively, in Group Ser–E and .50 and .51 in Group Sim–E.

Figure 9.10 shows the primary data of Experiment 4, the results of the summation test. Consistent with the results of other investigators (e.g., DeVito, 1980; Pearce et al. 1982; Rescorla, 1982b; Zimmer–Hart & Rescorla, 1974), we found no evidence in a summation test for the extinction of inhibition after nonreinforced presentations of an inhibitor established using conventional, simultaneous feature negative procedures. In fact, significant facilitation of inhibitory power was observed in Group Sim–E relative to Group Sim–N, extending Rescorla's (1982b) and Zimmer–Hart and Rescorla's (1974) reports of nonsignificant facilitation after nonreinforced feature presentations (DeVito, personal communication, has recently found a similar facilitation effect in rat lick suppression conditioning). Although the present experiments do not bear directly on the origin of this facilitation, one reasonable account assumes that the feature's true inhibitory power is normally masked somewhat by its tendency to evoke second-order conditioned responses (Cunningham, 1981; Rescorla, 1982b), acquired as a result of feature–excitor pairings. Nonreinforced feature presentations would destroy the feature–excitor relation, extinguishing the second-order conditioned responses that previously interfered with the display of all of the feature's inhibitory capabilities. Alternately, presentations of the inhibitor during the extinction phase may have forestalled the decay or forgetting of inhibition that might otherwise occur over time (e.g., Henderson, 1978).

However, nonreinforced presentations of the inhibitor resulted in a small but significant loss in its inhibitory powers if it had been established using serial feature negative procedures (Group Ser–E vs Group Ser–N). This observation indicates, as Ross suggested, that occasion-setting properties of feature stimuli are at least partially susceptible to simple extinction procedures, in both feature positive and feature negative procedures. We have conducted other related experiments, but remaining unresolved are a number of issues concerning this extinction effect, such as whether it is the feature–common element relation that must be maintained to maintain negative occasion setting or the three-term relation among feature, common element, and nonreinforcement.

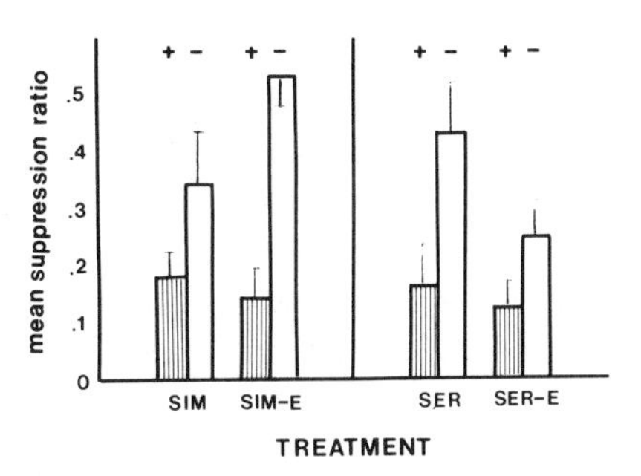

FIG. 9.10. Summation test responding in Experiment 4. + and – refer to responding to the excitor presented alone and within the feature–excitor compound, respectively. Brackets show the standard errors of the means.

Mechanisms of Solution of Feature Relevant Discriminations

Occasion Setting. All the data described previously suggest that quite different mechanisms are involved in the solution of serial and simultaneous feature-relevant discriminations. When stimuli are arranged to occur simultaneously, rats simply acquire associative strength to the feature stimulus as anticipated by most recent conditioning theories, accruing excitation if the feature is a predictor of reinforcement and inhibition if it is a predictor of nonreinforcement. However, if the feature precedes the common element, it acquires a conditional or occasion-setting function distinct from simple associative strength. This occasion setting or conditional function of serial features apparently acts on the association between the common element and the US, rather than on simply the US, facilitating or inhibiting the action of those associations as appropriate. Further, the occasion setting and simple associative functions of stimuli are apparently relatively independent, to the point that they may be of opposite valence, as in Experiments 3a and 3b in which a feature that signaled the occurrence of shock simultaneously signaled the cancellation of a common element–shock relation. We plan further experiments to investigate the various interactions among simple excitation and inhibition and positive and negative occasion-setting functions of stimuli.

A "Configural" Alternative to Occasion Setting. An alternative to occasion setting in dealing with the various data from serial feature-relevant discriminations is the notion of configural or unique compound cues (e.g., Kehoe & Gormezano, 1980; Rescorla, 1972); that is, perhaps subjects configure the two cues in serial discriminations, learning excitatory associations to one cue but not to the configured pair. For example, in the experiments reported here rats may have acquired excitation to the noise, and no association to the light-then-noise compound stimulus. Thus, responding to the noise on compound trials is withheld not because of inhibitory powers of the light but rather because the light–noise pair is discriminated from the noise presented alone. Consequently, acquisition of excitation to the light would not be retarded, nor would variations in the light's associative strength affect responding to the light-then-tone compound, except to the extent that the light alone and that compound generalized to each other. Similarly, a serially trained negative feature would not be expected to reduce responding to an excitor other than its training partner excitor, unless the feature + new excitor compound cue generalized to the feature + old excitor cue (Holland & Lamarre, in press).

Although it is difficult to see how the configural approach could be entirely ruled out, I favor the occasion-setting approach described earlier (see Jenkins, this volume, for other arguments against the configural alternative). First, it seems implausible that rats would be more likely to abstract a configural compound cue when the stimuli are presented serially than when they are presented simultaneously. Similarly, although it seems reasonable that configuring would be more likely with intramodal stimuli than with intermodal stimuli (e.g., Hull, 1945), Holland (1983) found more clear-cut occasion setting when intermodal stimuli were used within the compounds. Second, in Experiment 3 it seems likely that excitatory conditioning to the light alone would generalize more to the light trained within the light-then-noise compound than to the light trained within a simultaneous compound. Third, other data from my laboratory (e.g., Holland, 1983; Holland & Block, 1983; Ross & Holland, 1982) show that conditioned responding in situations in which configuring was explicitly encouraged differs in several ways from conditioned responding observed in the serial feature-relevant discriminations under consideration here. For instance, Holland and Block (1983) encouraged rats to abstract a configural light + tone cue by reinforcing presentations of that simultaneous compound and nonreinforcing each of the elements presented separately (positive patterning.) We noted the emergence of a unique response topography to that simultaneous compound. Other rats that received reinforced compound presentations but no nonreinforced element trials showed responding typical of each of the compound's elements alone but failed to acquire the unique response pattern observed in the positive-patterning subjects. However, Ross and Holland (1982) found no evidence of a unique response topography in serial positive-patterning discriminations: Subjects instead apparently used the first element to set the occasion for responding to the second element in the manner appropriate to the latter stimulus, just as do subjects in serial feature positive discriminations. Thus, although we found evidence for a configural process in simultaneous positive patterning, we found no such evidence in a comparable serial discrimination.

Conditions for Establishing Occasion Setting. An important question that remains is the conditions under which an occasion setting rather than a simple association mechanism is put into action. The most obvious condition seems to be that of temporal arrangement: Occasion setting was observed when the more predictive stimulus was less contiguous with reinforcement or nonreinforcement than the common element. Perhaps such a conflict between validity and contiguity engenders the strategy shift. Or perhaps occasion setting is induced whenever a perceptual discontinuity in the stream of events occurs (Estes, 1979). Simultaneous presentation of a tone and light followed by food might encourage the subject to abstract tone–food, light–food, and "tone+light"–food relations, whereas serial light-then-tone-then-food presentations might encourage the encoding of the relation between the light and the tone–food relation. The insertion of trace intervals or distractor stimuli might further segment the sequence of events. In fact, it is worth noting that Ross and I (Ross & Holland, 1981) found that the introduction of a trace interval between feature and common element within reinforced compounds enhanced positive occasion setting. Further, recent unpublished experiments from my laboratory suggest that the occurrence of a gap is more important than particular stimulus durations in generating positive occasion setting. Similarly, Jenny Lamarre and I (Lamarre & Holland, in press) found that the rate of acquisition of serial feature negative discrimination performance, thought to involve negative occasion setting, was more influenced by the occurrence of a trace interval between feature and common element than the particular intervals involved. Rats that received nonreinforced compound trials comprising a 30-sec light feature followed by a 30-sec empty trace interval followed by a 60-sec noise excitor acquired the feature negative discrimination considerably more slowly than rats that received 60-sec light then 60-sec noise compounds, but no more rapidly than rats that received compounds consisting of a 60-sec light separated from the 60-sec noise by a 60-sec trace interval. We are undertaking experiments isolating the contributions of the feature–common element, feature–US, and common element–US intervals and their ratios in an effort to piece together the importance of timing functions in the generation of the hierarchical event representations apparently reflected in occasion setting.

Of course, discontinuities need not all be temporal. Moore, Newman, and Glasgow (1969) reported that a related occasion-setting phenomenon in human eyelid conditioning was considerably more robust when the occasion-setting and common element (eliciting) stimuli were of a different modality. Similarly, Memmott and Reberg (1977) found rapid acquisition of serial feature-relevant discriminations in rat-conditioned suppression when the feature was a shock and the common element tones or lights, but Reberg (1978) was unsuccessful in training serial feature negative discriminations when both stimuli were tones or lights. Finally, Holland (1983) noted faster acquisition at occasion-setting in serial feature-positive discriminations when the feature and common element were of different stimulus modalities than when they were of the same modality. Other variations in stimulus type, such as that of diffuse and long-lasting stimuli versus discrete, punctate ones (as in Rescorla's work, this volume) may similarly encourage the engaging of more hierarchical coding of information.

One important stimulus type variable that may frequently engage such a coding system is the contrast of discrete stimuli with everpresent "contextual" stimuli (e.g., Balsam & Tomie, in press; Nadel & Willner, 1980; Rescorla, Durlach, & Grau, in press). For instance, Rescorla et al. (in press) have attempted to model the action of contextual cues using conditioning procedures similar to those described here. Further, Miller and his colleagues (e.g., Balaz, Capra, Hartl, & Miller, 1981) have noted that contexts apparently acquire potentiating functions apart from simple associative ones even in the absence of explicit context discrimination training. A related contrast is that of place with event: A variety of spatial tasks have been described by some investigators (e.g.,

Hirsh, 1980; Nadel & Willner, 1980; Olton, Becker, & Handelmann, 1979) as involving analogous conditional control functions. Finally, spatial discontinuities such as those encountered in foraging over a variety of food-source locations or traversing multiple-arm mazes might also engage hierarchical representation of events and occasion-setting strategies as well (Olton et al., 1979).

Brain Mechanisms in Occasion Setting. Some researchers have suggested that performance in many of the situations just described may involve different brain mechanisms than performance on simpler conditioning tasks. Recently, attention has focused on the hippocampus: Animals with hippocampal lesions often show severe deficits in the solution of spatial problems, context discriminations, and a variety of other tasks that might be conveniently described as involving occasion setting (see O'Keefe & Nadel, 1978 and Olton et al., 1979, for recent reviews). Ross and I speculated that hippocampal mechanisms might similarly be involved in the solution of serial feature-relevant discriminations but not in simple discrimination and conditioning tasks.

Together with William Orr and Dr. Theodore Berger, we recently examined the effects of bilateral hippocampectomy on the learning and retention of serial feature positive discriminations like those described earlier, simple nonconditional discriminations, simple CS–US associations, and within-compound associations (Ross, Orr, Holland, & Berger, 1984). We found that hippocampectomy prevented the acquisition of serial feature positive discriminations and eliminated discriminated responding in rats that had been trained on those discriminations prior to surgery but had no effect (relative to control subjects that received lesions of overlying neocortical areas) on the acquisition or retention of simple discriminations, CS–US associations, or within-compound associations. It is worth noting that in these experiments the same feature stimulus controlled both occasion setting and simple associative response tendencies. Thus, the disruption of only the occasion-setting function further supports the claim that occasion setting and simple associative functions of stimuli can be independent.

Ross et al.'s (1984) findings are consistent with the notion that a major function of the hippocampal function is that of dealing with conditional operations (e.g., Hirsh, 1980) or relations among stimuli. It is conceivable that occasion setting is a pervasive conditioning phenomenon involved in many of the tasks that show sensitivity to hippocampal damage. Considering that much of the data often claimed in support of a hippocampal role in conditional operations are obtained from rather complex, incompletely analyzed preparations, Pavlovian serial feature-relevant procedures may prove useful in the analysis of brain function, as well as of behavioral conditioning.

Generality of Occasion Setting. A final comment concerns the generality of occasion-setting phenomena. Clearly we need to examine the involvement of occasion setting across species and behavior systems. For example, although Rescorla's data from the pigeon autoshaping preparation and our rat-conditioned suppression data are similar in many ways, there are notable differences, such as the effects of nonreinforced feature presentations, and the occurrence of across-CS transfer. It is not clear whether these differences reflect species, response systems, or procedural differences. For instance, as Jenkins (this volume) points out, we do not know how much some instances of purported transfer of inhibition are dependent on generalization between the trained and transferred excitors (e.g., Gokey & Collins', 1980, finding of transfer in serial feature negative discrimination training in pigeon autoshaping). It is also worth noting that both Jenkins (this volume) and Lamarre and I tested transfer of inhibition with an unambiguous excitor, that is, one that had never been nonreinforced, whereas Rescorla's transferred excitors had themselves been trained with feature negative procedures. Lamarre has collected preliminary data in our conditioned suppression situation that show that inhibition to serially established features that did not transfer to an unambiguous excitor did in fact transfer to new excitors that had been trained within other serial feature negative discriminations.

Similarly, we have been unable to find evidence for occasion setting in serial feature-relevant procedures in rat flavor-aversion conditioning. Although our attempts have so far been rather feeble, it is worth pointing out the consistency of that failure with Garcia's (e.g., Garcia, Kovner, & Green, 1970) notion that the flavor-aversion paradigm involves different learning mechanisms at some levels.

Perhaps equally important is the extension of our analyses to more complex conditional discrimination procedures. Considerable current research, especially that directed toward memory processes, makes use of a variety of such procedures. But there has been relatively little consideration of the mechanisms involved in the performance of those tasks. The current research on occasion setting in feature-relevant discriminations may provide a link between analyses of simple compound conditioning phenomena and of more complex paradigms, which are perhaps more characteristic of organisms' natural experience.

Modern learning theory has made great progress assuming only a minimum of basic conditioning constructs, such as excitation, inhibition, association, and attention. Data such as those presented here (see also Soltysik, this volume) however suggest that relatively simple manipulations (such as the temporal arrangement of elements within a compound CS) can affect drastically the *content* as well as amount of learning. Under some circumstances, conditioned stimuli can acquire higher order functions, like occasion setting. There has been little consideration of such stimulus functions. I suspect that one of the more exciting areas of behavior theory in the next few years will be the investigation of the nature of functions like occasion setting, their determining and boundary conditions, and their interaction with other stimulus functions in the control of behavior. Several of the chapters in this volume, such as those of Jenkins and Rescorla, reflect the beginnings of such an endeavor.

ACKNOWLEDGMENTS

The research described here was supported in part by grants from the National Science Foundation and the National Institute of Mental Health. The discussions presented here profited greatly from my conversations with Jennifer Lamarre, to whom I am also indebted for collecting much of the data described here. Experiments 1a and 1b were reported at the April, 1983 meeting of the Eastern Psychological Association.

REFERENCES

Annau, Z., & Kamin, L. J. (1961). The conditioned emotional response as a function of intensity of the US. *Journal of Comparative and Physiological Psychology, 54,* 428–432.

Balaz, M. A., Capra, S., Hartl, P., & Miller, R. R. (1981). Contextual potentiation of acquired behavior after devaluing direct context-US associations. *Learning and Motivation, 12,* 383–397.

Balsam, P. D., & Tomie, A. (Eds.). (in press). *Context and learning.* Hillsdale, NJ: Lawrence Erlbaum Associates.

Cunningham, C. (1981). Association between the elements of a bivalent compound stimulus. *Journal of Experimental Psychology: Animal Behavior Processes, 7,* 425–436.

DeVito, P. L. (1980). *The extinction of a conditioned fear inhibitor and a conditioned fear excitor.* Unpublished Doctoral thesis, University of Pittsburgh.

Estes, W. K. (1979). Cognitive processes in conditioning. In A. Dickinson & R. A. Boakes (Eds.), *Mechanisms of learning and motivation.* Hillsdale, NJ: Lawrence Erlbaum Associates.

Garcia, J., Kovner, R., & Green, K. S. (1970). Cue properties versus palatability of flavors in avoidance learning. *Psychonomic Science, 20,* 313–314.

Gokey, D. S., & Collins, R. L. (1980). Conditioned inhibition in feature negative discrimination learning with pigeons. *Animal Learning & Behavior, 2,* 231–236.

Hearst, E. (1972). Some persistent problems in the analysis of conditioned inhibition. In R. A. Boakes & M. S. Halliday (Eds.), *Inhibition and learning.* London: Academic Press.

Henderson, R. W. (1978). Forgetting of conditioned fear inhibition. *Learning and Motivation, 9,* 16–30.

Hirsh, R. (1980). The hippocampus, conditional operations, and cognition. *Physiological Psychology, 8,* 175–182.

Holland, P. C. (1977). Conditioned stimulus as a determinant of the form of the Pavlovian conditioned response. *Journal of Experimental Psychology: Animal Behavior Processes, 3,* 77–104.

Holland, P. C. (1979). The effects of qualitative and quantitative variation in the US on individual components of Pavlovian appetitive conditioned behavior in rats. *Animal Learning & Behavior, 7,* 424–432.

Holland, P. C. (1983). Occasion-setting in Pavlovian feature positive discriminations. In M. L. Commons, R. J. Herrnstein, & A. R. Wagner (Eds.), *Quantitative analyses of behavior: Discrimination processes* (Vol. 4). New York: Ballinger.

Holland, P. C. (in press). Differential effects of reinforcement of an inhibitory feature after serial and simultaneous feature negative discrimination training. *Journal of Experimental Psychology: Animal Behavior Processes.*

Holland, P. C., & Block, H. (1983). Evidence for a unique cue in positive patterning. *Bulletin of the Psychonomic Society, 21,* 297–300.

Holland, P. C., & Lamarre, J. (in press). Transfer of inhibition after serial and simultaneous feature negative discrimination training. *Learning and Motivation.*

Hull, C. L. (1945). Discrimination of stimulus configurations and the hypothesis of afferent neural interaction. *Psychological Review, 52,* 133–142.

Jenkins, H. M., & Sainsbury, R. S. (1969). The development of stimulus control through differential reinforcement. In N. J. Mackintosh & W. K. Honig (Eds.), *Fundamental issues in associative learning.* Halifax: Dalhousie University Press.

Kehoe, E. J., & Gormezano, I. (1980). Configuration and combination laws in conditioning with compound stimuli. *Psychological Bulletin, 87,* 351–387.

Konorski, J. (1948). *Conditioned reflexes and neuron organization.* Cambridge: University Press.

Lamarre, J., & Holland, P. C. (in press). Acquisition of serial feature negative discrimination in rats' conditioned suppression. *Bulletin of the Psychonomic Society.*

Lubow, R. E., Schnur, P., & Rifkin, B. (1976). Latent inhibition and conditioned attention theory. *Journal of Experimental Psychology: Animal Behavior Processes, 2,* 163–174.

Marchant, H. G., Mis, F. W., & Moore, J. W. (1972). Conditioned inhibition of the rabbit's nictitating membrane response. *Journal of Experimental Psychology, 95,* 408–411.

Memmott, J., & Reberg, J. (1977). Differential conditioned suppression in the Konorski–Lawicka paradigm. *Animal Learning & Behavior, 5,* 124–128.

Moore, J. W., Newman, F. L., & Glasgow, B. (1969). Intertrial cues as discriminative stimuli in human eyelid conditioning. *Journal of Experimental Psychology, 79,* 319–326.

Nadel, L., & Willner, J. (1980). Context and conditioning: A place for space. *Physiological Psychology, 8,* 218–228.

O'Keefe, J., & Nadel, L. (1978). *The hippocampus as a cognitive map.* Oxford: Oxford University Press.

Olton, D., Becker, J. T., & Handelmann, G. E. (1979). Hippocampus, space, and memory. *The Behavioral and Brain Sciences, 2,* 313–365.

Pavlov, I. P. (1927). *Conditioned reflexes.* London: Oxford University Press.

Pearce, J. M., Nicholas, D. J. & Dickinson, A. (1982). Loss of associability by a conditioned inhibitor. *Quarterly Journal of Experimental Psychology, 33B,* 149–162.

Reberg, D. (1978). The feature positive effect in conditioned suppression: Effects of simultaneous and sequential compound stimuli. *Psychological Record, 28,* 595–604.

Rescorla, R. A. (1969). Pavlovian conditioned inhibition. *Psychological Bulletin, 72,* 77–94.

Rescorla, R. A. (1972). "Configural" conditioning in discrete-trial bar pressing. *Journal of Comparative and Physiological Psychology, 79,* 307–317.

Rescorla, R. A. (1973). Second-order conditioning: Implications for theories of learning. In F. J. McGuigan & D. B. Lumsden (Eds.), *Contemporary approaches to conditioning and learning.* Washington, DC: Winston.

Rescorla, R. A. (1974). A model of Pavlovian conditioning. In V. S. Rusinov (Ed.), *Mechanisms of formation and inhibition of conditional reflex.* Moscow: Academy of Sciences of the USSR.

Rescorla, R. A. (1975). Pavlovian excitatory and inhibitory conditioning. In W. K. Estes (Ed.), *Handbook of learning and cognitive processes,* (Vol. 2). New York: Lawrence Erlbaum Associates.

Rescorla, R. A. (1980). *Second-order conditioning.* Hillsdale, NJ: Lawrence Erlbaum Associates.

Rescorla, R. A. (1981). Within-signal learning in autoshaping. *Animal Learning & Behavior, 9,* 245–252.

Rescorla, R. A. (1982a). Simultaneous second-order conditioning produces S–S learning in conditioned suppression. *Journal of Experimental Psychology: Animal Behavior Processes, 8,* 23–32.

Rescorla, R. A. (1982b). Some consequences of associations between the excitor and the inhibitor in a conditioned inhibition paradigm. *Journal of Experimental Psychology: Animal Behavior Processes, 8,* 288–298.

Rescorla, R. A., Durlach, P. J., & Grau, J. W. (in press). Contextual learning in Pavlovian conditioning. In P. D. Balsam & A. Tomie (Eds.), *Context and learning.* Hillsdale, NJ: Lawrence Erlbaum Associates.

Rescorla, R. A., & Holland, P. C. (1977). Associations in Pavlovian conditioned inhibition. *Learning and Motivation, 8,* 429–447.

Ross, R. T. (1983). Relationships between the determinants of performance in serial feature positive discriminations. *Journal of Experimental Psychology: Animal Behavior Processes, 9,* 349–373.

Ross, R. T., & Holland, P. C. (1981). Conditioning of simultaneous and serial feature-positive discriminations. *Animal Learning & Behavior, 9,* 293–303.

Ross, R. T., & Holland, P. C. (1982). Serial positive patterning: Implications for "occasion-setting." *Bulletin of the Psychomomic Society, 19,* 159–162.

Ross, R. T., Orr, W. B., Holland, P. C., & Berger, T. W. (1984). Hippocampectomy disrupts acquisition and retention of learned conditional responding. *Behavioral Neuroscience, 98,* 211–225.

Skinner, B. F. (1938). *The behavior of organisms.* New York: Appleton–Century.

Zimmer–Hart, C. L., & Rescorla, R. A. (1974). Extinction of Pavlovian conditioned inhibition. *Journal of Comparative and Physiological Psychology, 86,* 837–845.

10 Conditioned Inhibition and Facilitation

Robert A. Rescorla
University of Pennsylvania

INTRODUCTION

Modern thinking about Pavlovian conditioned inhibition has been importantly influenced by the notion that inhibitors and excitors are opposites. This influence is nowhere more evident than in our choice of procedures for the production and measurement of inhibitors. The currently popular assessment procedures, summation and retardation (e.g., LoLordo, this volume; Rescorla, 1969), are clearly founded on an assumption of oppositeness. Moreover, descriptions of the circumstances that produce inhibition, such as negative correlation between a CS and US or the arrangement of a negative discrepancy between expected and obtained US, are explicitly intended as opposites to those that produce excitation, such as positive correlation between CS and US or a positive discrepancy between expected and obtained US. Historically, the identification of inhibitors as the equal opposites of excitors was important in achieving their acceptance in the psychology of learning. However, the notion that a stimulus might have an inhibitory, as well as an excitatory, function now seems well accepted. Consequently, we can more freely consider the possibility that inhibition is not always the opposite of excitation.

Some sense of the alternative interpretations of inhibition can be gained by consideration of one of the most popular procedures for endowing a stimulus with inhibitory power, what Pavlov called the conditioned inhibition paradigm. In that paradigm, the organism is exposed to two kinds of trials in intermixed fashion: those on which one stimulus (A) is followed by a US and others on which A is accompanied by another stimulus (B) and the US is omitted. The most common result is that the organism shows a response to A when it is presented alone but displays little responding to A when it is accompanied by B. The response to A is taken to mean that A is a conditioned excitor by virtue of its association with the US. The lower response to the AB compound is commonly interpreted as B's having inhibitory power. But as several authors (e.g., Rescorla, 1975) have noted, this A+/AB− paradigm offers a variety of modes by which B might exhibit its inhibitory action. One possible interpretation is that indeed there is an inhibitory B–US association that opposes the presumed A–US excitatory association. But there are elements in this paradigm other than the US with which B might be associated. For instance, B might have an inhibitory association with the excitatory A, enabling it to reduce A's power. Alternatively, B might have a direct association with the response to A, enabling it to prevent that response. These alternatives can be evaluated by assessing the degree to which B's inhibitory power transfers to other excitors that are discriminably different from A and that evoke a response other than that which A controls. Unfortunately, although these alternatives are frequently considered unlikely, in many Pavlovian settings the essential evidence is quite sparce (see LoLordo, this volume).

The present chapter is motivated by yet a third, less widely discussed, view of conditioned inhibition: that B acquires the ability to regulate the A–US association. It is possible that some of the inhibition controlled by B results not from its association with any *individual* event but rather from B's ability to modulate learned *relations* among those events. In this conception the organism learns that B is a stimulus in the presence of which the A–US relation is not in force. Such a view seems natural in many expositional circumstances, such as when one attempts to describe an A+/AB− paridigm to a novice. The stimulus A is readily described as predicting the US when it is presented alone but not when it is accompanied by B; that is, in casual discourse, it is natural to emphasize the hierarchical relation that one stimulus has, such that it modulates the strength of the association between two other stimuli. However, this possibility has received relatively little formal discussion.

This chapter is intended to begin that discussion by examining two elementary implications of this view of inhibition. The first is that conditioned inhibition may not be unique in having this kind of modulatory role; there may be other instances in which one stimulus is informative about the relation between other stimuli. Some inhibitors may best be thought of as members of a general class of modulators rather than as the opposite to excitors. In the first part of this chapter we examine one other instance of modulation, "facilitation," and note its parallels to inhibition. The second implication is that, because excitation concerns the relation of a stimulus to the US, whereas inhibition can involve the modulation of the relation of another stimulus to the US, the two may not always be opposites. Indeed, in principle they might be independent of each other, such that the same stimulus could concurrently control both. In the second part of the chapter we explore that implication.

The intention of this chapter is not to claim that all examples of conditioned inhibition are of this modulatory sort. Rather our point is that there is value in thinking about one stimulus as modulating the relation among others, and that some instances of inhibition may belong to a general class of stimuli having that function.

FACILITATION

Two Demonstrations

In this section we explore a procedure that is the analogue to the conditioned inhibition paradigm. That procedure generates a stimulus that promotes, rather than interferes with, responding to an excitor. If one describes the A+/AB− conditioned inhibition procedure as giving B the ability to signal the *absence* of the relation between A and the US, it is natural to inquire whether the converse procedure, A−/AB+, allows B to signal the *presence* of that relation. Would a B so treated take on the ability to modulate the response to A in a facilitative fashion? A similar question was raised for instrumental learning in an elegant paper by Brown and Jenkins (1967) but has received little subsequent analysis.

Of course, current theories of conditioning anticipate that the organism can solve such an A−/AB+ discrimination, eventually coming to respond in the presence of AB but not in the presence of A. Indeed, the solution to this problem, compared with that of the A+/AB− problem, has generated a literature of its own, describing the "feature-positive effect" (e.g., Hearst, 1978). According to many theories (e.g., Mackintosh, 1975; Rescorla & Wagner, 1972), the solution to the A−/AB+ problem is accomplished simply by B's becoming a conditioned excitor and A's eventually becoming relatively neutral. Such theoretical accounts require no appeal to B's having any modulatory function; they only demand that B become associated with the US. There is little doubt that this sort of learning occurs and may in many cases provide a complete account of such discriminations (see Ross & Holland, 1981).

However, these discrimination paradigms may also permit the development of a modulatory role for B. Under some circumstances, B may develop the ability to modulate A's excitatory strength, either in parallel with or instead of B's developing its own excitatory association with the US. Indeed, it is often quite difficult to separate the two alternatives. When we observe a response to AB it is often difficult to tell whether the organism is responding to B because of its association with the US or to A because the presence of B signals that the A–US association is in force. Should we attribute the response to A, to B, or perhaps to some mixture of the two?

We have recently been attempting to detect the presence of a modulatory role for B in such paradigms. In order to separate B's potential modulating function from its simple association with the US, we have adopted two tactics. The first tactic picks an A and B that display different, but compatible, responses when they develop excitatory associations with a US. Under those circumstances, one can separate B's potential functions in terms of the nature of the response exhibited in the presence of AB. A modulatory role would be detected in the form of B's presence enhancing the response normally observed when A is associated with the US; a B–US association would be detected as a response like that normally observed when B is associated with the US. The second tactic separates A and B in time, so that the response to A and B are temporally discriminable. We can then ask independently about the ability of B to evoke a response in its presence and to modulate the response to the subsequent A. Both of these tactics follow comparable procedures that have been used for similar purposes by Holland and his collaborators (e.g., Holland, in press; Ross & Holland, 1981).

Our first experiment illustrates the first approach. In this experiment pigeons were trained in an autoshaping paradigm in which various CSs are paired with food. When a key light is used as a CS, the resulting CS-food association is exhibited as directed pecking at the key light. However, placing a diffuse auditory CS in the same excitatory relation to food does not enable *that* CS to produce keypecking. This is true even when that diffuse CS is accompanied by an illuminated response key (An illustration of that point is given later, in Fig. 10.7). This is not to say that tone–food relations are not learned; of course they are, but that learning must be detected in ways other than directed keypecking. For instance, direct observational procedures indicate increased general activity during an auditory signal for food; moreover, such a CS has the ability to serve as a conditioned reinforcer for key-light CSs (see for example, Nairne & Rescorla, 1981).

In the present experiment we exploited this difference in response form to study a possible modulatory role of the tone. One group of 8 birds was exposed to an A−/AB+ procedure in which an 1800-Hz tone played the role of B and a black key light "X" on a white background played the role of A. On each day the birds received 20 nonreinforced 5-sec presentations of X and 20 reinforced presentations of the tone-X compound spaced with a 1-min mean intertrial interval. On compound trials the tone duration was 15 sec, the last 5 sec of which were accompanied by X. The intention of this temporal asynchrony was to encourage processing of the tone prior to the occurrence of the key light, as well as to discourage strong tone–food associations. Our simple question was whether the pigeon would come to exhibit a higher response rate to the X stimulus when it was accompanied by the tone, indicating that the tone facilitated responding to X. For purposes of comparison, a second group of eight birds received the identical treatment but with the reinforcement contingencies reversed so as to produce a standard conditioned inhibition paradigm, with the tone playing the role of the inhibitor.

Figure 10.1 shows the rates of keypecking on the two trial types. Because no keypecking occurred in either group when the diffuse B was the only stimulus present, that figure only shows responding when A was present. Consider first the animals exposed to the conditioned inhibition treatment (open symbols). These animals initially responded on both A+ and AB− trials but rapidly came to confine their responding primarily to the reinforced A trials. This result would normally be taken as evidence that the auditory B had developed inhibitory power. More interesting here is the fact that the animals exposed to the A−/AB+ paradigm also came to respond differentially, pecking primarily on the AB+ trials. That the discrimination was solved is no surprise; but the fact that its solution took the form of enhanced pecking at A is of interest. Because tones do not evoke pecking when simply paired with food, these results suggest that the present solution involved something other than simple excitatory training of the tone. Instead the directed form of the response suggests that the key light controlled responding on the compound trials. The role of the tone seems to be to facilitate responding to that key light. Moreover, although the response rate was substantially lower in the A−/AB+ treatment, the course of learning is remarkably similar to that observed in the A+/AB− (conditioned inhibition) treatment, raising the possibility of a shared basis.

In another attempt to detect facilitation independent of direct excitatory associations with the US, we employed key lights in the roles of both A and B but separated the events in time. In this experiment, pigeons were exposed to two concurrent facilitation procedures, A−/BA+ and A′−/B′A′+. All four stimuli were 5-sec key light. To ensure low levels of direct conditioning of B, a 5-sec.

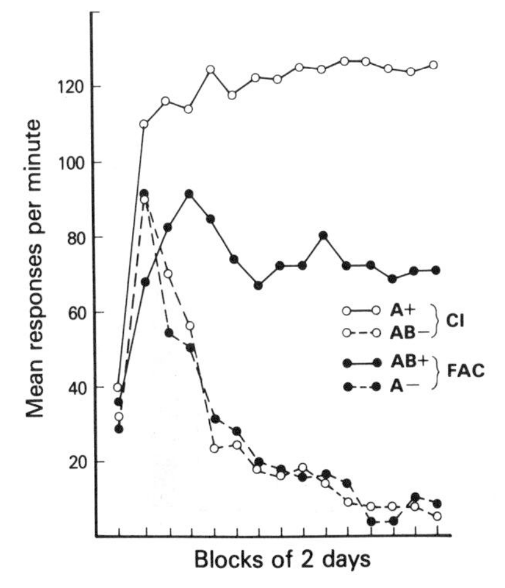

FIG. 10.1. Responding during the element and compound trials for animals given conditioned inhibition (A+/AB−) or conditioned facilitation (A−/AB+) treatment. The response measure is keypecking directed at the key-light A in the presence and absence of a diffuse auditory B. Responding to B itself was near zero in both conditions.

blank key intervened between the B and A stimuli on the compound trials. Two of the stimuli were compounds consisting of a color on the left-hand side of the key and a grid on the right-hand side: red-horizontal and green-vertical; the other two were a blue key light and a black X on a white background. Half the animals received X and blue as the A stimuli and the color-orientation compounds as the B stimuli; half received the reverse assignments. All 16 animals received 12 presentations each of A−, BA+, A′−, and B′A′+ during each 48-min session on each of 30 days. The question of interest was whether pecking would develop differentially to A and A′ as a function of whether or not they were preceded by their respective B and B′ facilitators.

The left-hand side of Fig. 10.2 shows the course of responding, collapsed across the two instances of facilitation treatment. The results parallel those of Fig. 10.1, in that initially there was a good deal of responding to the target key light (A) both when presented alone and when preceded by the facilitator (B); however, responding to A alone eventually dropped to a lower level. The initial low level of responding to B itself rapidly dropped to near zero. Thus, although there was little responding during B, there was a substantial enhancement of responding to A when it was preceded by B. B's prior occurrence facilitated responding to A.

These results are in substantial agreement with those previously reported for rat subjects by Holland (in press). Rats learning associations between food and diffuse auditory and visual stimuli exhibit those associations in quite different response topographies. When those stimuli are presented in the sort of sequential facilitation paradigm described previously, responding to the second stimulus in the sequence is promoted by the first. Moreover, as in the aforementioned studies, the form and timing of that responding is characteristic of the second stimulus itself. Holland has described B as "setting the occasion" on which A is reinforced.

These results thus suggest that one stimulus can acquire the power to facilitate the response to another stimulus if the converse to a conditioned inhibition

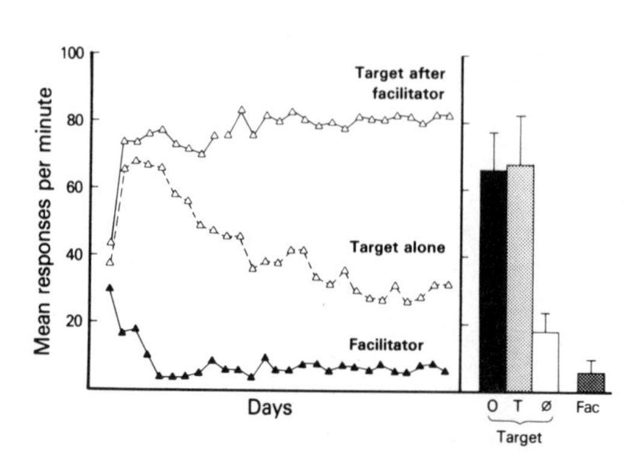

FIG. 10.2. Performance during a facilitation paradigm involving key-light stimuli. The left-hand side shows pecking to target and facilitator key-light stimuli on trials when the target was presented alone (nonreinforced) and presented after the facilitator (reinforced). The right-hand side shows pecking to the target during a test when it was preceded by its original (O) facilitator, a transfer (T) facilitator, or nothing (ø). Also shown is responding to the facilitators (Fac).

paradigm is arranged. The next two sections provide some analysis of that "facilitation." First we examine the kinds of stimuli on which facilitators act positively. Second, we discuss in more detail the relation between facilitation and excitation. Our primary intention is to provide for facilitators the same sorts of data that are standardly available for conditioned inhibitors. But the experiments reported in the following sections were also done with an eye to evaluating two obvious modes in which faciltators might function. In gaining acceptance for the notion of conditioned inhibition it was important to rule out two alternatives: that an inhibitor converts the excitatory A into configural AB stimulus and that it is only a neutral stimulus. Similarly, it is important to evaluate the comparable alternatives for conditioned facilitators, that they create a configural AB stimulus or that they are only excitors.

Transfer Tests: Stimuli on which Facilitators Act

As noted earlier, transfer tests have historically been very useful in the analysis of the nature of conditioned inhibitors. Similarly, one can learn a good deal about facilitators by examining the kinds of stimuli to which they promote responding. In this section we use transfer tests to examine four aspects of facilitators: the issue of configural cues, the role of excitation conditioned to the target stimulus, the role of inhibition conditioned to that stimulus, and the role of the response it controls.

Facilitation and Configural Cues. One may view an A+/AB− conditioned inhibition paradigm as requiring a discrimination between two stimuli, A and AB. Similarly, the facilitation paradigm is easily conceptualized as the reverse discrimination. It is an easy step from that observation to the suggestion that the organism simply treats AB as a new stimulus, different from A. In that view, AB is not well described as A plus B but rather as some new stimulus. In the case of the A+/AB− paradigm the further conclusion has sometimes been drawn that the concept of "inhibition" is not needed (e.g., Skinner, 1938); instead, the excitatory association controlled by A is simply absent in the AB compound. This view amounts to the claim that AB is a configuration that is treated differently from its component parts.

The standard procedure for evaluating this interpretation of inhibition is to conduct transfer tests (e.g., Brown & Jenkins, 1967; Rescorla, 1969). The idea is that if B is able to inhibit not only the response to A but also the response to other excitors similarly paired with the US, B must have some power of its own, beyond simply being a component of the AB compound. There are now a sufficient number of examples to give one confidence that sometimes B can be endowed with separate inhibitory power (see LoLordo, this volume).

Similarly, transfer tests are central to any claim that B might have separate facilitory power. Perhaps the most straightforward question is whether one facilitator promotes responding to a target stimulus that has been trained in conjunction with another facilitator. If one trains two facilitators, each with its own target excitor, to what degree are those facilitators interchangeable?

The right-hand side of Fig. 10.2 suggests that there is considerable interchangeability. After their previously described facilitation training was complete, those animals were tested for their response to each target key light, presented alone and preceded by both its original (O) and transfer (T) facilitator. Each target was presented alone 6 times and preceded by each facilitator 3 times; all trials were nonreinforced. The data indicate that the facilitators and target stimuli each continued to show little response when they were presented singly. However, responding was enhanced to a target stimulus when it was preceded by either facilitator. Moreover, the magnitude of facilitation was comparable for both the original and transfer facilitators.

Figure 10.3 shows the data from another transfer experiment that employed diffuse auditory and visual facilitators. In this experiment 16 birds received concurrent training with two facilitators, each 15 sec long. One facilitator was a white noise (N), the other was the flashing of the normally illuminated houselight (L). For the final 5 sec of one facilitator, a yellow (Y) stimulus was projected on the right-hand side of a response key and followed by food; during the final 5 sec of the other, a horizontal stripe pattern (H) was projected on the left-hand side of the key and reinforced. Both types of trials were presented 12 times in each 60-minute session. In addition, the birds received 12 nonreinforced presentations each of Y and H given in the absence of any facilitator. Consequently, N and L each served as signals that a particular key light would be reinforced. In order to prevent the absence of diffuse stimuli from becoming a general predictor of nonreinforcement, all animals additionally received 12 reinforced presentations of a blue key light in each session. The left-hand side of Fig. 10.3 shows the course of learning under this regime. The results are quite similar to those shown in the left-hand side of Fig. 10.2. The birds initially came to peck at the key

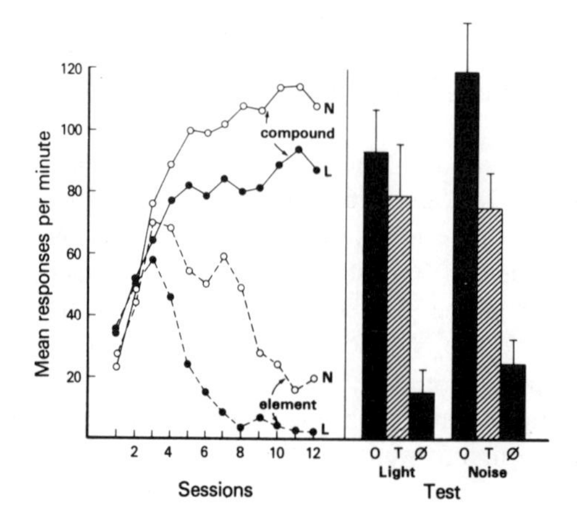

FIG. 10.3. Performance during a facilitation paradigm using diffuse light (L) and noise (N) facilitators. The left-hand side shows pecking to the key-light stimuli when they were presented as elements and in compound with the facilitators. The right-hand side shows test responding to the key-light when it was presented alone (ø), in compound with its original (O) facilitator, and in compound with a transfer (T) facilitator, separated according to whether the facilitator was a noise or light.

lights in both the presence and absence of the facilitators but then confined responding primarily to their presence. Although overall responding was greater to the element whose facilitator was a noise, clearly the diffuse houselight also could be trained as a facilitator.

The right-hand panel shows responding during a transfer test in which each key light element was presented 6 times alone (ø)and 6 times in combination with each facilitator. The results have been presented separately for the original training (O) and transfer (T) compounds. It is clear that the ability of each diffuse stimulus to facilitate performance transferred to the other key light. Despite the fact that the key lights were selected to be easily distinguishable, a hue and a line pattern on different parts of the key, and the facilitators were chosen to be in different modalities, there was substantial transfer of facilitation to novel combinations. Nevertheless, in this instance transfer was incomplete, suggesting some learning about the particular combinations used in training.[1]

It is worth noting that this experiment also contained a control group treated in the same way as the experimental animals with one exception: The key light that was reinforced during one of the facilitators was also reinforced when presented alone. That group can be described as being exposed to an A−/AB+ procedure with one diffuse stimulus and to an A′+/A′B′+ procedure with the other. The facilitation effect developed normally to B. But of course the animal responded to A′ in both the presence and absence of B′. The feature to note is that both B and B′ have key lights reinforced in their presence, and in the presence of both keypecking occurs. But for this group B′ is not informative about whether or not its key light will be reinforced. When those animals received a transfer test, the mean rates of responding during A, AB, and AB′ were 7, 90.5, and 15 responses per minute, respectively. The lower level of responding to A in the presence of B′ in this group than under analogous conditions in the experimental group indicates that the training history of the diffuse stimuli is relevant. A diffuse stimulus does not become a facilitator simply by virtue of having a key-light stimulus reinforced in its presence; it must signal differential reinforcement of that key light.

These results are analogous to those of many similar tests with conditioned inhibitors (see LoLordo, this volume; Rescorla, 1969). The experimental group showed substantial, but incomplete, transfer to other excitatory stimuli. A control group failed to show comparable transfer. As is common in the case of conditioned inhibition, we take these data to suggest that a configural cue cannot provide a complete account of facilitation procedures.

The Role of Target Excitation. One important feature of the current view is that facilitators and inhibitors act to modulate the excitatory association of other

---

[1]Other experiments in our laboratory have identified the presence of associations between the facilitator and its target CS. It is possible that such associations form the basis of the specificity observed here.

CSs. An implication of that view is that their modulating effect should not be observed with all target stimuli, but only with ones having some excitatory strength. That implication is well accepted in the case of conditioned inhibition, where the ability to detect inhibition is heavily dependent on the presence of an excitor. But it is of interest to explore its validity for the case of facilitation.

Consequently, in one recent experiment we examined the ability of a facilitator to induce responding to various types of target stimuli. Sixteen pigeons were given facilitation training like that used in the previous experiment, with a white noise (N) and a flashing houselight (L) serving as facilitators. But four key lights received different treatments intended to make them somewhat different kinds of targets: facilitation training with the houselight, facilitation training with the noise, simple excitatory training followed by extinction, and simple presentation without reinforcement. During the initial 12 days of the experiment, the birds received training of the form LA+, A−, NB+, B−, C+, D−, where the four key-light stimuli are indicated by A−D. Then the birds received an additional 6 days of the same training except that C was nonreinforced. This resulted in A and B being trained as targets of facilitators, C having been trained and extinguished, and D having never been trained. Ten trials of each type were given per day with a mean intertrial interval of 1 min. The four key lights were blue, yellow, and two grids oriented 45° and −45° from the vertical, assigned to the roles of A−D in a balanced fasion. Then all animals were tested for the ability of L and N to promote responding to each stimulus. Over the course of 2 test sessions they received 12 presentations each of A−D alone, 6 presentations of each preceded by N, and 6 presentations of each preceded by L.

Figure 10.4 shows the results of those tests. That figure plots separately the responding to each target stimulus when it was presented alone (ø) and in conjunction with a facilitator (F). As intended by the treatment conditions, responding was low and equivalent to the separately presented targets. The two pairs of bars to the left show responding to the facilitators presented in conjunction with their original and transfer targets. Those results replicate the findings shown in

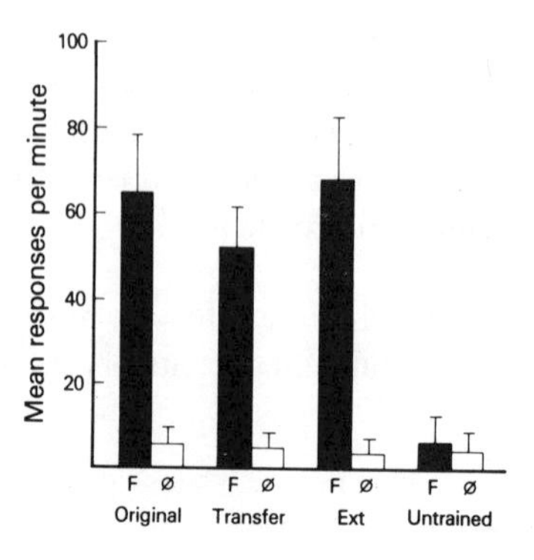

FIG. 10.4. Keypecking to various target key lights when presented alone and in the presence of a diffuse facilitator (F). The facilitator was tested in conjunction with its original target, a transfer target trained with another facilitator, a stimulus that had been trained and extinguished, and an untrained stimulus.

Fig. 10.3: A facilitator transfers its effect to a key light that has been trained in another facilitation paradigm. The third pair of bars indicates that facilitators also transfer to previously trained and extinguished stimuli, even if they have never been embedded in a facilitation paradigm. But the final set of data suggest that this facilitation does not extend to stimuli that have never been trained at all. If a stimulus has no history of conditioned excitation, a facilitator apparently cannot promote responding to it. That suggests that indeed a facilitator requires that the stimulus have some level of excitation in order to enhance performance to it. This agrees with the proposition that a facilitator acts to modulate the effectiveness of the A–US association.

The Role of Target Inhibition. All the stimuli on which a facilitator has been shown to act share a history of excitatory conditioning. But they also share an extinction treatment. They were all once excitatory but then lost the power to evoke a response when presented alone. We have tested facilitators against such stimuli for an important practical reason: Facilitation is shown by the enhancement of responding to a stimulus; if that stimulus is a strong excitor, it will evoke substantial responding on its own and one could not anticipate detecting enhancement. But it is then natural to ask whether a facilitator would promote responding to a stimulus that is only a moderate excitor by virtue of having a minimal history of conditioning. Such a stimulus would have excitation without having a history of extinction. Does a facilitator enhance responding to any excitor or does it restore previous levels of excitation that are currently depressed?

In order to address that question, we carried out excitatory training with the "untrained" stimulus of the previous experiment. Our intention was to monitor continuously the ability of a facilitator to act at various excitatory levels of a target stimulus. But we did not want explicitly to subject the stimulus to facilitation training. Consequently, we reinforced the target stimulus *both* when it was presented alone and when it occurred during the previously established facilitator. On each of 2 days each bird received four reinforced presentations each of the untrained stimulus alone and in conjunction with one of the facilitators. This allowed us to trace facilitation during initial acquisition of excitation. Then we gave nonreinforced presentations of the key light both alone and in conjunction with the facilitator, in order to monitor facilitation as the target CS underwent extinction.

The results of these treatments are shown in Fig. 10.5. Responding was rapidly acquired to the key-light element. Moreover, throughout the course of this acquisition there was a small (but nonsignificant) facilitation effect; responding was somewhat greater to the compound of the key light and the facilitator. However, during the course of extinction the facilitation effect became quite large. In order to evaluate the magnitude of facilitation, compare the second and third blocks of trials in both acquisition and extinction, at which points respond-

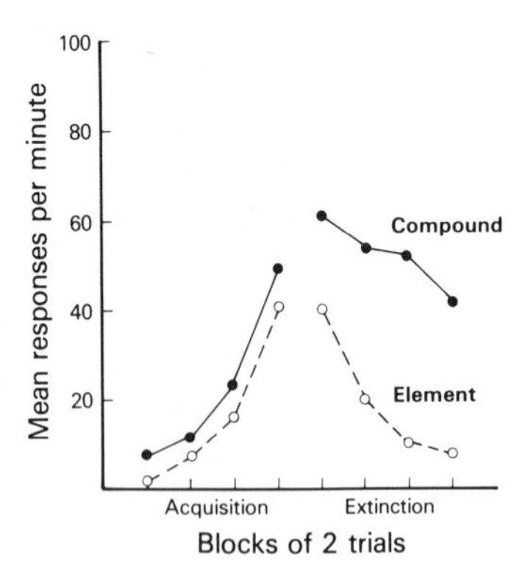

FIG. 10.5. Keypecking to a key-light element when presented as a single element and in compound with a facilitator. Performance is shown as the key light underwent acquisition and extinction.

ing to the key light alone was comparable. The facilitator produced a substantially greater augmentation from that level during extinction, compared with acquisition. That result suggests that facilitators are especially powerful when acting on previously trained and extinguished stimuli. Perhaps "ambiguous" stimuli are particularly susceptible to facilitation (see Bolles, this volume).

This conclusion is supported by two recent results that I just mention briefly. First, another way to produce a weak excitor is to have the CS bear a trace relation to the US. We have recently found that the responding to a weak excitor generated in this way is not augmented by a facilitator. Second, we have recently compared the action of a facilitator on two stimulus compounds, neither of which evoked substantial responding: one compound constructed of an excitor and an inhibitor and another compound constructed from two neutral stimuli. Only the former compound showed augmented responding in the presence of a facilitator. If one views extinction as superimposing inhibition on an excitor, that result agrees with those described previously. Apparently stimuli having both excitatory and inhibitory histories are especially subject to facilitation. This pattern of results is sensible if one views facilitation as the opposite of inhibition.

The Role of the Response to the Target Stimulus. Although we have found facilitators to transfer across stimuli that are perceptually quite distinct, all the preceding evidence comes from stimuli that evoke a common response form, namely directed pecking. It is then of interest to ask about the response specificity of facilitation; will it transfer across stimuli that evoke different excitatory responses? This is an issue that has been addressed only infrequently in the conditioned inhibition literature, but it is quite important. Knowing whether or not inhibition transfers to new excitors paired with the original US but evoking a different response gives important information on the locus of action of inhibition. The little evidence available suggests that inhibition can make that transfer (see Rescorla & Holland, 1977).

We have only a little evidence for the case of facilitation, but it is positive. Recently, we conducted an experiment in which birds were given training with key lights as facilitators. However, these facilitators differed in the target stimuli with which they were used. The trial types administered may be described as BA+, A−, B′V+, V−, C− − +, where B, B′, and C refer to different key lights, all of which were 5 sec long and 10 sec distant from a food reinforcer. B and B′ provided information that two stimuli occurring 5 sec later, A and V, respectively, would be reinforced; however, V was a visual stimulus, a key light that controls directed pecking, whereas A was a diffuse auditory stimulus during which no pecking occurs. The C was a control stimulus with the same temporal relation to food as did B and B′ but that lacked an intervening target CS. Each of 16 animals received 12 trials of each type on each of 12 days. The question of interest was how much facilitation B, B′, and C would promote to a common target. For this purpose, we trained and extinguished yet another key light and tested it 6 times alone and in conjunction with each of these three stimuli.

Figure 10.6 shows the results of that test. The left-most bar shows that responding was low to the extinguished stimulus when tested alone (ø). The far right-hand bar indicates that the control stimulus, trained in a trace relation to food, produced a small increment in responding. Substantially greater facilitation was produced by the key light that had been trained in conjunction with the visual target stimulus. The finding of most interest, however, is shown in the bar labeled "A": A facilitator trained with a diffuse auditory target, to which it did not promote the pecking response, also acted very effectively on the extinguished target. This data pattern suggests that if a stimulus is informative about the relation between another stimulus and food, it develops facilitatory power that transfers to other stimuli that evoke a different response. Apparently facilitation does not simply involve elevation of a particular response; a facilitator during which pecking had not previously occurred nevertheless augmented responding to a key-light target.

These transfer results only examine a small subset of the possible stimuli to which transfer might be made. Many questions about the stimulus domain on

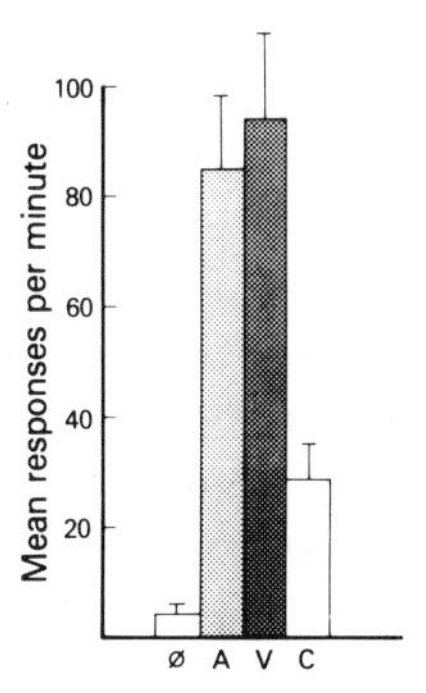

FIG. 10.6. Pecking to a target key light when it was presented alone (ø) and following key lights trained as a facilitator with an auditory (A) target, as a facilitator with a visual (V) target, or given a control (C) treatment.

which facilitators act remain to be addressed. But the available data seem to indicate two points. First, they discourage a view of facilitation in terms of configural conditioning. A facilitator seems to do more than form a unique perceptual combination with its target. It appears to have a power of its own that it can use to promote responding to other excitators. Second, they suggest a considerable parallel between facilitation and inhibition in the stimuli to which they transfer. Like an inhibitor, a facilitator demands the presence of an excitatory association but not a response of a particular form. It is not clear, however, whether or not the special sensitivity of an extinguished target to the action of facilitation also extends to the action of an inhibitor.

Facilitation and Excitation

We have argued on the basis of response form and timing that facilitators are other than simple Pavlovian excitors. Nevertheless, one might wish some more direct evidence on the relation between these two types of stimuli. In this section we look at that relation in three ways. First, we directly compare the ability of stimuli that have been trained in the two procedures to serve as excitors and facilitators. Second, we examine the different roles that the excitatory target and the facilitator play on the AB trials of a facilitation paradigm. Third, we observe the interaction of facilitation and excitation when training regimes appropriate to each are conducted with the same stimulus. To anticipate the results, we find that excitors and facilitators do not have equivalent augmenting and eliciting effects, they are not interchangeable in their roles on a trial, and the two kinds of learning are relatively independent when the same stimulus is treated in both ways.

Comparison of Excitatory and Facilitory Power. First, we directly compared stimuli trained according to the two paradigms, measuring the facilitative and excitatory powers of each. Sixteen birds were given the same 15-sec white noise and light as used in previous experiments; those stimuli were counterbalanced such that one (B) was embedded in a facilitation design with a 5-sec orange key light (A), whereas the other (B′) received simple excitatory training in which it was directly paired with food. Each bird received 12 trials each of A−, AB+, and B′+, on each of 16 days. As a result, B and B′ were both reinforced with the same frequency, but only B had an otherwise nonreinforced key light present on those reinforced trials. We then asked two questions of B and B′: First, to what degree does B′, trained as a simple excitor, also act as a facilitator for responding to A? To answer that question, we presented B′ in conjunction with the orange key light. All animals received a single test session that contained 6 nonreinforced presentations each of A, AB, and AB′. Second, to what extent does B, trained as a facilitator, also act as an excitor? Answering that question requires a measure of excitation for a diffuse stimulus. For this purpose we examined the reinforcing power of these stimuli, asking both B and B′ to serve as reinforcers for key-light CSs.

The left-hand side of Fig. 10.7 shows the course of acquisition. As in previous experiments, responding to the key light rose in the presence of the facilitator and fell in its absence. Responding to the diffuse B′ stimulus, which was simply trained as a Pavlovian excitor, was at zero throughout this stage and consequently its results are not plotted. The middle panel of Fig. 10.7 shows the results of the facilitation test session, during which the animals received the orange key light alone and in conjunction with both B and B′, each presented 6 times. It is clear that there was little augmentation of pecking to the key light during its presentation with the excitor B′. That result suggests that excitatory training does not bestow facilitatory power on a stimulus.

The right-hand panel of Fig. 10.7 shows the results of a second-order conditioning test of the reinforcing power of B and B′. That panel displays the results of responding to two key lights (a dot pattern and a magenta color) when one was paired with B and one with B′. This training was conducted in a single session during which each type of second-order trial was presented 6 times with a 2-min intertrial interval. It is clear that by this test the stimulus simply paired with the food was a substantially stronger excitor than was the stimulus employed in a facilitation design.

It should not be thought, however, that B is without effect on its antecedent key light. Subsequent experiments have shown that a key light that signals a facilitator itself acquires the ability to facilitate. Consequently, a facilitator and an excitor each condition something of their own (different) powers to antecedent stimuli.

These results thus indicate that the two paradigms produce two quite different powers on the part of a diffuse stimulus. Pairing directly with food generates an excitor; pairing another stimulus with food in its presence endows a stimulus

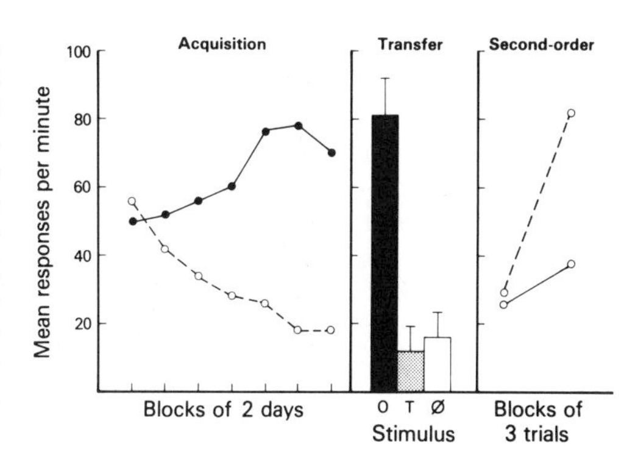

FIG. 10.7. Comparison of facilitation and second-order conditioning for a diffuse stimulus trained either as a facilitator or an excitor. The left-hand panel shows acquisition of facilitation in an A−/AB+ procedure. Keypecking is shown to a key-light A when presented alone (open circles) and during B (filled circles). The middle panel shows responding to A when it was presented during its original (O) facilitator, in a transfer test with a stimulus trained as an excitor (T), and alone (ø). The right-hand panel shows acquisition of second-order conditioning to two new key lights, one paired with the facilitator (solid line) and one paired with the excitor (dotted line).

with facilitatory power. Neither procedure produced a stimulus with the other's properties.

Interchangeability. We next asked about the degree to which stimuli trained as facilitators and target excitors could be interchanged in their roles. This test was conducted with those animals for which the primary results are shown in Fig. 10.2. Those animals had been trained with key lights as both facilitators and excitors, the former preceding the latter by 5 sec on reinforced trials. After they had been tested for the ability of each facilitator to act on the excitor appropriate to the other (shown in the right in Fig. 10.2), they received a day of retraining and then a second test. In that second test they were given each target stimulus 6 times preceded by three events: nothing, the facilitator appropriate to the other target, and the other target itself. They were also given each facilitator preceded by nothing, the other facilitator, or the target appropriate to the other facilitator. One can then compare the ability of a facilitator and a target each to promote responding to a subsequent facilitator and target.

The results of responding to the second stimulus of each test sequence are shown in Fig. 10.8. To the left is responding to a target stimulus when preceded by each of the three events. Responding to the target was low when preceded by nothing (ø) but high when preceded by the facilitator (F); those results replicate those of Fig. 10.2. More interesting is the failure of responding to be promoted by prior presentation of the other target (T). To the right is responding to the facilitator when it was preceded by nothing, the other facilitator, or the other target. It is clear than in none of those three cases was responding high. These results indicate that facilitators and target excitors have their separate roles. A facilitator can promote responding to a target but is not itself augmented either by another facilitator or by a target. Similarly, a target can be augmented by a facilitator but can augment responding neither to another target nor to a facili-

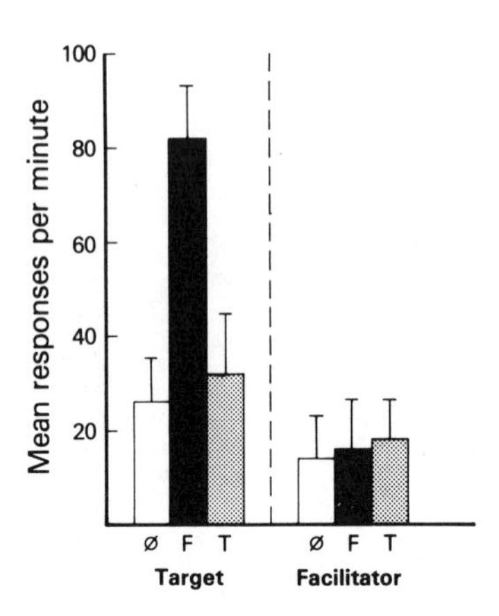

FIG. 10.8. Responding to a target and facilitator key light when each was preceded by nothing (ø), another facilitator (F), or another target (T).

tator. This training regime has resulted in asymmetrical consequences for the various stimuli.

Training with the Same Stimulus. Finally, we can look at how the excitatory and facilitatory functions interact when the same stimulus is given both kinds of training. Our strategy here has been to subject previously established facilitators to manipulations intended to change their level of conditioned excitation. Figure 10.9 shows one sample piece of data. In that experiment we trained naive pigeons in a procedure using both white noise and flashing houselights as 15-sec facilitators for the same 5-sec key light (a black X on a white background). On each of 12 days the birds received 24 nonreinforced presentations of X and 12 reinforced presentations each of NX and LX. Then each bird received extinction of either N or L. On each of 6 days the birds received 24 nonreinforced presentations of one facilitator. The intention was to extinguish any excitation that the facilitator might have. After extinction, each bird received a test session consisting of 12 nonreinforced presentations each of NX and LX, as well as 24 nonreinforced presentations of X alone. The left-hand panel of Fig. 10.9 shows the results of that test. They are easily summarized: Both facilitators continued to promote responding to X without any evidence that the nonreinforced presentations had had an effect. Carrying out an extensive extinction procedure did not disrupt facilitation.

We next examined the consequences of conducting deliberate excitatory conditioning of a facilitator. After several days of retraining with the facilitation procedure, half of each of the preceding subgroups received simple Pavlovian conditioning of N and half received conditioning of L. On each of 6 days the birds received 24 reinforced presentations of 1 facilitator, a training procedure that other experiments indicate is quite adequate to establish substantial excitatory conditioning. Then they were retested in the same manner. The right-hand panel of Fig. 10.9 shows that again the conditioning manipulation was without

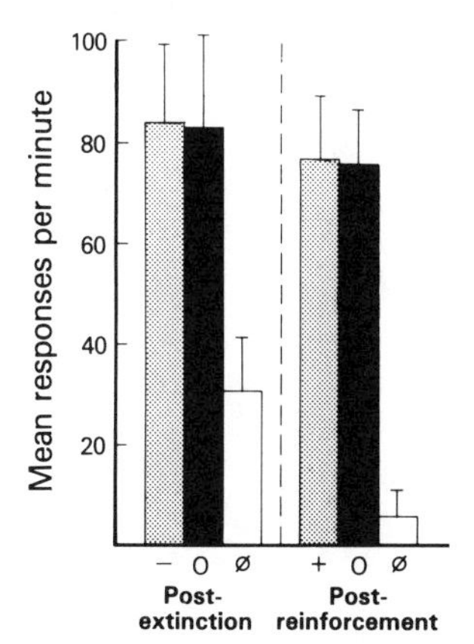

FIG. 10.9. Responding to a key-light stimulus when presented alone (ø), and in the presence of facilitators given Pavlovian extinction (−), Pavlovian training (+), or no treatment (O).

detectable effect. Converting a facilitator into an excitor neither augmented nor depressed its facilitative power.

This independence of the facilitative power of a stimulus from its level of conditioned excitation was also observed in the animals whose transfer data are shown in Fig. 10.8. After those animals had received that test, they received 3 days of retraining and then were subjected to nonreinforced presentation of one of their facilitators. On each of 6 days they received 24 nonreinforced presentations of one facilitator. They were then retested. The left-hand panel of Fig. 10.10 shows the results of that retest. Nonreinforcement (−) of one facilitator reduced responding during its own presentation (although this was not a significant effect). However, it had only a small impact on its ability to facilitate responding to the target stimulus. Next, half of each extinction group was given excitatory training (+) on one facilitator and nonreinforced presentation (−) of the other. On each of 6 days they received 24 presentations of one facilitator paired with food and 24 nonreinforced presentations of the other. Then they were retested, with the results shown in the right-hand side of Fig. 10.10. It is clear that differential conditioning of the facilitators had a substantial effect on responding in their presence; the facilitators evoked quite different levels of responding. However, those profoundly different levels had very little impact on the facilitatory power of the key lights. Again both facilitators continued to promote responding to the targets without an appreciable difference between them. These results suggest that manipulations of excitatory value of a stimulus have very little effect on its ability to serve as a facilitator.

This last experiment has another implication worth mentioning. As was the case previously (see Fig. 10.2), the transfer of a facilitator to a new target was complete during this test, so complete that we have not separately displayed responding for the original and transfer excitors. Note, however, that whenever two stimuli are equivalent in their inhibitory or facilitative impact on a target, the issue arises as to whether the animal discriminates those stimuli. We have deliberately picked as facilitators stimuli that we know that the bird readily discrimi-

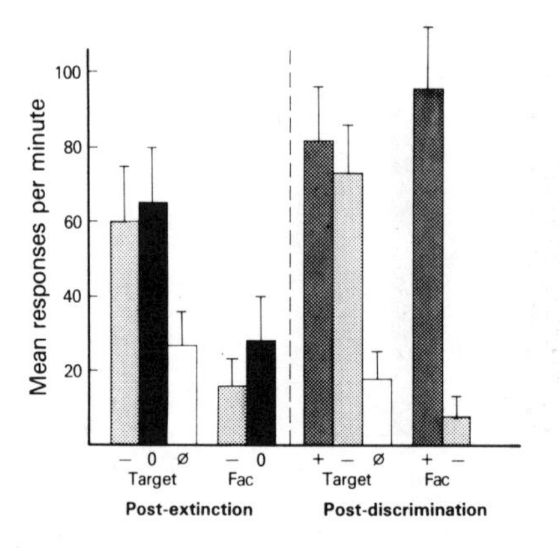

FIG. 10.10. Responding to a target key-light stimulus when presented alone (ø) and following key-light facilitators given Pavlovian extinction (−), Pavlovian training (+), or no treatment (O). The left-hand panel shows the results after simple extinction of one facilitator; the right-hand panel shows the results after discriminative conditioning of two facilitators. Within each panel responding to the facilitators is shown at the right.

nates in other settings. But the very success of transfer deprives one of evidence that they are discriminated when they act as facilitators. The results shown in the right-hand side of Fig. 10.10 demonstrate more directly that the birds discriminate the facilitators on the very trials that they show transfer. It is more difficult to be sure that the animal discriminates between the target key lights. However, an internal analysis provides some information indicating a consistency in the animals' response to the targets, whichever facilitator precedes them. In particular, a superiority of pecking rate at target A over target A′ when one facilitator was used was an excellent predictor of greater responding to A than to A′ when the other facilitator was used. That outcome could only occur if the birds detected the difference between A and A′ whatever the facilitator. Consequently, this experiment contains evidence of substantial transfer at a time when we know that the birds discriminate both between the facilitators and between the targets. That appears to rule out the more elementary configural interpretations.

Finally, it should not be thought that once a facilitator has been established there is nothing that can change its power. If one changes the treatment of the target stimulus in the presence of a facilitator, a substantial modification can be produced. In a recent experiment we trained two targets (a yellow key light and a horizontal grid) reinforcing each only in the presence of two facilitators (the same noise and houselight). Then each animal received 144 nonreinforced presentations of one target in the presence of one facilitator. Subsequent testing showed that facilitator to have suffered a loss in its ability to enhance responding to the other target. Those results suggest that the facilitating power of a stimulus can be modified if one manipulates the reinforcement of another stimulus in its presence.

Summary of Facilitation Results

The picture that emerges from these experiments is that a facilitator cannot easily be thought of either as a simple excitor or as a generator of a configural cue. In terms of the circumstances that produce a facilitator, those that reduce its power, and of the functions that it appears to play in behavior, it seems to be a modulator of the excitatory strength of other stimuli. Moreover, it is along just those dimensions that a facilitator appears analogous to a stimulus treated in a conditioned inhibition paradigm. Both appear to develop because of the information they give about the reinforcement of other stimuli. Neither is easily disrupted by changing its own relation to reinforcement (see later for more evidence on inhibition). Both facilitation and inhibition appear to transfer to a range of other excitatory stimuli. Consequently, it may be that the natural opposite to inhibition is facilitation, rather than excitation. In any case, the existence of facilitation is consistent with the proposition that some examples of conditioned inhibition may involve modulation of associations among other events.

CONDITIONED INHIBITION

The previous exploration of facilitators was motivated by the notion that conditioned inhibitors might have counterparts that enhance responding. Such stimuli were indeed identified and found to play a modulatory role in governing responding to conditioned excitors. We now return to inhibitors themselves to explore whether they too can play such a modulatory role. As we have noted earlier, many of the properties of facilitators are parallel to those previously observed for inhibitors. But there is one finding that may be especially important for evaluating whether inhibitors act by modulating excitatory associations controlled by other stimuli: its independence of its own excitatory strength. Like a facilitator, an inhibitor that acts through modulation should do so independently of its own excitatory association with the US.

There have been several reports of attempts to influence an inhibitor's association with the US by one of the manipulations used previously, simple nonreinforcement. Perhaps not surprisingly, that manipulation apparently leaves an inhibitor largely intact (Rescorla, 1982; Zimmer–Hart & Rescorla, 1974). More interesting would be an attempt to influence an inhibitor's power by giving it excitatory training. Of course, it is well known that such a treatment undermines the ability of an inhibitor to reduce responding in a summation test. But it is not clear whether this occurs because the inhibitor has lost its power to modulate responding to another stimulus or because the inhibitor itself has conditioned excitation that evokes a response; that is, the inhibition may be still intact but masked by conditioned excitation.

The problem here is the same that we faced in discussing facilitators: When the animal now responds to the AB compound, is it because the formerly inhibitory B has lost its power to modify the response to A, or is it because B still has that power but is also now an excitor capable of evoking its own response? Maybe B does depress the response to A but replaces that with its own response. We have adopted the same solution to this problem: Use as B and A stimuli for which the excitatory responses can be separated, either because they are separated in time or because they have different topographies.

Figure 10.11 shows the results of one experiment that attempted such a separation by temporal sequencing. In this study 16 female Carneaux pigeons were given 2 excitors, a white key and a black X on a white background. Each was reinforced on 12 of its daily presentations. The birds also received 3 colored key-light stimuli, red, green, and blue, 2 of which were treated as inhibitors and 1 as a control stimulus. This training was accomplished by presenting each color 12 times without reinforcement. One color was followed immediately by W, 1 was followed by X, and 1 was presented alone. Consequently, the treatment may be schematized as an X+, AX−, W+, BW−, C− procedure. The left-hand side of Fig. 10.11 shows the course of responding to the excitors when presented alone and when preceded by a color. It is clear that initially the excitors evoked

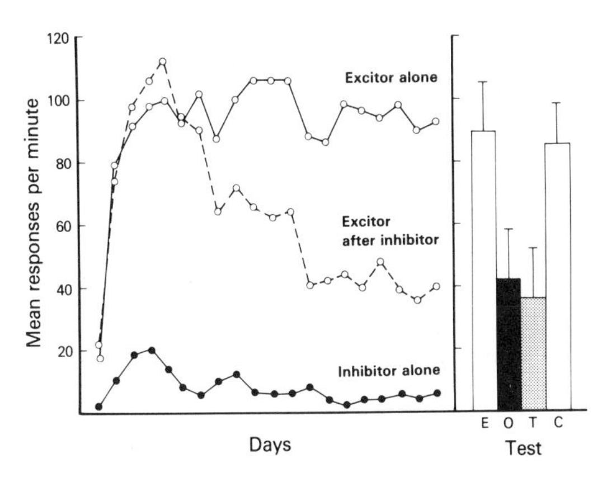

FIG. 10.11. Pavlovian conditioned inhibition in which a key-light excitor was reinforced when presented alone but not when it followed a key-light inhibitor. The left-hand panel shows acquisition of inhibition. The right-hand side shows responding to the excitor when presented as an element (E), after its original (O) inhibitor, after a transfer (T) inhibitor, and after a control (C) stimulus.

responding on both kinds of trials, but eventually the colors took on the ability to reduce responding to the excitors. Responding to the colors themselves was low throughout the experiment.

The right-hand panel of Fig. 10.11 shows the results of a transfer test in which each excitor was preceded by each color. The first bar shows the responding to the excitors alone (E). The second shows responding to each excitor when preceded by its origianl training inhibitor (O). The difference in those levels of responding accords quite well with that observed at the end of training. The third bar shows responding to the excitor when it was preceded by the inhibitor trained with the other excitor. Transfer of that inhibition was clearly substantial. Finally, the fourth bar shows responding to the excitor when it was preceded by the control (C) color that was simply repeatedly presented without reinforcement. The high level of responding in this case suggests that simple nonreinforcement did not generate an inhibitor. These data document that conditioned inhibition was obtained, transferred to a new excitor, and demanded a particular training procedure for its occurrence.

Next each animal received a discrimination procedure in which one inhibitor was reinforced and the other was not. Over 3 days they received a total of 60 reinforced and 60 nonreinforced trials. By the end of that training, the mean response level to the reinforced inhibitor was 84 responses per minute; that to the nonreinforced inhibitor was 6 responses per minute. Finally, all animals received a second test like the first.

The results of that test are shown in the left-hand panel of Fig. 10.12, which displays separately responding to the excitor and inhibitors presented alone and in compound. Responding to the excitor alone (E) remained high. Responding to the reinforced (+) inhibitor was substantially greater than that to the nonreinforced (−) inhibitor, verifying the success of their differential training. More interesting are the results of responding to the excitor in compound (C) with each inhibitor. The middle set of bars suggests that both inhibitors continued to reduce

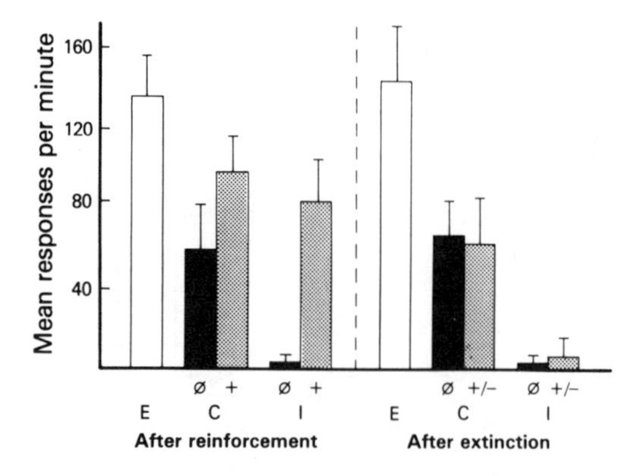

FIG. 10.12. Testing of Pavlovian conditioned inhibition after reinforcement or extinction of the inhibitor. The left-hand panel shows responding to the excitor when presented as an element (E) or in compound (C) with an inhibitor that had been reinforced (+) or given no treatment (0). The right-hand side shows performance after both inhibitors were repeatedly nonreinforced. Within each panel, responding to the inhibitors is shown at the right.

responding to the excitor. Even though the newly trained inhibitor produced substantial excitatory responding of its own, it continued to act as an inhibitor for responding to the W and X stimuli. That result suggests that the same stimulus can be excitatory and maintain some of its inhibitory control over another excitor.

Nevertheless, reinforcing an inhibitor apparently reduced its ability to attenuate responding to another excitor. One interpretation of that finding is that sequential stimulus presentation incompletely separates the excitatory conditioning of a stimulus from its ability to modulate responding to other stimuli. For instance, the color may leave an excitatory aftereffect that summates with the value of the target excitor. It is then of interest to ask whether reinforcement of the inhibitor resulted in reduction in its inhibition or some superimposed conditioning of excitation. To address this question, we next exposed these animals to repeated nonreinforced presentations of each inhibitor. On each of 6 days they received 24 nonreinforced presentations of each inhibitory color. The intention was to extinguish the excitation conditioned to the inhibitor so as again to permit unambiguous evaluation of the residual inhibitory control. The results of that test, shown in the right-hand frame of Fig. 10.12, were quite clear. After that extinction treatment, neither the untreated (ø) nor the trained and extinguished (+/−) inhibitor showed much evidence of excitation. More interestingly, both colors showed substantial and equivalent inhibitory control. That suggests that reinforcement of the inhibitor had not removed its power but only temporarily masked it. In this situation reinforcing a conditioned inhibitor does not first remove its inhibition and then produce excitation; rather that excitation is superimposed on the inhibition, apparently leaving it intact. This observation agrees with the proposition that inhibition of this type is independent of conditioned excitation.

These conclusions are further supported by a parallel experiment that we have conducted with diffuse auditory inhibitors. In this experiment 16 naive pigeons were given excitatory autoshaping with two key lights, a red color (R) and a horizontal grid (H). In addition, they received nonreinforced presentations of one of those key lights in the presence of a white noise. On each of 22 days the birds received 6 reinforced presentations each of H and R as well as 24 nonreinforced presentations of either H or R at the end of a 15-sec noise. Next, half the birds received 24 reinforced presentations of the noise on each of 6 days; the other half received no treatment. Finally, all animals were given 3 test sessions each containing nonreinforced presentations of each excitor alone and in compound with the noise. The question of interest is how much inhibition the noise will show with both its trained and a transfer excitor as a function of whether or not the noise had received excitatory conditioning.

Figure 10.13 shows the results of those test sessions, separated according to the noise treatment. To the left are the results from those animals receiving no excitatory conditioning of the noise (ø). They showed substantial responding to both the excitor (E) used in original training of the inhibitor (O) and that used for the transfer test (T). They also showed reduction in responding to the original excitor when it was given in compound (C) with the noise. Furthermore, they showed transfer of that inhibitory control to the other excitor, but that transfer was only partial. To the right are the comparable results from those animals who received excitatory training with the noise. Their data pattern was essentially the same; despite that training, the noise continued to inhibit responding to the original excitor and to show a moderate level of transfer to the other excitor.

These results thus indicate for inhibitors a result parallel to that we found with facilitators. The ability to show inhibitory modulation of responding to another CS can be independent of the excitation that a stimulus itself controls. That observation agrees with the proposition that under some circumstances inhibitors are not well thought of as opposites of excitors based on their having another relation with the US. Perhaps they are sometimes the opposites of facilitators based on their signaling of when other stimuli bear various relations to the US.

COMMENTS ON MODULATION

The two kinds of results described here support the conclusion that a stimulus can have a modulatory role independently of its own association with the US. It is

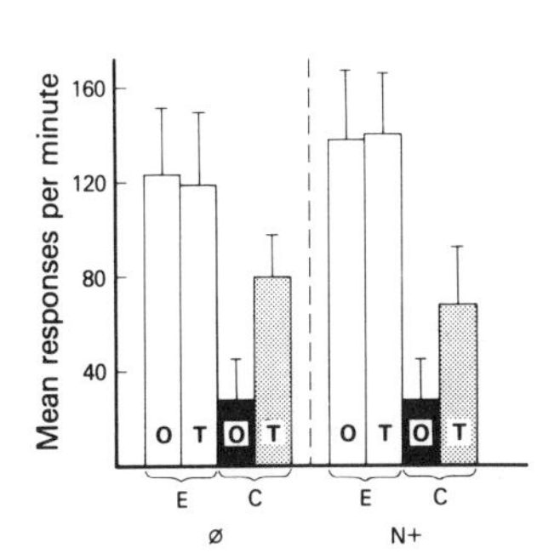

FIG. 10.13. Testing of Pavlovian conditioned inhibition after reinforcement of a diffuse white noise (N) inhibitor. Each panel shows responding to a key-light stimulus with which a diffuse inhibitor was originally trained (O) or to which it was transfered (T) when each excitor is presented as an element (E) or in compound (C) with the inhibitor. The inhibitor in the left-hand panel had no additional treatment (ø). That in the right-hand panel had been given Pavlovian excitatory training (N+).

then of interest to consider two further questions: (1) What is the character of that modulatory role and (2) where does it function?

There are multiple ways in which a stimulus might modulate responding to a conditioned excitor. As mentioned earlier, in the case of inhibitors there are four plausible loci of action: on the processing of the CS, on the processing of the US, on the production of the response, and on the functioning of the association itself. It seems quite likely that different modes of action will be responsible for different instances of conditioned inhibition and facilitation (see Holland, this volume, and Jenkins, this volume). However, decisions among the alternatives can be made in terms of transfer tests.

Although very few data are presently available to aid in choosing among those alternatives for the instances of conditioned inhibition and facilitation discussed here, it is worth noting their implications. For instance, action on the production of the response seems like the least attractive alternative for the present cases. There are several demonstrations that conditioned inhibitors transfer across excitors that control different responses (e.g., Brown & Jenkins, 1967; Rescorla & Holland, 1977). Moreover, I have here described an instance in which such transfer was also successful for a facilitator. Consequently, although response-inhibition or response-facilitation are certainly possible, the cases discussed here do not seem of that sort.

Neither do the available data provide support for the processing of the CS as the locus of action. The same results that indicate successful transfer across excitatory responses imply successful transfer across CSs that the organism acknowledges to be different. We have emphasized previously the negative implications of that finding for one form of impact on CS-processing, perceptual modification to generate a configural cue. But that finding is incompatable with any manner in which modulation might affect the processing of a particular CS. On the other hand, there is little in the data to preclude an interpretation in terms of modulators generally changing the degree of processing of all stimuli. They might function by means of a general alerting or depressing process. Only the failure of facilitators to enhance responding to a minimally trained target seems difficult for this view.

A third possibility is modulation by action on the representation of the US. The occasional evidence that conditioned inhibitors fail to transfer their power to excitors paired with other USs favors such a possibility. However, the appropriate experiments have seldom been done for inhibitors and are lacking altogether for facilitators. If facilitators do act on the US representation, it seems clear that this cannot take the form of activation in the manner of a conditioned excitor. The previously described failure of a simple Pavlovian excitor to facilitate responding to another Pavlovian excitor, together with the fact that facilitation promotes behavior directed at the excitor, suggests that something other than simple summation of excitors is involved. More plausible is a suggestion that several authors have made for conditioned inhibitors: that they affect the threshold for activation of the US (Konorski, 1948; Rescorla, 1979). According to this view, a conditioned inhibitor may adversely affect responding to an excitatory A by raising the threshold for activation of the US representation. We have taken as evidence for that position the observation that inhibitors do not typically evoke measurable behavior on their own and the fact that separately presented inhibitors do not extinguish. Facilitators may act in a comparable fashion, to lower the threshold. In that way they would make it more possible for an excitor to activate responding of the form it would usually evoke. But facilitators by themselves would not be able to evoke behavior. Moreover, it would not be surprising if facilitators had their primary action on stimuli that had a history of inhibitory training. It is just such stimuli, which presumably have elevated thresholds, that would be most vulnerable to stimuli capable of lowering the threshold. As we have previously noted (e.g., Rescorla, 1979), conceptualizing conditioned inhibition as shifting a threshold would not force substantial alterations in one popular model of the circumstances that produce conditioning, the Rescorla–Wagner model. It would simply force that model to acknowledge the separate preservation of excitatory and inhibitory tendencies, the latter being represented in threshold shifts. In the same way, a facilitator could be represented as the opposite of inhibition. In fact, one might consider the possibility that once an excitatory association has been formed it remains intact. Subsequent decrements and improvements in performance might be largely the result of temporary shifts in the threshold of activation of the US.

Finally, one might more frankly localize the effect of facilitators and inhibitors on the operation of the association itself (see Holland, this volume). For instance, stimuli trained in these paradigms might act as amplifiers, increasing or attenuating the ability of an association to be used. This sort of suggestion was made by Konorski (1967) when he discussed contextual stimuli. A facilitator is a stimulus in the presence of which another stimulus is reinforced and so undergoes substantial increases in associative strength; an inhibitor is one in the presence of which another stimulus is nonreinforced and so undergoes decreases in strength. It is then possible that those associative changes in other stimuli act as reinforcers for the facilitator or inhibitor, such that they take on the ability themselves to temporarily change effective associative strength. Because a change takes place in their presence, these stimuli become able to produce a similar change.

There is obviously not a great deal of evidence to help one select among alternatives such as these. But the present data seem sufficient to suggest that we need to begin acknowledging modulatory functions and to make that selection.

This need becomes more acute when it is observed that modulatory functions like those described here may very well play a role in a variety of other conditioning settings. Several familiar paradigms have logical relations among events that are formally analogous to those investigated here. Of course, the familiar feature positive and feature negative procedures are essentially the same as the present procedures. Hearst (1978), Jenkins and Sainsbury (1969), and others have extensively investigated responding under circumstances in which a compound stimulus containing two features, A and B, receives one reinforcement contingency and another compound lacking one of those features receives another contingency. A good deal is known about the circumstances under which those discriminations are learned. Whether these discriminations involve the sort of modulatory power discussed here remains to be seen. There is some evidence from Holland's laboratory to indicate that such details of the presentation mode as temporal relations can have an important impact on whether or not a modulatory role develops (e.g., Holland, this volume; Ross, 1983; Ross & Holland, 1981).

Somewhat further afield are delayed discriminations and delayed conditional discriminations, in which an initial stimulus specifies whether or not a subsequent stimulus will be reinforced. Primary interest in such paradigms has focused on how long the organism holds information about the first stimulus and just what information it holds. Less attention has been paid to the issue of concern here: How does that information interact with the presentation of subsequent stimuli to control responding?

Yet further afield are three examples of stimuli that are widely acknowledged to play an important role in behavior but for which detailed analysis is not available: contextual stimuli, drive stimuli, and operant discriminative stimuli. Each of those can be described as an event in the presence of which one, rather than another, relationship holds.

For instance, as Holland (in press) has noted, an operant discriminative stimulus is the formal analogue to the present facilitation paradigm: In its presence a response-reinforcer relation holds. With the fall from favor of an S–R theory of instrumental learning, there has been little formal description of the precise role that a discriminative stimulus plays. It seems possible that one might use the present Pavlovian analogues to model that role and better investigate its nature. Indeed, one might entertain the possibility that a stimulus trained as a facilitator (or an inhibitor) might function in an operant paradigm as an S^D (or an S-delta). It may be noted that the kind of amplification function about which we speculated earlier was envisioned by Konorski (1967) for operant discriminative stimuli. Despite the insistance of two-process theories, there is evidence that excitatory CSs and S^Ds are not identical (e.g., Holman & Mackintosh, 1981). Moreover, there are data to suggest that manipulations that change the conditioned excitatory value that sometimes develops to an S^D do not change its ability to control responding (Wilson, Sherman, & Holman, 1981). Those data seem parallel to the results reported here for facilitators and encourage consideration of the analogy.

Contextual stimuli also frequently take the form of being present when a relation applies. We have elsewhere noted that contextual stimuli sometimes develop direct associations with the US and with the CS (Rescorla, Durlach, &

Grau, in press). But we have also indicated that contextual stimuli can play a modulatory role. The most obvious example is the so-called "switching" experiment in which contextual stimuli inform the organism about which of two discrete CSs is reinforced.

Finally, we may note that drive stimuli also can be thought of as potential facilitators. When an animal has an opportunity to respond for the same reinforcement under different drive states, the drive often controls the effectiveness of that reinforcement.

These examples suggest that the kind of modulatory role here discussed could be pervasive in otherwise elementary learning paradigms. Perhaps they provide a way for us to consider constructing more hierarchical descriptions of behavior based on circumstances controlling the operation of associations. But at least they suggest that conditioned inhibition can be profitably thought of in the context of stimuli that play a modulatory role.

ACKNOWLEDGMENTS

This research was supported by National Science Foundation grants BNS-78-02752 and BNS-83-08176. I would like to thank Ruth Colwill, Paula Durlach, and Jim Grau for helpful discussions. Pamela Hill assisted in data collection.

REFERENCES

Brown, P. L., & Jenkins, H. M. (1967). Conditioned inhibition and excitation in operant discrimination learning. *Journal of Experimental Psychology, 75,* 255–266.

Jenkins, H. M., & Sainsbury, R. S. (1969). The development of stimulus control through differential reinforcement. In N. J. Mackintosh & W. K. Honig (Eds.), *Fundamental issues in associative learning.* Halifax: Dalhousie University Press.

Hearst, E. (1978). Stimulus relationships and feature selection in learning and behavior. In S. H. Hulse, H. Fowler, & W. K. Honig (Eds.), *Cognitive processes in animal behavior.* Hillsdale, NJ: Lawrence Erlbaum Associates.

Holland, P. C. (in press). "Occasion-setting" in conditional discriminations. In M. L. Commons, R. J. Herrnstein, & A. R. Wagner (Eds.), *Quantitative analyses of behavior: Discrimination processes.* (Vol. 4). New York: Ballinger.

Holman, J. G., & Mackintosh, N. J. (1981). The control of appetitive instrumental responding does not depend on classical conditioning of the discriminative stimulus. *Quarterly Journal of Experimental Psychology, 33*B, 21–31.

Konorski, J. (1948). *Conditioned Reflexes and neuron organization.* Cambridge: Cambridge University Press.

Konorski, J. (1967). *Integrative activity of the brain.* Chicago: University of Chicago Press.

Mackintosh, N. J. (1975). A theory of attention: Variations in the associability of stimuli with reinforcement. *Psychological Review, 82,* 276–298.

Nairne, J. S., & Rescorla, R. A. (1981). Second-order conditioning with diffuse auditory reinforcers in the pigeon. *Learning and Motivation, 12,* 65–91.

Rescorla, R. A. (1969). Pavlovian conditioned inhibition. *Psychological Bulletin, 72,* 77–94.

Rescorla, R. A. (1975). Pavlovian excitatory and inhibitory conditioning. In W. K. Estes (Ed.), *Handbook of learning and cognitive processes* (Vol. 2). Hillsdale, NJ: Lawrence Erlbaum Associates.

Rescorla, R. A. (1979). Conditioned inhibition and extinction. In A. Dickinson & R. A. Boakes (Eds.), *Mechanisms of learning and motivation.* Hillsdale, NJ: Lawrence Erlbaum Associates.

Rescorla, R. A. (1982). Some consequences of associations between the excitor and the inhibitor in a conditioned inhibition paradigm. *Journal of Experimental Psychology: Animal Behavior Processes, 8,* 288–298.

Rescorla, R. A., Durlach, P. J., & Grau, J. W. (in press). Contextual learning in Pavlovian conditioning, In P. D. Balsam & A. Tomie (Eds.), *Context and learning.* Hillsdale, NJ: Lawrence Erlbaum Associates.

Rescorla, R. A., & Holland, P. C. (1977). Associations in Pavlovian conditioned inhibition. *Learning and Motivation, 8,* 429–447.

Rescorla, R. A., & Wagner, A. R. (1972). A theory of Pavlovian conditioning: Variations in the effectiveness of reinforcement and nonreinforcement. In A. Black & W. F. Prokasy (Eds.), *Classical conditioning II.* New York: Appleton–Century–Crofts.

Ross, R. T. (1983). Relationships between the determinants of performance in serial feature positive discriminations. *Journal of Experimental Psychology: Animal Behavior Processes, 9,* 349–373.

Ross, R. T., & Holland, P. C. (1981). Conditioning of simultaneous and serial feature-positive discriminations. *Animal Learning and Behavior, 9,* 293–303.

Skinner, B. F. (1938). *Behavior of organisms.* New York: Appleton–Century–Crofts.

Wilson, C. L., Sherman, J. E., & Holman, E. W. (1981). Aversion to the reinforcer differentially affects conditioned reinforcement and instrumental responding. *Journal of Experimental Psychology: Animal Behavior Processes, 7,* 165–174.

Zimmer–Hart, C. L., & Rescorla, R. A. (1974). Extinction of Pavlovian conditioned inhibition. *Journal of Comparative and Physiological Psychology, 86,* 837–845.

11 Conditioned Inhibition of Keypecking in the Pigeon

H. M. Jenkins
McMaster University, Hamilton, Ontario

In recent years considerable attention has been devoted to the task of identifying the mode of action of a stimulus that has become an inhibitor through the training procedure that Pavlov (1927) termed *conditioned inhibition*. In this procedure, reinforced trials with CS alone are interspersed with nonreinforced trials on which the CS is accompanied by another stimulus, the conditioned inhibitor, which gradually acquires the ability to suppress the response that is otherwise evoked by the CS. The formulation of possible modes of action of the inhibitor depends to some extent on how the excitatory process itself is conceived. Rescorla and Holland (1977) take as their starting point a view of the excitatory process similar to Konorski's (1967). Their view is that excitatory conditioning occurs when the CS has developed an excitatory association with an internal representation of the US, the reactivation of which results in a CR. Within this framework they list four possible modes for the inhibitor's action. First, the inhibitor might act on the CR. Second, it might act to countermand the excitatory effect of the CS against which the inhibitor was trained, presumably by suppressing its ability to arouse a representation of the US. Third, it might act to suppress the activation of the US representation generally for any CS that is otherwise capable of arousing it. Finally, the inhibitor might act on the CS–Us relation rather than on either the CS or the US taken singly.

The principal means for evaluating these possibilities has been to examine the transfer of the inhibitor's action to other conditioned behaviors. The assumption is that the inhibitor's action is specific to the element or elements of the excitatory process on which the inhibitor has acted during conditioning. Conversely, the inhibitor does not act upon other elements except insofar as these elements are coupled by the nature of the excitatory conditioning process itself to the element or elements on which the inhibitor acted. Accordingly, if the action is on the CR, the inhibitory effect would be expected to decrease when tested with CR's of increasing dissimilarity to the first-trained CR. At the same time, inhibition should show transfer on combined-cue or summation tests carried out with dissimilar CSs and, moreover, be independent of the similarity between the first-trained CS and the transfer-test CS except insofar as different CSs might be coupled to different CRs through the nature of the excitatory conditioning process (see Holland, 1977). If the inhibitor acts on the CR, transfer to different USs would generally be expected to fail because of the coupling between CR and US in classical conditioning. If the action were on the US, transfer across dissimilar CSs, but not USs, would be expected.

Application of the transfer test logic to the hypothesis that the inhibitor acts on the CR, the CS or the US is straightforward. Demonstrating an effect on the CS–US relation is, however, more complex. If the inhibitor acts on the CS–US relation, one would expect transfer to be dependent on the similarity of both the first-trained CS and US to the test CS and US; changing either should result in a loss of inhibitory effect on test. Such a result, whereas consistent with the view that inhibition acts on the CS–US relation, is also consistent with inhibitory action focused separately on the CS and US. To show that the inhibitory action is specific to the relation, and not just to the terms that enter the relation, it would be necessary to test for transfer that is specific to the particular CS–US combination. Accordingly, one might carry out conditioned inhibitory training on the pairs, CS_1–US_1 and CS_2–US_2, followed by excitatory training and a test for inhibitory transfer on CS_1–US_2 and CS_2–US_1 (see discussion by Asratian, 1972). To my knowledge, a test of this kind has not be carried out.

Rescorla and Holland (1977) cite several previous experiments that have shown successful transfer of conditioned inhibition across different excitatory CSs. Their own results on the inhibition of fear conditioning (conditioned suppression of food-reinforced bar pressing in rats using shock as the US) and of appetitive conditioning (conditioned reactions to diffuse auditory or visual CSs paired with food) also showed transfer across excitatory CSs. Moreover, in the appetitive case, transfer to different CSs occurred despite the fact that the CRs to trained and tested CSs were distinctively different. On the other hand, the inhibitory effect did not transfer to conditioned responses based on a US different from the US on which the inhibitory training was carried out (food/shock). Transfer was not observed even when the inhibitor was presented in conjunction with the same CS with which it was first trained when that CS now signaled a different US. These results led to the conclusion that a conditioned inhibitor does not act directly on the CR, nor does it countermand the action of the CS, nor does it act on the CS–US relation. The results support the view that the locus of inhibitory action is on the representation of the US. In other words, the inhibitor appears to convey the message that the specific US (specific at least to an appetitive US in contrast with an aversive US) will not be forthcoming. It conveys that message independently of the CS that signals the US and independently of the CR evoked by different CSs as long as they signal the same US. The conclusions reached by Rescorla and Holland (1977) on the locus of action of a conditioned inhibitor have contributed to the prevailing view, which is that an inhibitor signals the absence, or dimunition, of the US (see Dickinson & Mackintosh, 1978).

Although cautions are sometimes expressed to the effect that the mode of inhibitory action might depend on "different preparations and paradigms" (Rescorla, 1979), the question of mode of action has commonly been approached at a level of abstraction that encourages the assumption of generality. The experiments reported by Peter Holland (this volume) show that the caution is well advised. He has shown, using conditioned suppression of bar pressing with rats, that whether the action of a conditioned inhibitor transfers across CSs or is specific to the CS on which the inhibitor was trained depends critically on whether in training the presentation of the inhibitor was temporally separated from the excitatory CS (serial procedure) or was simultaneous with it. Serial presentation resulted in CS-specific inhibition, whereas the simultaneous procedure led to inhibitory transfer across CSs.

The present experiments are concerned with the mode of action of an inhibitory stimulus that acquires its ability to suppress keypecking in pigeons through training similar to the Pavlovian paradigm of conditioned inhibition, which has also been referred to as the feature negative discrimination procedure (Holland, this volume). Although the procedure in the present experiments involved the simultaneous presentation of the inhibitor (noise) with the excitatory stimulus (key light), the inhibitory action of the noise shows two properties that Holland (this volume) has found in the case of serial, as contrasted with simultaneous, feature negative procedures. Those properties are: (1) specificity of inhibitory action to the excitatory stimulus with which the inhibitor was originally trained; and (2) a high degree of independence of the inhibtor's action from a direct, or first-order relation of the inhibitor to the US. These properties lead me to the same view of the inhibitor's action in the present experiments that Holland takes for the case of the serial presentations: namely, that the inhibitor acts as a conditional, or gating stimulus that modulates the signal value of the key light.

That the form of inhibitory action depends on the nature of the behavior system appears to be an inescapable conclusion. We know little, however, of the features of a behavior system that affect the mode of inhibitory action. We do not know, for example, the circumstances in which one stimulus will modulate the signal value of another. The experiments I describe do not constitute a direct experimental approach to such questions. Nevertheless, I believe the groundwork for a direct approach needs to be laid by more explicit recognition of readily observable features of the behavior system involved in a given experiment that might differentiate it from other commonly studied systems. Accordingly, the present experiments on the inhibition of keypecking in the pigeon are introduced by a discussion of the distinctive relation between peck initiating stimuli and the pecking response that characterizes this behavior system. Two types of initiating stimuli are identified, a remote stimulus and a target stimulus. One purpose of the experiments is to explore the action of inhibitors when keypecking is initiated by each of these types of initiating stimuli.

Initiating and Target Stimuli for the Keypeck

Experimental analysis of the mode of inhibitory action has so far been based almost entirely on classical conditioning preparations in which the measured CR is an undirected reaction to the CS, as is the case in conditioned suppression, salivary conditioning, conditioning of the nictitating membrane, and appetitively conditioned responses to diffuse visual and auditory stimuli. The pigeon's keypeck, on the other hand, is a response involving approach and contact with the stimulus object afforded by the key. When keypecking is engendered by classical pairings of food with a stimulus presented on the key, which corresponds to the CS in other classical conditioning preparations, the pigeon's response is typically directed to the CS. Hearst and Jenkins (1974) used the term *sign tracking* to capture the signal-centered character of the autoshaped keypeck. It would not be surprising if the inhibition of an action system involving sign tracking showed some properties that differ from the inhibition of undirected conditioned responses.

The focus in the present experiments is on the inhibition of the keypeck when that response is initiated in each of two quite different ways. It is useful to develop a terminology for these cases. The distinctions I wish to make cut across the more commonly made distinctions based on Pavlovian versus instrumental contingencies of reinforcement.

Keypecks are directed to the stimulus object afforded by the key. The key may be in a prevailing stimulus condition (e.g., constantly lighted or unlighted), or it may display momentary stimulus changes as in the usual autoshaping experiment. In either case, I refer to the stimulus object contacted by pecking and followed directly by a feeding as the *terminal target stimulus*.

In the experimental arrangements considered here the keypeck to the terminal target stimulus comes under the control of a momentary initiating stimulus. Two broad types of initiating stimuli, and some subtypes, need to be considered. The initiating stimulus may be the same stimulus that serves as the terminal target stimulus, or it may be a stimulus different from the terminal target stimulus. In autoshaping, the initiating stimulus (e.g., lighting the key) and the terminal target stimulus are one and the same. This is also true of discrete-trial experiments involving a peck-feeding contingency in which the lighting of the key initiates the response (e.g., Jenkins & Sainsbury, 1970). An initiating stimulus different from the terminal target stimulus is often provided by a diffuse stimulus, such as an unlocalized sound or overall change of illumination (see Rescorla, this volume).

If the terminal target stimulus is a prevailing or constant stimulus, peck-contingent feedings will generally be necessary to establish the keypeck and bring it under the control of an initiating stimulus. It is possible, however, to use a momentary target stimulus, and to establish an initiating stimulus without the use of a peck-feeding contingency. If, for example, the momentary lighting of the key is followed by a feeding when accompanied by a diffuse stimulus and not otherwise, responses to the terminal target stimulus will come under the control of the diffuse stimulus that then serves to initiate the peck to the terminal target stimulus. The presence of the terminal target stimulus continues to be necessary to the occurrence of the keypeck even when, as the result of the differential training just indicated, responses occur only in the presence of the diffuse initiating stimulus (see, for example, Rescorla, this volume).

Initiating stimuli that are separate from the terminal target stimulus need not, of course, be diffuse. Pecks directed at the terminal target stimulus can be initiated by a localized visual stimulus similar in kind to the terminal target stimulus. Depending on its timing in relation to food and its spatial location, a localized visual stimulus that initiates pecks to a terminal target stimulus may itself become a target of pecking. If the terminal target stimulus is continuously available, a response-reinforcer contingency will be necessary in order to establish control by the initiating stimulus over-pecks to the terminal target stimulus. If the initiating and terminal target stimuli are presented together prior to food delivery, a response-reinforcer contingency may be necessary in order to direct the peck to the terminal target stimulus rather than, or in addition to, pecking directed to the initiating stimulus itself.

In summary, an initiating stimulus may be the same as or different from a terminal target stimulus. The target stimulus must be localized, but it may be momentary or prevailing. The initiating stimulus may be diffuse or localized. If localized, it may itself be pecked. These cases make different demands for a response-reinforcer contingency.

When the initiating stimulus is other than the terminal target stimulus, it appears to play a role not unlike the role assigned to the discriminative stimulus by Skinner (1938), which was to set the occasion for a response. In the present context the initiating stimulus appears to set the occasion for pecks directed to the terminal target stimulus without substituting for, or replacing, the function of the terminal target stimulus. We do not, unfortunately, have a commonly agreed upon label for such stimuli. The term *discriminative stimulus* has an historical association with response-reinforcer or operant paradigms, which argues against its general application to any initiating stimulus that is separate from a target stimulus. Rescorla (this volume) and Rescorla, Durlach, and Grau (in press) use the term *facilitator,* and Holland (this volume) uses the expression *occasion setter,* whereas others have used such terms as *conditional stimulus, gating stimulus*, or *modulator stimulus* when referring to a stimulus that appears to control responses to another stimulus. In the present report I refer to an initiating stimulus that is distinct from the terminal target stimulus as a *remote initiating stimulus*.

The present experiments examine the mode of action of a conditioned inhibitor when keypecking is initiated by the terminal target stimulus (Experiment 1), and when it is initiated by a remote initiating stimulus (Experiment 2). As in many previous experiments, transfer summation tests were used to examine the specificity of inhibitory action to the initiating stimulus. The second method used to assess the mode of action of the inhibitor is one also used by Holland (this volume) to explore inhibitory action and by Rescorla (this volume) to explore the action of a facilitator. It involved converting the conditioned inhibitor into a signal for food (Pavlovian CS) or, in another case, into a remote initiating stimulus for food-reinforced keypecking, and then observing the effect of these treatments on its previously trained inhibitory function.

Experiment I, Inhibition of Keypecking When the Initiating Stimulus is the Terminal Target Stimulus. Part 1: Transfer of Conditioned Inhibition Across Target Stimuli

In the first set of experiments, the terminal target stimulus was a small backlighted dot on the key that was used in an autoshaping-like procedure. Presentation of this stimulus initiated the sequence of approach and pecking, whereas the stimulus itself served as the target of the peck. Although the conditioning procedure has much in common with autoshaping, a short run of pecks on positive trials terminated the stimulus and produced food so that a response-reinforcer contingency as well as a stimulus-reinforcer contingency was in effect.

The essentials of the procedure were these. A square pecking key divided into four smaller squares 1.65 cm on a side by narrow metal strips (0.2 cm wide) was used. From trial to trial the dot that served as the target stimulus appeared at the center of a randomly selected quadrant. Pecks were recorded by quadrant. Following the usual tray training, birds received 12 positive (food reinforced) and 12 negative (unreinforced) trials per session in a restricted random order. On positive trials a small lighted dot (0.4 cm in diameter), red for six subjects and green for another six, appeared until terminated either by the passage of 6 sec or by four pecks, whichever occurred first. A 4-sec feeding was made available at the moment the dot went off whether it did so as the result of the fourth peck or the passage of 6 sec. It may be noted that prior to the occurrence of four pecks on a positive trial, acquisition was brought about entirely through an autoshaping procedure. As responding increased, the response contingency on positive trials came into play. On negative trials the dot was accompanied by a 90 dB white noise, and both were presented for a fixed period of 6 sec without regard to responses. Noise onset and termination coincided with onset and termination of the dot.

After 30 training sessions, 12 reinforced trials per session with the second dot color were added. Training continued for 10 sessions. During the next 10 sessions, 12 negative trials on which the noise accompanied the second dot color were added. Accordingly, these sessions contained 48 trials; 24 positive trials, 12 on each of the first- and second-trained dot colors, and 24 negative trials with the noise present, 12 with each of the first- and second-trained dot colors.

Shown in Fig. 11.1 is responding on various trial types throughout the several phases of training. With the exception of but one of the 12 animals, responding was almost completely confined to the quadrant containing the dot. Because the responses on positive trials were limited to a maximum of 4 by the response contingency, the data do not allow an estimate of the degree of inhibition achieved. They do show, however, that inhibitory training was effective. The median number of responses per 6-sec negative trial was reduced from a peak value of about 9 responses to a value of between 1 and 2 responses over the later sessions of training. Over sessions 21 through 30, subjects trained with the green dot made a median of 1.1 responses per negative trial with a range from 0 to 2.7. For those trained with the red dot, the median was 1.6 with a range from 0.9 to 13.6 responses per trial. The highest responder in this group was an extreme outlier on the distribution, since the next highest score was 2.0 responses per trial.

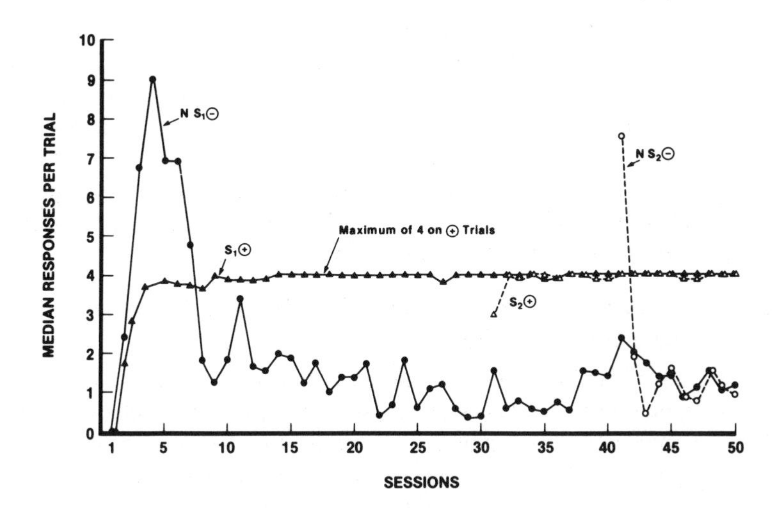

FIG. 11.1. Median responses per trial on positive trials with dot alone, and on negative trials with noise plus dot. Positive trials with a dot of another color (red or green) were introduced beginning with session 31. Negative trials with noise plus the second-trained dot were introduced in session 41.

There was obviously considerable transfer from positive training with the first-trained dot color to the second, although in the first session in which the second dot was introduced (session 31) responding was noticeably below the level of responding to the first-trained dot. Evidence on the transfer of inhibitory control from one target stimulus to the other was obtained beginning with session 41. During this session there were far more responses to the second-trained dot when accompanied by the inhibitor than to the first-trained dot when accompanied by the inhibitor. In subsequent sessions, continued inhibitory training on the new dot resulted in a very much more rapid development of response suppression than occurred during inhibitory training on the first stimulus. It is therefore clear that there was some transfer of inhibitory control from the first- to the second-trained target stimulus, but the greater part, perhaps all, of that transfer was in the rate of learning rather than on the initial level of responding.

Given the fact of stimulus generalization based on similarity, complete specificity of inhibition is hardly possible. The critical question is whether the initial level of suppression to a test stimulus shows transfer when that stimulus is sufficiently dissimilar to the original stimulus to be beyond the range of stimulus generalization. If it does not, the inhibitory action can properly be viewed as completely dependent on the stimulus with which it was trained. Although the present experiment does not provide a directly comparable test of responding to the second-trained stimulus with and without the inhibitor, it does provide data that suggest that there was in fact no initial transfer. The following comparison was made. The session with the highest overall mean rate of response on noise-plus-dot trials in the first phase of inhibitory training was identified for each subject individually. For the first three noise-plus-dot trials of this peak-response session, a mean of 13.1 responses per 6-sec trial was obtained. For the first three trials with noise plus the second-trained dot in session 41, the comparable figure was 13.9 responses, indicating no reduction in responding due to transfer. We may conclude from the present results that the initial action of the inhibitor was highly dependent, probably completely dependent, on the similarity between the original stimulus and the transfer test stimulus.

Experiment I, Part 2: Effect of Converting the Inhibitory Stimulus into a Signal for Food

If the conditioned inhibitor were acting as a signal of no food, and thereby causing a suppression of keypecking to initiating stimuli, its inhibitory effect should be eliminated by converting the inhibitory stimulus into a signal of food. In order to test this implication three groups were formed and given the following treatments. Group NS_1+ received reinforced trials in which noise accompanied the first-trained stimulus, S_1. This amounts to a reversal of the inhibitory training previously received on S_1, and it would be expected to cause a loss of inhibitory action of noise on this stimulus along with some loss of its inhibitory effect on

S_2, the second-trained stimulus, due to generalization. Group N+ received presentations of the noise alone followed by a feeding; the key was unlit. Finally, Group Control received no presentations of noise during the treatment phase. The critical comparison is between Group N+ and the other groups. If the inhibitory action of noise is on the signal value of the initiating stimulus, and not on the representation of the US, the pairing of noise with food would not be expected to attenuate the inhibitory action of noise. Accordingly, the results for Group N+ should not differ from those of the control group.

The details of the procedure were as follows. Each of the three groups contained two subjects first trained on the red dot and two first trained on the green dot. The groups were matched as closely as possible on the level of responding on negative trials during the last five sessions of the previous phase. There were 10 sessions in the treatment phase. In each session, Group NS_1+ received 12 positive trials in which noise accompanied S_1, and 12 positive trials on each of S_1 and S_2 (the latter were received by all groups). Group N+ received 12 positive trials in which only noise was presented. Although pecking was not expected to emerge from the noise–food pairings, the reinforcer contingency on these trials was the same as for the other types of positive trials: A feeding was programmed to occur after the passage of 6 sec or four pecks, whichever occurred first. Group Control received only the 12 positive trials on each of the first- and second-trained stimuli; noise was not presented.

Following these treatments, all groups were returned to the inhibitory training procedure previously in effect: 12 positive trials on each of S_1 and S_2 and 12 negative trials in which noise accompanied S_1 and S_2.

The results are shown in Fig. 11.2. As would be expected, the reversal of inhibitory training in Group NS_1+ caused a marked loss of response suppression to S_1. Suppression did not return to its former level until about the 12th session of retraining. There was also a marked loss in this group of the suppressive effect of noise on the second-trained stimulus, S_2. The mean number of responses per trial with noise and S_2 for sessions 43–50 was 1.2. The comparable mean for the first posttreatment session was 7.4, $t(3) = 4.17, p < .025$. The most important result was the absence of a substantial change in the suppressive effect of noise in Group N+ despite the intervening treatment in which noise was repeatedly paired with feedings. The pretreatment mean on the noise-plus-S_2 trial was 1.0 responses. The posttreatment mean was 3.2 responses. The comparable values for the noise-plus-S_1 trial in this group were 2.3 and 2.2 responses. Neither of these differences approached significance by t test. The results for Group Control also showed no significant difference between the pre and posttreatment means.

Three of the 4 subjects in Group N+ made no responses to the unlit key during presentation of noise with food, but 1 subject did show some acquisition of key pecking during these trials, averaging about 2 responses per trial over the last 4 of the 10 treatment sessions. Reinforcements were earned by pecks on about 10%

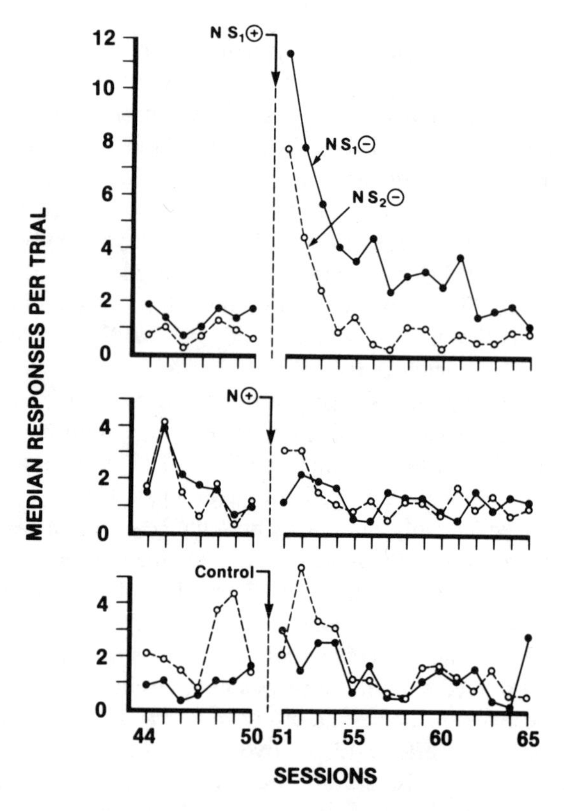

FIG. 11.2. Median responses per trial on negative trials with noise plus the dot for the last seven sessions of the previous phase of training and for sessions 51–65 following treatment sessions. Top panel shows results for Group NS_1+, middle panel for Group N+, and bottom panel for Group Control.

of the noise trials. This bird showed, however, no loss of the suppressive effect of noise on keypecking.

The persistence of the inhibitory effect of noise on keypecking despite interpolated training with noise as a signal of food implies that the mode of action of the inhibitor is not as a signal for the absence of food. The result is consistent with the view that the action of the conditioned inhibitor on keypecking is tied to the initiating-target stimulus.

Experiment I, Part 3: Effect of Signaling Food with the Inhibitory Stimulus on Concurrent Conditioned Inhibitory Training

The previous results showed that the suppressive effect of noise on pecking to an initiating-target stimulus was not appreciably diminished by a series of training sessions in which noise alone was used to signal food. It was of interest to learn whether a similar result would obtain when the noise–food trials were presented concurrently with conditioned inhibitory training.

The three groups of the previous phase were maintained intact. They received added trials of the type previously received during the treatment phase, but these trials were now intermixed with inhibitory training. Specifically, the composition of each session for each group was as follows. All groups received 12 negative trials in which noise accompanied S_2 and 12 positive trials on each of S_1 and S_2. In addition, Group NS_1+ received 12 positive trials per session with S_1 accompanied by noise. Group N+ received 12 positive trials with noise alone. Group Control received no additional trials. There were 14 sessions of training under these conditions. Extinction sessions were given at the end of this training. A session consisted of 48 trials, 12 on each of S_1, S_2, noise plus S_1, and noise plus S_2. All trials were 6 sec in duration regardless of responding.

The results for training are shown on the left side of Fig. 11.3. Responding on negative trials prior to the introduction of intermixed positive trials (sessions 60–65 of the previous phase) is shown at the far left for purposes of comparison. In Group N+, the intermixture of positive noise-alone trials had no effect on the inhibitory action of noise on responding to S_2. The overall median number of responses on negative trials with noise plus S_2 for the six sessions prior to the introduction of positive noise-alone trials was 0.9 responses per trial. The comparable figure for the first six sessions after the introduction of these trials was 0.6 responses per trial. The values for the control group were 0.7 and 1.0 responses per trial. In Group NS+ there was some loss of inhibitory effect on responding to S_2 due to the intermixed positive trials with noise plus S_1. For the six sessions prior to the introduction of these trials the overall median number of responses per negative trial with noise plus S_2 was 0.8 responses. For the six sessions following their introduction the comparable value was 2.1 responses. Each animal in this group showed an increase and each responded at a higher level during the first session with these intermixed positive trials than during any of the six sessions immediately prior to their introduction.

Results for the first extinction test session are shown in the right side of Fig. 11.3. All groups showed strong inhibition by noise of responding to S_2. Leaving aside one extremely high responder in Group Control, responses per trial for the remaining 11 subjects ranged from 3.8 to 17.5 when S_2 was presented alone and from 0 to 0.3 when S_2 was presented with noise. The groups were statistically indistinguishable in the amount of suppression in responding to S_2 attributable to the noise. It is evident that in Group NS_1+ the pairing of noise with the reinforcement of pecking to a stimulus of one color did not prevent the noise from exercising a strong inhibitory effect on responding to the stimulus of a different color. The extinction test corroborates previous findings by showing that the pairing of noise with feedings does not interfere with the inhibitory effect of

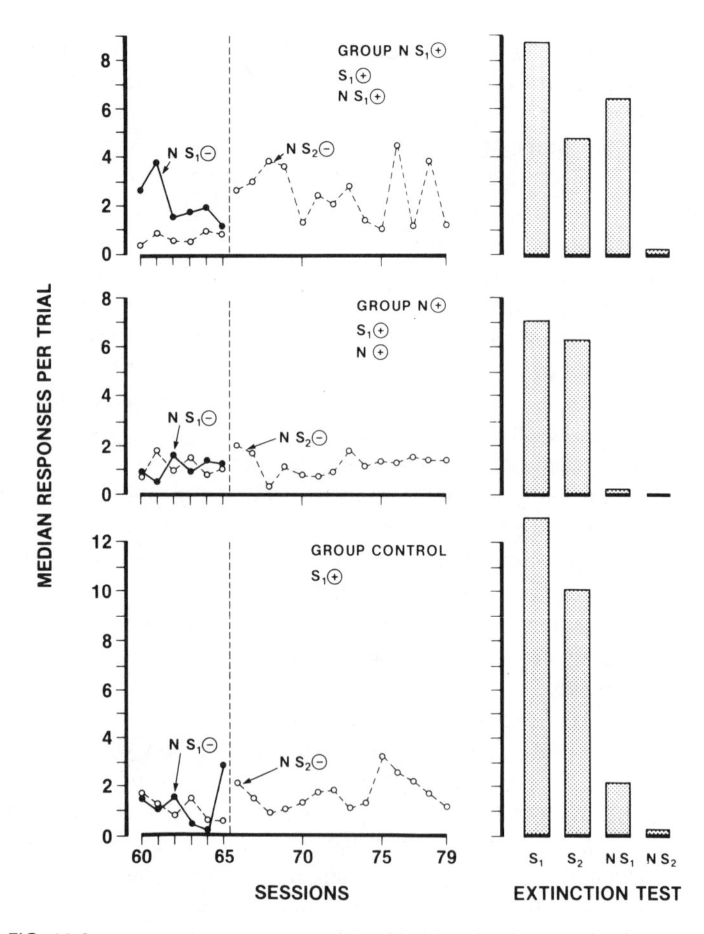

FIG. 11.3. Responding on negative trials with noise plus the second-trained dot during sessions 66–79 is shown following the vertical dotted line. Other trial types received concurrently during these sessions are shown under the Group designations. Responding on negative trials in the previous phase, in which noise was not reinforced, is shown at the far left (sessions 60–65). Median responses per trial during the first extinction test session is shown for different trial types in bar graphs.

noise on the initiating stimuli. The noise was no less inhibitory in Group N+ than it was in Group Control.

The results show that noise can act to inhibit the response to an initiating-target stimulus while concurrently being paired with food, and even while concurrently being paired with another initiating-target stimulus and food. The results are consistent with the view that in this behavior system the inhibitor acts specifically to modulate the signal value of the initiating-target stimulus. It does not appear to suppress the representation of the US. In other words, its inhibitory action does not depend on being a signal for the absence of food.

Experiment I, Part 4: Effect of Converting the Inhibitory Stimulus into a Peck-Initiating Stimulus

The absence of an effect on inhibition when the inhibitor is converted into a signal of food is consistent with the specificity of inhibitory control to the initiating stimulus. This pattern of results suggests that noise should maintain its inhibitory action on an initiating stimulus even after the noise is trained to initiate pecking. In the next phase of the experiment the effect of training the noise to function as a remote initiating stimulus for pecking a constant target stimulus afforded by an unlit key was examined. The question is whether such training destroys the previously established inhibitory effect of the noise on pecking to the initiating-target stimulus afforded by the presentation of the dot on the same key.

There is a suggestion from an incidental result previously mentioned that the inhibitory action will not be destroyed by converting the noise into an initiating stimulus for pecking. Recall that one subject in Group N+, which received positive trials on noise alone, began pecking the key during these trials yet showed no loss of the inhibitory effect of noise on pecking to the dot.

The procedure was as follows. Subjects were retrained for 9 sessions with the first-trained dot, S_1, on positive trials, and noise plus S_1 on negative trials. There were 12 positive and 12 negative trials per session. The procedure was as in original training. At the end of training, 2 groups of 5 subjects were formed. For the purposes of matching, an extremely high-rate responder and 1 other subject from the initial set of 12 subjects were not run. Each group contained 3 subjects originally trained with the green dot and 2 originally trained with the red dot. Each group also contained at least 1 subject from each of the 3 experimental groups of the previous phase of training, and the groups were matched as closely as possible on the level of responding on negative trials during the final session of retraining. The experimental group was trained to peck the unlighted key when the noise was presented. The control group received no treatment during this phase.

In the experimental group, training the noise as a remote initiating stimulus for keypecking began with shaping of the keypeck response by reinforcing successive approximations to pecking in the presence of noise. The requirement for reinforcement was gradually increased to 4 responses, and an intertrial interval of 30 sec with noise off was put into effect. Birds were then placed on automatically programmed sessions consisting of 40 positive noise-on trials. A feeding was delivered on the completion of 4 pecks or after the passage of 60 sec; whichever occurred first. A fixed 2-minute intertrial interval was used. From 5 to 6 training sessions were run under these conditions. The retention of inhibitory control over pecking to S_1 was tested in an extinction session in which there were 16 trials with S_1 alone, 16 trials with S_1 plus noise, and 16 trials with noise alone. All trials had a fixed duration of 6 sec.

In the last session of training with noise as a remote initiating stimulus each animal completed all trials. The mean time to make the four required pecks was 6.0 sec (note that maximum duration of these trials was 60 sec). The mean number of responses per 2-min intertrial period was only 1.3. Noise was, therefore, strongly established as an initiating stimulus for keypecking by the end of training.

The results for the first extinction-test session are shown for individual subjects in Fig. 11.4. Responding on the trial type that received the greatest number of responses was rescaled to 1.0. The mean number of responses per trial for the trial type that evoked the maximum number of responses is shown at the upper end of the appropriate bar. Responding to the other trial types is shown as a proportion of this maximum. In the experimental group, four of the five subjects made their greatest number of responses to the dot alone, responded least when the dot was presented with noise, and responded at an intermediate level to the unlighted key when the noise alone was presented. For the remaining subject, subject 59, noise retained none of its inhibitory effect since the maximum response occurred to the combination of noise with the dot stimulus.

Each of the control subjects showed clear inhibitory effects of the noise and, of course, no responding to noise alone. Comparison of the individual results in the experimental and control groups shows a considerable overlap in the strength

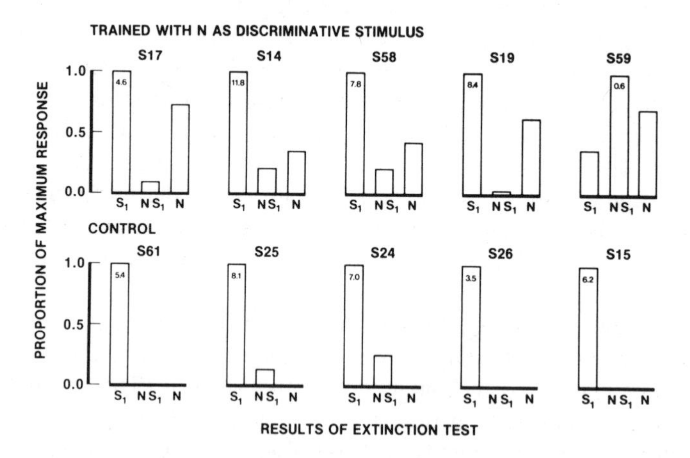

FIG. 11.4. Proportion of maximum response to test stimuli for individual subjects trained with noise as a discriminative stimulus (remote initiating stimulus) shown in the top half of the figure, and for control subjects shown in the lower half. The mean number of responses per 6-sec trial is shown for the test stimulus that evoked maximal responding.

of the inhibitory action of noise. Leaving aside subject 59 in the experimental group, the results show that the inhibitory effect of noise on the dot as an initiating-target stimulus was not removed by converting the noise into a remote initiating stimulus for keypecking at a different terminal target stimulus (unlit key).

The odd result from subject 59 in the experimental group may contain a clue to the locus of action of the conditioned inhibitor. Unlike any of the other subjects, this bird made a substantial proportion, 64%, of its positive-trial responses to a blank quadrant of the key rather than to the dot during training. No other subject made more than 1% of such responses. During the extinction-test session, subject 59 made a total of 28 responses on trials with the noise plus S_1. All but four of these were made to the blank quadrant. If the inhibitory action of the noise were exerted on the terminal target stimulus, a loss of inhibition would be expected when the inhibitory stimulus in converted to a remote initiating stimulus for pecking at the *same* terminal target stimulus, as it was, inadvertently, for this subject. The possibility that the inhibitor acts on the control otherwise exerted by the terminal target stimulus was examined in Experiment II.

Experiment II, Inhibition of Keypecking when the Initiating Stimulus is Remote. Part 1: Effect of Converting the Inhibitory Stimulus into a Remote Initiating Stimulus

In the experiments reported so far, the peck to be inhibited was, with an unplanned exception, directed at the dot that also initiated pecking. When the initiating and target stimuli are the same, it obviously makes no sense to ask whether the inhibitor acts on the initiating or on the target functions of the stimulus. The question does make sense, however, when a remote initiating stimulus is used; that is, one that initiates pecking directed to a different stimulus object at a separate location.

In the present experiment birds were trained so that, when the colored dot appeared, they pecked an adjacent, but clearly separate, sector of the pecking key. This separate sector became the terminal target of the key peck. Noise was then established as a conditioned inhibitor by the usual differential conditioning procedure. If the inhibitory action of the noise is centered on, and is specific to, the remote initiating stimulus, the noise might maintain its inhibitory effect on responses initiated by that stimulus even after the noise had been established as a remote initiating stimulus for pecking the common terminal target stimulus. On the other hand, if the inhibitory effect of noise is focused on the terminal target stimulus itself, a complete loss of inhibition should occur following training with the noise as an initiating stimulus, because in this case the inhibitory and initiating functions of the noise would be converging on the same element.

The essentials of the procedure were as follows. A two-sector key was used. Each sector was a square 1.65 cm on a side. The left- and right-hand sectors were separated by a metal strip 0.2 cm wide. An illuminated red dot, 0.4 cm in diameter, was used as a remote initiating stimulus. It was always presented in the center of the left-hand key. I refer to the left key as the signal key. Pecks to the signal key were recorded but were without effect. To receive a feeding the birds were required to peck the right-hand, or terminal target key. At the center of this key was permanently affixed a yellow-paper dot, 0.65 cm in diameter. It was illuminated only by the ambient chamber light, and it was visible even though the terminal target key was otherwise unlighted throughout (unlighted means not backlighted).

In the first phase of the experiment the birds were trained, using response-contingent reinforcement, to make eight responses on the target key to receive a feeding. They were then trained to discriminate between the presence and absence of a red dot on the signal key by reinforcing target-key responses only when the red dot was present. By the end of this phase of training each session consisted of 20 positive trials in which the dot on the signal key appeared for a maximum of 8 sec. The completion of eight pecks to the target key before 8 sec had elapsed turned off the dot on the signal key and produced an immediate 4-sec feeding. If eight responses were not made before 8 sec, the trial ended without reinforcement. At all other times the signal key was unlighted and reinforcement was unavailable. For purposes of measuring control by the signaling dot, 20 dummy trials were presented in which responses were recorded as they were for actual trials. These dummy trials were indistinguishable within the chamber from intertrial periods. Positive trials and dummy trials were presented in a random sequence. A fixed 1-minute intertrial interval was used. These training parameters were reached by a process involving gradual changes from initial values. In some cases, where performance on positive trials faltered, a temporary reversion to less demanding parameter values was necessary. Training continued until each subject was regularly completing at least 90% of positive trials and not more than 10% of dummy trials. From 30 to 80 sessions were required for different subjects to reach this criterion.

The purpose of the second phase of training was to establish the noise as a conditioned inhibitor. Each training session consisted of 20 positive trials, 20 negative trials, and 10 dummy trials. A fixed 1-min intertrial interval was used. The positive trial and the dummy trial were as in the previous phase. On the negative trial, white noise simultaneously accompanied the presentation of the signaling dot. Negative trials were terminated by the completion of 8 pecks on the target key, or at the end of 8 sec, as were positive trials, but no food was delivered. Training sessions were continued until each animal was regularly completing at least 90% of positive trials and no more than 15% of negative trials. Dummy trials were almost never completed in this, or any subsequent phase of training. From 8 to 106 sessions were required to complete this training.

The very large number of sessions was required in the case of one subject that often failed to complete positive trials and required repeated adjustments of the training parameters before performing to criterion in the standard training sessions described previously.

The purpose of the next phase of training was to establish the noise as a remote initiating stimulus for pecking the target key. The target key was constantly illuminated by the ambient chamber light. It was not backlighted. The birds were shaped to peck the target key in the presence of noise and were then placed on automatically controlled sessions each consisting of 40 positive trials and 10 dummy trials. The fixed 1-min intertrial interval was retained. Trials had a maximum duration of 8 sec. They were terminated by 8 pecks. On positive trials the noise sounded and food was delivered if 8 pecks were completed before 8 sec had elapsed. Dummy trials were as previously described. From 9 to 16 sessions of training were required before each subject was regularly completing at least 90% of positive trials and not more than 10% of the dummy trials.

The purpose of the final phase of training was to learn whether the inhibitory action of noise survived the establishment of noise as a remote initiating stimulus for pecking the common target stimulus. The test was made during the resumption of conditioned inhibitory training, which now included positive trials in which noise was presented and pecks to the target key were reinforced (signal key unlighted). There were 12 such trials per session. In addition, each session contained 12 positive trials with the red dot presented on the signal key, 12 negative trials with the red dot on the signal key accompanied simultaneously by noise and 12 dummy trials. A fixed 1-min intertrial interval was used as before. The reinforcement contingencies for each trial type were as previously described. The trials were presented in 2 blocks of 24 trials. Each block contained 6 presentations of each trial type in a random order. Sixteen of these training sessions were run.

No attempt is made to describe in detail performance during the several training phases leading up to the final assessment phase. Seven subjects were destined for the treatments described earlier, one was discarded during the phase in which conditioned inhibitory training was first attempted because for long periods it repeatedly ceased to respond to any trial type. A second subject was discarded because it quickly ceased responding on all trials during the assessment phase. Complete data were obtained on five subjects. In every case these subjects met the performance criteria for one phase before going on to a later phase. Their terminal performances were therefore quite homogeneous when measured by the proportion of different trial-types completed. They did, however, differ widely in their tendency to peck the dot on the signal key. The mean percentage of total positive-trial responses to the signal key during the last session of the first phase of conditioned inhibitory training was 29% with a range from 0 to 69%. An overall mean of 2.3 responses were made to the signal key. Observation showed that responses to the signal key tended to occur at trial onset followed by a switch to the target key. There was a high degree of consistency in the level of responding to the signal key within individuals.

The principal results are shown in Fig. 11.5. The first four pairs of points show the mean percentage of trials completed on positive and negative trials during the last four sessions of conditioned inhibitory training. The next point shows performance during the last session of training with noise as a discriminative stimulus for pecking the target key. The remaining points represent performance during the resumption of conditioned inhibitory training with intermixed noise-positive trials.

The mean percentage of positive trials completed over the last four baseline training sessions ranged from 94 to 100%. The range for negative trials was 0 to 10%. In the first inhibitory training session following the treatment in which noise served as a signal for pecking, the percentage of positive trials (S+) completed ranged from 92 to 100%. The range for negative trials (NS−) was now 67 to 100%, showing virtually a complete loss of the inhibitory effect of noise. Over the last four sessions of training, the range for positive trials was 79 to 100%, whereas for negative trials it was 0 to 40%, showing reacquistion of the inhibitory effect of noise.

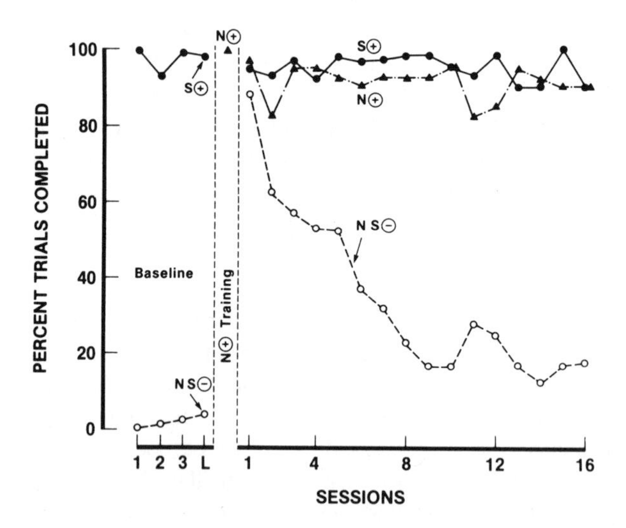

FIG. 11.5. The effect of establishing noise as a remote initiating stimulus (N+ training) for pecking a common target stimulus. Baseline performance at the end of inhibitory training is shown at the far left. Responding on reinforced noise-alone trials (signal key unlighted) at the end of training with noise as the initiating stimulus is shown by the single point between the vertical dotted lines. The remaining section of the figure shows performance when inhibitory training was resumed and reinforced noise-alone trials were intermixed with the inhibitory training trials.

It is clear that training the noise as a remote initiating stimulus for pecking the target stimulus eliminated all, or very nearly all, of its inhibitory action when presented together with the dot. As training continued, however, the inhibitory action of the noise was reestablished despite its concurrent role as a remote initiating stimulus for pecking the target key. Qualitatively the same pattern appeared in each animal with no apparent difference as a function of whether, or how frequently, the signal key was pecked.

The initial loss of the inhibitory effect of noise in this experiment contrasts with the maintenance of its inhibitory effect after similar training in the previous experiment. Although the experiments involve other differences in procedure that cannot be confidently put aside, the most direct interpretation of the divergent outcome is that in the previous experiment the noise served to inhibit pecks at one target stimulus, namely the dot, while it served as a remote initiating stimulus for pecking quite a different target stimulus; namely, a blank unlighted key. In the present experiment, in contrast, there was a common target stimulus for pecking initiated by the dot and for pecking initiated by the noise; namely, the permanently affixed yellow dot on the target key.

As previously discussed, a loss of inhibition under the present arrangement might mean that the locus of action of the inhibitory stimulus is on the stimulus that serves as the target of the peck and not on the initiating stimulus that directs the peck to the target. Although the present results are consistent with that interpretation, they are not adequate to establish it. It is possible that the inhibitory action was exerted on the red dot as a remote initiating stimulus, but reinforced training with noise alone overrode this inhibition by separately creating a tendency to peck the target stimulus. This question could be resolved by applying the transfer test design to remote initiating stimuli immediately following inhibitory training; i.e., without interpolated training in which the inhibitor serves as a positive stimulus.

The present results show that with a moderate amount of training it is possible to inhibit the action of a remote initiating stimulus selectively. That was shown by the ability of the noise to inhibit the initiating function of the dot while acting concurrently as an initiating stimulus for pecking the common target stimulus.

Experiment II, Part 2: Transfer of Conditioned Inhibition Across Remote Initiating Stimuli

Experiment I showed that the inhibitory action of noise was very specific to the properties of the target stimulus on which it was trained. The initial effect of a change of dot color from red to green, or vice versa, was a very marked, perhaps complete, loss of inhibitory effect. It is of interest to learn whether specificity also characterizes an inhibitor that is acting on a remote initiating stimulus. The procedure in effect for the last 16 sessions of training shown in Fig. 11.5 ensures that inhibitory action comes to be focused on the remote initiating stimulus and therefore provides an appropriate precondition for examining the specificity of inhibition once the inhibition has been focused on the remote initiating stimulus.

In order to test for the transfer of inhibitory control to other remote initiating stimuli, training was given on four new visual stimuli of varying degrees of similarity to the small red dot (0.4 cm in diameter) that was the first-trained stimulus. The new stimuli were: a large red or green dot (0.8 cm in diameter) and a large red or green square (1.5 cm on a side).

Training sessions on these four new stimuli consisted of 10 positive trials on each stimulus plus 10 dummy trials. The number of responses on the target key required for reinforcement was increased from 4 to 6 to 8 as training continued. A total of from 8 to 10 sessions were run, with at least 3 sessions at the 8-response requirement. The reinforcement contingency was as before. These training sessions were alternated with conditioned inhibitory training on the first-trained stimulus as in the previous phase. Inhibitory training sessions continued to include noise-positive trials.

The specificity of inhibitory action was assessed during continued training sessions in which each of the five visual stimuli, the first-trained stimulus and the four new stimuli, were reinforced when presented alone and nonreinforced when presented with noise. In addition, there were reinforced trials with noise alone, and nonreinforced dummy trials. Accordingly, there were 12 trial types.

Sessions were divided into two blocks of 24 trials. In each block there were 2 presentations of each of the 12 trial types. Within these blocks the order of presentation was random. A 1-min fixed intertrial interval was used. Reinforcement contingencies for each of the trial types were as previously described for the training sessions in this experiment. From 15 to 45 training sessions were run with the 5 subjects that completed the preceding training phases.

During these sessions responses to dummy trials were very infrequent, and responding on the target key was strongly controlled by the initiating stimuli. As training continued, inhibitory control by noise over responding to each of the visual stimuli developed gradually. The results for the first five training sessions, which contained a total of 20 presentations of each trial type, show the early profile of transfer across the remote initiating stimuli. They are shown in Fig. 11.6.

It is evident that the transfer of inhibitory control was dependent on the similarity of the first-trained stimulus to the transfer-test stimuli. Each of the five subjects showed a progressive loss of inhibitory control with increasing dissimilarity of the excitatory stimuli from the one on which the original inhibitory training had been carried out. For the most dissimilar stimulus (large green square) no animal responded appreciably less in the presence of noise than in its absence. The dependence on similarity may be said to be complete. The dependence of inhibitory action on similarity that was previously found for the case in which the initiating stimulus is also the target stimulus (Experiment 1) has its

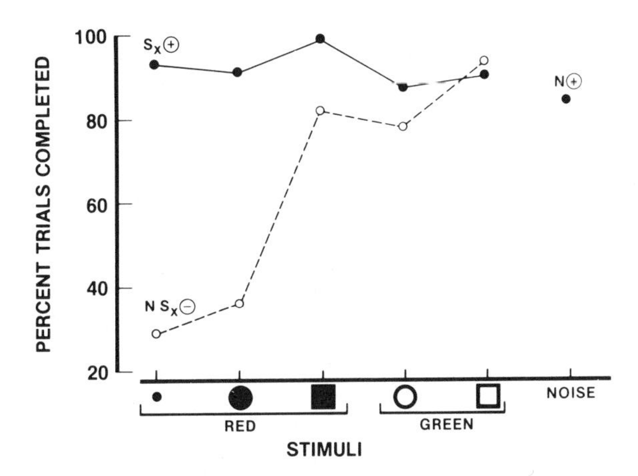

FIG. 11.6. Transfer of inhibition across remote initiating stimuli, S_x, which vary in similarity from the first-trained stimulus (small red dot). Data based on a total of 20 presentations of each stimulus over 5 training sessions. Positive trials with noise, N+, were intermixed with the other training stimuli.

parallel when, through appropriate training, inhibitory action is focused on a remote initiating stimulus.

Summary of Results

The principal results of the present experiments may be summarized as follows: (1) The initial transfer of the inhibitory action of noise from one initiating-target stimulus to another is highly, perhaps completely, dependent on the similarity of the initiating-target stimuli; (2) the inhibitory action of noise on pecking at an initiating-target stimulus is not eliminated, and may be unaltered, by the subsequent conversion of noise into a signal for food through noise–food pairing; (3) the inhibitory action of noise on an initiating-target stimulus is also not eliminated, and may be unaltered, by the subsequent conversion of the noise into a remote initiating stimulus for pecking a target stimulus that is visually distinct from the inhibited initiating-target stimulus; (4) the inhibitory effect of noise when the initiating stimulus is remote is severely weakened (perhaps eliminated) when noise is subsequently converted into a remote initiating stimulus for pecking a common target stimulus; (5) by means of appropriate training, the inhibitory effect of noise can, however, be focused on a remote initiating stimulus as distinct from the target stimulus. This is evidenced by the fact that noise can serve to inhibit keypecking initiated by a remote stimulus, while noise functions concurrently as a remote initiating stimulus for pecking a common target stimulus; (6) when the inhibition has been focused on the remote initiator, the initial transfer of the inhibitory action to other remote initiating stimuli is completely dependent on their similarity to the remote initiator on which inhibitory training was initially carried out.

Discussion

Conditions Under Which Inhibition is CS-Specific and Independent of the First-Order Relation of the Inhibitor to the US. The properties of CS-specificity, and independence of inhibitory action from the first-order relation of the inhibitor to the US, which we have found in the present experiments with simultaneous presentations of the inhibiting and initiating stimuli, were not found in Holland's experiments on the inhibition of bar pressing with the simultaneous procedure. They were, however, found by Holland with the serial procedure. With the benefit of hindsight, this does not seem too surprising. The critical factor appears to be whether the common stimulus in the feature negative training procedure maintains an active status on nonreinforced trials, or, on the other hand, becomes irrelevant. With diffuse stimuli and an undirected response, such as Holland used in his experiments, serial presentation may be necessary to maintain an active role for the common stimulus on negative trials. However, with the keypecking response directed at the lighted key, the target stimulus maintains control even under simultaneous presentations. It is interesting to note that in a feature positive discrimination with noise presented as the distinctive feature on positive trials, pecks continue to be directed to the common stimulus of the lighted key. Rescorla (this volume) found that to be the case even when the noise, which in his experiment preceded the key light, remained on during the key-light presentation. In an unpublished experiment involving a feature positive discrimination with noise presented simultaneously with the key light on positive trials, I have also found that the key light maintains control in the presence of noise, since changes in the color of the key light eliminated responding to the key even though the noise was present. I am suggesting that the stimulus object to which the peck is directed is not necessarily made irrelevant by the presence of remote modulating stimuli, whether they be facilitators or inhibitors, presented simultaneously with the stimulus object. Moreover, it appears that if the common stimulus maintains an active status, the modulation will be relatively independent of the first-order relation of the modulator to reinforcement or nonreinforcement. The results reported by Rescorla (this volume) and by Holland (this volume), as well as my own results, are in agreement on this point.

The results on the specificity of inhibitory action to the common stimulus with which the inhibitor was trained are, however, not so clear. Holland (this volume) reports specificity of inhibition to be associated with independence of inhibitory action from the first-order relation of inhibitor to reinforcement. My results also show both specificity and independence. Rescorla (this volume, reference to his p. 318), on the other hand, finds virtually complete transfer of inhibitory action to another excitor even though he demonstrates independence of inhibitory action from the inhibitor's first-order relation to reinforcement. However, he tested transfer to an excitor that had itself been subject to inhibition by another inhibitor in a feature negative discrimination. This is clearly a special test procedure that could well show transfer even though transfer might not be found to an excitor with a simple history of reinforcement. The results of another of Rescorla's experiments (see his Fig. 13 in Chapter 10, this volume) in which the conventional transfer summation test was made show much less transfer, perhaps no more than could be accounted for by generalization between excitors.

The Parallel Between Facilitation and Inhibition. Both Holland (this volume) and Rescorla (this volume) develop the parallel between inhibition and facilitation as modulators of the signal value of another stimulus. The action of facilitators on the signal value of a common stimulus shows a high degree of independence from the first-order relation of the facilitating stimulus to reinforcement or nonreinforcement. It also appears that the action of a facilitator is highly specific to the common stimulus with which it was trained when transfer is tested by combining the facilitator with a new stimulus that has no history of reinforcement. Rescorla (this volume) and Rescorla, Durlach, and Grau (in press) do report transfer of facilitation if the test stimulus has a history of excitation. It is quite possible that modulating effects of inhibitors, when the inhibitor is one that acts on the CS, and the modulating effects of facilitators, are each CS-specific provided that inhibition is tested with stimuli that have only an excitatory history, and that facilitation is tested on stimuli that have no excitatory history.

In the literature on conditioned inhibition, a number of authors have been concerned with the question of whether conditioned inhibition is the symmetrical opposite of conditioned excitation (see Baker, 1974, Kaplan & Hearst, in press; Konorski, 1972; Rescorla, 1979). The present results show that in the case of keypecking, the inhibitor modulates the signal value of the initiating stimulus. This does not parallel the operation of a conditioned excitor, because establishing a stimulus as a conditioned excitor for keypecking does not entail a second-order or modulating function. When the initiating and target stimuli are provided by the same event (e.g., lighting the key), second-order relations are not brought into play. Accordingly, the parallel to conditioned inhibition on the excitatory side arises with a remote initiating stimulus that appears to enable, or switch on, the target stimulus. In other words, the parallel is to what has been variously referred to as an "if-then" stimulus, a facilitator, a gating stimulus, a modulating stimulus, a conditional stimulus, an occasion setter, or a discriminative stimulus. The same view has been expressed by Holland (this volume), Rescorla (this volume), and by Rescorla, Durlach, and Grau (in press). The experiment on conditioned inhibition and excitation by Brown and Jenkins (1967) embodied this view as well.

Implications of Inhibitory Specificity for the Accepted Tests of Inhibition. The arguments advanced by Rescorla (1969) and by Hearst (1972) in support of using both a summation (or combined-cue test), and a test for retarded acquisition of excitatory conditioning to the inhibitor, have been widely accepted (see Dickinson & Mackintosh, 1978; Zimmer–Hart & Rescorla, 1974). Moreover, when inhibition is trained in the Pavlovian conditioned inhibition, or feature negative procedure, it has been generally accepted that the summation effect must be tested on a new excitatory stimulus with which the inhibitor has not previously been combined, or, in other words, by a transfer summation test. The transfer summation test was recognized by Pavlov (1927) as a critical test in order to show "that the phenomenon of conditioned inhibition is really in the nature of inhibition . . . , and is not merely a passive disappearance of the positive conditioned reflex owing to the compound stimulus remaining habitually unreinforced" (p. 75). The transfer summation test is designed to show that the inhibitor maintained its identity during the training when it was combined with an excitatory stimulus and did not interact with the excitor to form a unique configural stimulus.

What we have repeatedly referred to as the inhibiting stimulus in the present experiments passes neither the retardation test nor the transfer summation test. Inhibition was found to be dependent on the similarity of the transfer test excitors to the parent excitor with which the inhibitor was trained. But that is exactly what is to be expected if the inhibitor acts on the signal value of the parent CS rather than acting on the US representation. If we are to allow that inhibitors might act on the signal value of the CS, we cannot then insist on transfer of inhibition to other CSs independently of their similarity to the parent CS as a criterion for inhibition in general.

The retardation test also presupposes a particular conception of the inhibitor. If the inhibitor acts as a signal of no US, or acts to suppress the CR, retardation would be expected. On the other hand, when the inhibitor acts as a modulator of the signal value of the parent CS, it can continue to inhibit even though it is converted into a first-order excitatory signal of the US. In this case, the retardation test is simply precluded by the excitatory counterconditioning.

I find the evidence persuasive that, depending on details of the procedures and the nature of the behavior system involved, the Pavlovian conditioned inhibition paradigm can produce an inhibitory stimulus that acts as though it were a signal of no US by passing the transfer summation test and the retardation test, or it can produce an "inhibitory" stimulus that acts as though it were a signal that a particular CS will not be followed by the US by showing CS-specificity and independence of inhibitory action from the direct relation of the inhibitor to the

US. We could choose to call only the first type of stimulus a true inhibitor and give some other label to the second type. I suggest, however, that the two varieties of inhibition have far too much in common to warrant such treatment. Perhaps when an inhibitor acts as a signal for no US it might be termed a *first-order inhibitor*. When it acts on the signal value of an excitor it might be termed a *second-order inhibitor*.

Configural Conditioning as an Alternative to Second-Order Inhibition. Is it possible to distinguish between an inhibitory stimulus that acts specifically on the initiating function of the stimulus with which it was trained and one that forms a unique compound with that stimulus? Otherwise said, how can specific inhibitory action be distinguished from configural conditioning? The same issue arises for facilitation by a remote initiator whenever the facilitation proves specific to the target stimulus with which the remote initiator was trained.

It has been pointed out to me (Rescorla, personal communication) that the truth table for the proposition A implies (B implies C) is the same as for the proposition (A and B) implies C. Because the first of these propositions could be taken to represent the case in which noise (A) acts specifically to alter the status of the initiating stimulus (B) as a signal for food (C), and the second proposition to represent the case in which the compound of noise and initiating stimulus signals no food, it would appear that no distinction can be made at a formal level between stimulus-specific action of an inhibitor and the action of a compound stimulus.

I believe that a firm distinction between configural conditioning and CS-specific action of an inhibitor could in principle be made if we were able to obtain evidence on the conditions that influence the ease of learning based on configural patterning of stimuli. It would then be possible to compare those conditions with those that lead to specificity of inhibitory action (see Rudy & Wagner, 1975), and in that way to decide whether the phenomena are closely tied or separable. At present, we have only plausibility arguments. One of the more persuasive of these arguments is provided by Holland (this volume) when he notes that simultaneous presentation would be expected to favor configural conditioning when compared with serial presentation, yet it is the latter arrangement that produces CS-specific inhibition in the behavior system on which he has experimented.

Another approach to the problem is to search for variables that influence the degree of CS-specificity even though they involve no variation of the conditions of stimulus presentation during either the training or the transfer summation test. If, as now seems likely, it turns out that the transfer summation test can show either specificity or broad transfer depending on the history of reinforcement or nonreinforcement of the transfer test stimuli, it would be possible to reject configural conditioning as an explanation of CS specificity.

I believe it must be accepted that we do not at present have the evidence with which to distinguish CS-specific inhibition from subtle forms of configural conditioning. I trust, however, that this uncertainty will not impede the effort to learn more about how the nature of the behavior system that is being inhibited affects which of several possible modes of inhibitory action is brought into play.

ACKNOWLEDGMENTS

This research was supported by a grant to the author from the National Science and Engineering Research Council of Canada. I wish to express my appreciation to Miss. B. Royle for her painstaking execution of the experiments.

REFERENCES

Asratian, E. A. (1972). Genesis and localization of conditioned inhibition. In R. A. Boakes & M. S. Halliday (Eds.), *Inhibition and learning*. New York: Academic Press.

Baker, A. G. (1974). Conditioned inhibition is not the symmetrical opposite of conditioned excitation: A test of the Rescorla–Wagner model. *Learning and Motivation, 5,* 369–379.

Brown, P. L., & Jenkins, H. M. (1967). Conditioned inhibition and excitation in operant discrimination learning. *Journal of Experimental Psychology, 64,* 365–376.

Dickinson, A., & Mackintosh, N. J. (1978). Classical conditioning in animals. *Annual Review of Psychology, 29,* 587–612.

Hearst, E. (1972). Some persistent problems in the analysis of conditioned inhibition. In R. A. Boakes & M. S. Halliday (Eds.), *Inhibition and learning*. New York: Academic Press.

Hearst E., & Jenkins, H. M. (1974). *Sign-tracking: The stimulus-reinforcer relation and directed action.* Austin, TX: The Psychonomic Society.

Holland, P. C. (1977). Conditioned stimulus as a determinant of the form of the Pavlovian conditioned response. *Journal of Experimental Psychology: Animal Behavior Processes, 3,* 77–104.

Jenkins, H. M., & Sainsbury, R. S. (1970). Discrimination learning with the distinctive feature on positive or negative trials. In D. Mostofsky (Ed.), *Attention: Contemporary theory and analysis.* New York: Appleton–Century–Crofts.

Kaplan, P. S., & Hearst, E. (in press). Contextual control and excitatory vs. inhibitory learning: Studies of extinction, reinstatement, and interference. In P. D. Balsam & A. Tomie (Eds.), *Context and learning.* Hillsdale, NJ: Lawrence Erlbaum Associates.

Konorski, J. (1967). *Integrative activity of the brain.* Chicago: University of Chicago Press.

Konorski, J. (1972). Some ideas concerning physiological mechanisms of so-called internal inhibition. In R. A. Boakes & M. S. Halliday (Eds.), *Inhibition and learning.* New York: Academic Press.

Pavlov, I. P. (1927). *Conditioned reflexes.* London: Oxford University Press.

Rescorla, R. A. (1969). Pavlovian conditioned inhibition. *Psychological Bulletin, 72,* 77–94.

Rescorla, R. A. (1979). Conditioned inhibition and extinction. In A. Dickinson & R. A. Boakes (Eds.), *Mechanisms of learning and motivation: A memorial volume to Jerzy Konorski.* Hillsdale, NJ: Lawrence Erlbaum Associates.

Rescorla, R. A., Durlach, P. J., & Grau, J. (in press). Contextual learning in Pavlovian conditioning. In P. D. Balsam & A. Tomie (Eds.), *Context and learning.* Hillsdale, NJ: Lawrence Erlbaum Associates.

Rescorla, R. A., & Holland, P. C. (1977). Associations in Pavlovian conditioned inhibition. *Learning and Motivation, 8*, 429–447.
Rudy, J. W., & Wagner, A. R. (1975). Stimulus selection in associative learning. In W. K. Estes (Ed.), *Handbook of learning and cognitive processes* (Vol 2). Hillsdale, NJ: Lawrence Erlbaum Associates.
Skinner, B. F. (1938). *The behavior of organisms.* New York: Appleton–Century.
Zimmer–Hart, C. L., & Rescorla, R. A. (1974). Extinction of Pavlovian conditioned inhibition. *Journal of Comparative and Physiological Psychology, 86*, 837–845.

12 A Cognitive, Nonassociative View of Inhibition

Robert C. Bolles
University of Washington

When I teach undergraduate learning I sometimes ask my class, "how do we know that there is such a thing as inhibition?" My students usually say that the existence of inhibition is demonstrated by the spontaneous recovery effect and the disinhibition effect. That is the correct short answer. It is, after all, the answer Pavlov would have given. Some of my better students add reinstatement as a third phenomenon. The long answer to the question is that these three phenomena, all of which can be observed following extinction, indicate that during extinction the original excitatory association is still intact with at least some of its initial strength. The loss of the conditioned response does not reflect a corresponding loss in the underlying association. Therefore, it is logically required that there be some opposing process, some countermanding force, that operates to produce the loss in responding seen in extinction. We give the name inhibition to that logically required process. That long answer is also correct because it too is the answer Pavlov would have given. Since Pavlov's time there have been other developments, of course, and we consider some of these shortly. But let us turn first to consider one of the basic Pavlovian phenomena that originally inspired him to think about inhibition. And let us look at it from the rat's point of view.

Evidence of Inhibition

Initially, from the rat's point of view, it gets handled a good deal, and gets to know the experimenter, and learns that the experimenter is okay. One day it is put in a box it can explore. After a while it hears a tone and then discovers food.

After a number of such occasions the rat might begin to think that this is a good experiment. Whenever the tone comes on he can expect food. And, of course, he is very hungry. He is glad he volunteered to be in the experiment. But then one day, the rat discovers that when the tone comes on there is no more food. What is this? This is extinction—more tones, but no more food. At this point the rat might begin to wonder whether he should have volunteered for the study. On the next day, back in the box once again, the rat now has several things stored in long-term memory. One is that this has been a good experiment, and one is that this has been a bad experiment. Now he is not sure what to expect; the situation is obviously ambiguous because we have made it ambiguous. But the rat is very hungry, (we have arranged that also), and hungry rats are eternally optimistic. So, when the tone comes on again, it responds as though it expected food. Voila, we have spontaneous recovery. And notice that we have spontaneous recovery without having to suppose that there are any Freud-like warring forces trying to resolve themselves. There is no subterranean conflict, no associative competition. There is just accurate memory: We need only assume the animal has both the acquisition experience and the extinction experience stored in long-term memory. If the animal remembers both kinds of episodes, it will respond in the ambiguous test situation just as we would expect an uncertain but hungry rodent to respond: with some hesitant anticipation. And notice that as soon as we start to define what kind of an episode the current one is, the rat quickly assimilates that new information. Thus, if we give a series of reinforced trials we will see very rapid reacquisition; if we give nonreinforced trials we will see very rapid further extinction. What happens today can quickly resolve the ambiguity created by incongruous events in the past.

I am describing here a concept that has been developed by Mark Bouton and myself and described in more detail elsewhere (Bouton & Bolles, in press). Our basic notion is that when you condition some response in an animal, what you are really doing is conveying information about the environment: This CS signals that US. And typically the animal learns just that and remembers that. Then, when you carry out the extinction of that response, you are conveying a quite different message: This CS means no US, which the animal also learns and which it also remembers. How the animal will respond next is a little uncertain, as indeed it should be, because it has to act on ambiguous information. We assume that when the CS itself is ambiguous, the expression of behavior is quite likely to be governed by other stimuli, such as contextual cues. For example, if extinction is carried out in a different apparatus from where acquisition occurred, the animal's behavior is going to be very largely determined by which of the contexts it is tested in. If testing is carried out where acquisition occurred, we will see anticipation of the US. But if we test the animal in the extinction context, we see little responding. Contexts can have, we assume, that kind of disambiguating effect. "Disambiguate" is the key word. (Pavlov gave us 'disinhibition'; now I give you 'disambiguate.') One may suppose that any kind of setting can help to resolve ambiguity. It might be temporal, it might be spatial, or it might involve apparatus cues, or some sort of pattern such as alternation of positive or negative trials.

Bouton developed this point of view from his research on the reinstatement effect (Bouton & Bolles, 1979). Let me outline the reinstatement procedure as it was applied to the conditioned suppression situation by Rescorla and Heth (1975). From the animal's point of view, it first learns about being handled, about the experimenter, and so on. Then it learns about bar pressing for food. Then in the first experimental episode (acquisition) the animal is given a number of trials in which tone signals shock. Then in the next episode (extinction) there are a number of trials in which the tone occurs alone and there is no shock. Then in the third kind of episode (reinstatement), shock occurs alone and there are no tones. Surely the rat must be perplexed as to what it should expect when in the next episode the tone comes on once more. What happens, of course, is that the rat plays it safe. The frightened rat is eternally pessimistic, so it freezes or quits bar pressing or stops whatever it was doing. And so we see the reinstatement effect: Shock alone following extinction reinstates the extinguished response. Have we witnessed a memory being recaptured, as Rescorla and Heth suggested? Do we see the forces of excitation and inhibition battling each other, to be won in this case by excitation? No, I do not believe that anything like that as happened. I think that just as the excitatory association is still there in long-term memory and is still accessible by proper retrieval cues, so the inhibitory association corresponding to the negative message is also in permanent storage. Thus, the reinstatement effect does not seem to depend on associative strength, but rather on some sort of selection from among associations that are still in memory. This conclusion is based on Bouton's findings of how important contextual cues are in determining whether or not the reinstatement effect is obtained (Bouton & Bolles, 1979). Thus, it was found that reinstatement shocks given in the apparatus where the CS was to be tested did result in renewed fear, but equivalent shocks given in a different apparatus did not. Note that I am referring here not to the fear elicited by contextual stimuli, but to fear elicited by the CS. Thus, it appears that the context can disambiguate a CS that has a history of giving the animal ambiguous and conflicting messages.

It is important to recognize that all the classical evidence for inhibition is found following extinction. Spontaneous recovery, disinhibition, and reinstatement are all seen after a series of extinction trials has followed a series of acquisition trials. In other words, all the classical evidence for inhibition comes from experimental procedures in which the experimenter has made the CS inherently ambiguous. It is a remarkable fact that, as far as Bouton and I were able to determine, all the evidence for contextual effects also occurs with experimental procedures that create ambiguity; that is, if the CS consistently predicts the occurrence of the US, the animal seems to ignore contextual stimuli. It is only when there is a history of mixed messages from the CS that the animal begins to rely on contexts to resolve ambiguity. For example, context is very important in the reinstatement effect; it matters a great deal whether the reinstatement shocks are given in the original conditioning box or in the subsequent extinction box. And then it matters again which of the boxes the animal is tested in. The inflation effect is quite different, however. With Rescorla's (1974) procedure, fear conditioning is carried out with mild shocks, and then the animal is given a few trials with strong shocks that are not paired with the CS. These animals subsequently show heightened ("inflated") fear of the CS. Bouton (1984) has recently found that it makes no difference at all how many different boxes are used to carry out the procedure or in which one the animal is tested.

The importance of the context in the reinstatement effect indicates that the effect is not a matter of the animal forgetting the US in extinction and then having the US-memory brought back by the reinstatement shocks. The animal evidently does not forget the US in extinction. Nor, as we have already seen, does it seem likely that the CS–US association is materially weakened in extinction. Rather, I suggest that in extinction the animal is simply learning something about its current world, namely, that the US is not going to occur. I would suggest as an alternative to the "loss" theory of extinction, and as a basic working principle, that the animal remembers virtually everything that happens to it in a conditioning experiment. Any uncertainty it may have at the time of testing derives entirely from the ambiguity of the CS, and not from a loss of information.

The Problem of Behavior

If we start with the assumption that essentially all associative input is stored in long-term memory, there arises the problem of how this material is selected for behavioral expression. This is a question we have to worry about. We have always been inclined to take changes in behavior as direct evidence for changes in associative strength. We all tend to make that mistake; it is part of our psychological heritage. In the early days of behavior theory, everything was S–R; there was nothing but S–R associations. Therefore, changes in behavior necessarily had to reflect a change in associative strength. And *that* is why Pavolov had to have inhibition. He had no other way to account for the response decrement, the loss of salivation that occurred in extinction, even though the excitatory association still had considerable strength. Something had to intervene between that association and behavior, so Pavlov let a new kind of negative association intervene. And ever since then we have been trying to figure out just what kind of thing it is.

What happened next historically was that motivation was discovered. Here was a new class of intervening concepts that could lie between the associative structure and behavior. For example, drives could motivate the whole system, turning associations on and off to provide control over behavior. The hungry dog would salivate to the tone, but the sated one would not. Pavlov knew that, of course, but he failed to attach any importance to it.

Then as time went by, the drive concept began to fail us, and we turned to more flexible and more interesting motivators. One of the most important people to work on this problem was Kenneth Spence. He and his students proposed that between the associative structure (which of course still consisted of S–R connections) and behavior there stood a variety of little motivational mediators, little responses, such as r_G, which were classically conditioned. The essential thing about the hypothetical r_G and its several cousins was that they were motivational responses. Spence himself was reluctant to call these things emotions, but he credited them with doing all sorts of motivational work. They motivated defensive behavior and avoidance. They provided incentive motivation for appetitive behavior. And they were certainly important mediators of behavior in frustrating situations. This was a very attractive model. It was in the old, noble S–R tradition, and it provided a much needed behavioristic account of some of the cognitive-appearing results that had been reported from Berkeley and elsewhere. For example, the latent learning phenomenon could be easily dealt with by using r_G.

So the primary associative system was modulated and activated by little emotion-like responses. That worked pretty well for awhile. But then things began to go wrong. One problem was that no one could actually detect r_G. Another problem was that, whereas Spence had made great contributions to the concept of inhibition as it applied to discrimination learning, he had not thought of inhibition as it might apply to these little emotions. Thus, r_G and all its relatives were excitatory. The ironic problem was that no one had been reading Pavlov very carefully, so no one understood that there had to be an inhibitory as well as an excitatory side to all these little conditioned emotional responses. There had to be a r_G inhibitor.

There were two fellows, however, who apparently had been reading Pavlov; they were Rescorla and LoLordo (1965). These fellows found it, it was there—a fear inhibitor. A tone, established as a CS+ off the baseline, had the effect of accelerating ongoing Sidman-avoidance behavior. Another tone, established discriminatively as a CS− off the baseline, decelerated avoidance behavior. This was a totally new thing. There had been discriminative conditioning before, and there had been prior discriminative control studies of instrumental behavior, but this was the first report of turning off a r_G-like mediational motivator. It was the first explicit account of demotivation with a Pavlovian procedure, although the behavioral effect of such a thing had no doubt been seen before, for example, in extinction. The study of inhibition suddenly came to life, transfer of control studies began to blossom here and there, and all at once the whole experimental world of conditioning looked like springtime.

Meanwhile, Rescorla and Solomon (1967) were developing the theory. There were excitors and inhibitors. And they pertained either to positive USs such as food, or negative ones such as shock. Therefore, it was possible, conceptually, either to turn on or turn off appetite or to turn on or turn off aversion (in effect, fear). Thus, they had a neat 2 by 2 format indicating the different kinds of Pavlovian cues that were potentially capable of motivating or demotivating instrumental behavior. And because these Pavlovian CSs could be applied to either appetitive or defensive instrumental behavior, we ended up with an 8-cell table of mediational possibilities (see Table 12.1). For example, an excitatory conditioned stimulus predicting the occurrence of shock should have a negative effect upon, or demotivate, appetitive instrumental behavior. And that particular cell is well filled by a number of conditioned suppression studies. That part of the 8-cell table looks very solid.

Suddenly, we could see that r_G had more cousins than anyone had anticipated. With eight of these little mediators, there were too many for them all to have names. But it was easy to see where Mowrer's (1960) hope and fear fit into the overall scheme of things. And the fear cell was well filled by all the work on conditioned suppression. Rescorla and LoLordo (1965) had filled in two other cells in one great bold stroke. So all the rest of us had a clear invitation, and a very inviting invitation, to get to work and fill in the other cells of the 8-part table.

Rescorla and Solomon's Table

The invitation was irresistible, and a lot of people got to work trying to fill in the cells. Let us now briefly survey some of this work, and let us start where, historically, it all started, that is, with the report of Rescorla and LoLordo (1965). When I first heard of their work, I was very excited about it. At that point in my own research career, I was deeply involved in avoidance behavior and all set to do avoidance studies, and so my students and I set to work to replicate

TABLE 12.1
Effects of Different Kinds of Pavlovian Mediators on Instrumental Behavior

		Instrumental Behavior	
	Pavlovian Mediator	*Appetitive*	*Defensive*
Appetitive	Excitatory	+	–
	Inhibitory	–	+
Aversive	Excitatory	–	+
	Inhibitory	+	–

Rescorla and LoLordo's effect using rats (they had reported their results with dogs). We had a lot of trouble getting it. We could usually get accelerated avoidance performance with a CS+, a danger signal. But we found it was not so easy to get the deceleration effect with a CS−, a safety signal. Finally, however, we were able to report something that at least looked right with rats in a shuttle box (Grossen & Bolles, 1968). There were problems with this study, though, that kept it from being a very satisfying demonstration. We had used a procedure in which the CS− was probed on the baseline very shortly after an avoidance response had occurred. Thus, the possible demotivational effect of the CS− was confounded with its potential reinforcing effect. Worse than that, the CS− should have been probed on the baseline randomly in time so as to average out whatever dependence the demotivation or deceleration effect may have on such parameters as time since the last response. It is well known that the probability of responding on a Sidman schedule depends importantly on the time since the last response, and we may expect that the effects of Pavlovian cues do too. If the timing of the probe test is synchronized with respect to where the animal is in a trial, the effect of the cue may be specific to that temporal location and have little generality as an overall effect. The Grossen and Bolles study should be discounted.

Hendersen (1973) reported a reliable demotivating effect with rats in a shuttlebox. But in addition he found that when a novel tone stimulus, one that was never presented in the Pavlovian phase of the experiment, was probed on the baseline, it too produced marked weakening of the avoidance response. The novel stimulus was evidently inhibiting fear just as effectively as the CS−. Then Libby (1976) reported just the opposite baseline reference problem. Her pretest of the then novel test stimulus produced an unconditioned acceleration of behavior, and her CS− demotivation effect was obtained only relative to this elevated baseline. These two studies demonstrate that great care must be exercised to separate the possible CS− fear inhibition effects of the stimulus from its other, nonassociative effects.

The comparable effects reported earlier by Weisman and Litner (1969) in the wheel-turn apparatus appear to be large enough and durable enough to override all baseline problems. And the replication by Morris (1974), using the same apparatus, gives further credence to the idea that the CS− effect does prevail in the paddle-wheel apparatus (in which the rat moves a small wheel to avoid shock). One might wonder why Weisman and Litner's animals failed to show extinction, or why their reinforcement effect was so huge. But even if we put such skeptical opinions aside—even if we grant that rats show the CS− effect in the paddle wheel—there remains the more important question of why they do not show it in other avoidance situations.

In part because my whole approach to avoidance behavior was at that time concerned with the learning of different sorts of avoidance responses, I proceeded to look at CS+ and CS− effects with different responses. Most of the results we found were rather distressing. We tried to replicate the Rescorla and LoLordo results with runway avoidance, a running wheel, the Skinner box, and the shuttlebox. In each case we got strong CS+ acceleration, but everything else in the world except CS− deceleration. The more we pursued the problem, the worse it got. The Rescorla and LoLordo phenomenon began to look like it had very little response generality. Dogs showed it in the shuttlebox and rats showed it in the paddle-wheel situation, but in other situations it simply was not there. Even the accelerative effect of CS+ had become suspect. Hurwitz had found (e.g., Hurwitz & Roberts, 1969) that a CS+ could suppress bar-press avoidance, at least under some conditions. Thus, how one obtains the expected effects appears to be quite uncertain, and rather precarious. In the Skinner box, for example, the acceleration effect now appears to depend on shock intensity, CS duration, and the number of shocks per session. Overall, the effect appears to depend on what the animal is (dogs seem to behave according to the theory, but rats surely do not), and what the avoidance response requirement is (the dog's success in these matters may be fortuitous consequence of its exclusive deployment to the shuttle box, whereas the rat's only reasonable claim to success is seen in the paddle-wheel apparatus). The best thing I can say about the two cells of the table that were originally laid claim to by Rescorla and LoLordo is that there are serious problems, grave questions of generality, in both of those cells.

Probably the simplest and most straightforward cells of the table are those where a food-relevant cue is superimposed on food-getting behavior. We might have an animal bar pressing for food, or running for food, and we could present a CS+ or a CS− for food that had been established elsewhere. That kind of study was easy to do, and several people did it, and the results came out uniformly disappointing. The effect simply is not there. This negative finding is somewhat surprising because we know it is quite possible to establish stimulus control in a simple instrumental situation. If we make food available in the presence of a light but not in its absence, we soon see nearly all responses restricted to the light condition. We are inclined to think of the light as a classically conditioned incentive motivator, as an elicitor of r_G, or a food signal, or a CS+. That is what the theory in one form or another tells us. Why then does the CS+ established elsewhere, off the baseline, have no effect on the animal's behavior? Is it just because of a ceiling effect, or because the experimental CS+ is redundant with all the other predictors of food in the situation? No, I do not think so, because the CS− has no effect either. Rather, I think it is that the CS+ established elsewhere is just that: It was established elsewhere. It was conditioned in a different context, perhaps in a different apparatus, and surely in a different episode at a different time.

Consider this: There is substantial evidence reported by Bolles, Grossen, Hargrave, and Duncan (1970), Estes (1943), Marx and Murphy (1961), and others that, whereas an off-baseline mediator has little or no effect on the baseline behavior, it does have an effect when the baseline behavior is extinguished.

This is the key point. When the animal is carried into extinction, *then* the off-baseline predictors, both the CS+ and the CS−, do have their expected effects on the behavior. So now we find something quite unforseen. These little motivational mediators, or r_Gs, or emotions, only work following extinction, or following a history of ambiguous episodes. And it may not be out of place to observe that, for a variety of methodological reasons, Rescorla and LoLordo's dogs were also tested in extinction.

We have looked so far at just the four cells where simple results should be expected, the cells in which the predictive cues predict an event that is relevant in the baseline situation—the food predictor is relevant when the animal is working for food. And yet we find that these four cells are full of problems. Let us look briefly at the other four cells, some of which contain some rather funny studies. The conditioned suppression or CER cell is safe and secure, there can be little question about that. We can generally count on fear to inhibit appetitive behavior; that is, we can count on a predictor of shock disrupting food-getting behavior. There are a few exceptions (e.g., Amsel, 1950), but these are few and far between. I am not surprised at the robustness of this effect because I would assume that the defensive system in the rat should take priority over the food-getting system. But how about a safety signal? Even if we grant that a safety signal is a good thing in an avoidance situation or other fearsome place, we may still ask whether it is good only in those aversive contexts or whether it becomes a good thing in itself. Hammond (1966) reported what looks like the latter effect: He found that a safety signal enhanced bar pressing for food. However, the data are reported in terms of suppression ratios, which in this case were well above .50, implying that there was considerable residual fear in the situation. So it remains unclear whether the safety signal was really enhancing the hunger-motivated behavior or whether it was inhibiting defensive behavior. Two studies by Grossen (1971) and Hyde (1976) indicate quite clearly, I think, that the safety signal only enhances appetitive behavior in a context where there is residual fear. In a neutral situation, an unambiguous food context, it has no effect.

There is one curious cell that deals with the case of a food cue and what that does to avoidance behavior. The answer, which I cannot intellectualize but only report, is that the food cue inhibits avoidance behavior (Bull, 1970; Davis & Kreuter, 1972). This finding makes no sense to me; I would have thought than an animal busy defending itself would pay no attention to a food cue. Remember, defense is supposed to have priority over food getting. I have to confess, moreover, having contributed to this mystery myself, because we found the same thing in our lab (Grossen, Kostansek, & Bolles, 1969).

Perhaps there is a reasonable explanation. Davis and Kreuter found that the food cue did not actually alter the overall response rate of bar-press avoidance. What it did do was disrupt the temporal pattern of responding, resulting in the animal receiving more shocks (this is also what Hurwitz and Roberts (1969) found when they probed a CS+ for shock on avoidance behavior). The implication, then, is that the effect is not motivational. It is not a competition between r_G and one of its brothers; the competition simply occurs at a stimulus level. The food cue disrupted, literally, distracting the animal from its task that involved a subtle temporal discrimination. Perhaps it was no more than an external inhibitor.

The evidence taken altogether is disappointing. It was disappointing to me personally because I had wanted to believe in the picture Rescorla and Solomon (1967) had painted. But every cell of the table one looks at is full of trouble. Even the historically oldest body of data, that on the CER phenomenon, is not above suspicion. Thus, we really do not know, even after so many experiments, whether CER really involves motivational competition. Perhaps the CS+, the danger signal, is simply a more salient stimulus than all those subtle things that control appetitive bar pressing. Perhaps it is a matter of response competition. The animal is too busy freezing, or whatever, to be pressing a bar. Of course Rescorla and Solomon conceived of motivational competition in basic S–R terms, so that they were talking about a sort of response competition too, but there was a different sense. They were thinking something like "there is too much sympathetic activity to allow parasympathetically controlled behavior to occur."; that is, strictly speaking, response competition. But it is also one way to think of motivational competition. It is unfortunate that all this work, undertaken for the most part in the hope of establishing a nice behavioristic framework for understanding Pavlovian-instrumental interactions, has run into so much trouble. The attempt was to conceptualize a mechanism for translating the primary associative structure into action. The way to do that was through motivating emotions. Their concept had made excitation and inhibition symmetrical. They also had an elegant scheme for testing their proposal with transfer of control studies. But when the data came in it was disastrous. As early as 1974, Mackintosh was able to conclude that there are no grounds to justify the belief that the mediating mechanism involves anything like motivational competition. And at this time I see no reason to entertain a different conclusion. We appear to be in need of a new mechanism, new conceptual machinery, for translating the associative structure into behavior.

Conclusions

Let us look at the pieces and see if we cannot begin to fit them together. Part of the puzzle is that contextual stimuli, which are customarily assumed to enter into association with USs just as the primary CSs do, do not seem to behave like that. It begins to appear as though an animal responds mainly on the basis of the CS and does not rely on contextual cues *unless* there is uncertainty about the CS, perhaps because it has become ambiguous by being presented under both reinforced and nonreinforced circumstances. At that point, when the CS is ambiguous, the animal's behavior will reflect what contextual cues are present. Contexts are used to resolve ambiguity that has been experimentally produced in CSs (Bouton & Bolles, in press). Such a conclusion should not be surprising if we remember that the CS was the prime predictor of the US; it was deliberately made the primary predictor in the acquisition phase of the experiment. If it were not, we would probably not have any conditioning.

The role of Pavlovian cues established in a different context appears to be very much the same. Thus a predictor of food established elsewhere appears to have very little effect on an animal's ongoing food-getting behavior, *unless* the primary CS controlling the behavior has been made ambiguous by having been previously both reinforced and not reinforced. At that point, when the CS is ambiguous, the animal's behavior will reflect what out-of-context, off-baseline cues are present. The point is illustrated by many of the studies that have been cited and, in addition, a number of studies carried out by Trapold and his colleagues (see review by Bolles & Moot, 1972). Trapold's transfer of control studies were almost invariably of the discrimination-reversal type, which is a complicated and inherently ambiguous paradigm. For example, bar pressing is brought under the disciminative control of the houselight, and then the discrimination is reversed. Trapold (1966) found that animals learned the reversal more rapidly if reversal was immediately preceded by Pavlovian trials on which the new positive stimulus condition was used to signal noncontingent food in a Pavlovian manner. Why did Trapold et al. do their transfer of control studies in this curious complicated way? Because it was the only way they could get the effect (Trapold, personal communication)!

And so it seems that secondary stimuli, such as contextual cues and CSs conditioned off the baseline, do not enter into the determination of behavior just like any other stimuli. In particular, they have far less influence on behavior than primary stimuli, that is, those that have been experimentally selected to be the prime predictors in the situation (provided that the primary stimulus is unambiguous). Nor do secondary stimuli seem to play an omnibus role in behavior, such as providing motivation or demotivation for the baseline response, as Rescorla and Solomon (1967) proposed. Rather, it appears that such secondary cues enter into the associative matrix, along with everything else, and remain stored in long-term memory, but they play little or no part in the control of behavior until such time as the primary stimuli, such as the nominal CS, are made ambiguous. Then the animal attempts to resolve the primary ambiguity by drawing upon information provided by the secondary stimuli that are present.

It is a remarkable fact that all the classical evidence for inhibition, as well as the recent evidence provided by transfer of control studies, comes from inherently ambiguous situations. Thus we begin to see that conditioned inhibition is not a new kind of association, which "opposes" the old kind of association, as Pavlov first proposed. It is not just like excitation, but with a minus sign somewhere. It is not a negative center of some sort. Indeed, it appears to be hardly anything at all. The notion that it had to be an entity, or a device, or a process is,

I believe, an unfortunate historical accident that resulted from thinking that a change in behavior must reflect a change in the associative structure. But once that idea is put away, we may begin to consider the evidence for inhibition as reflecting not a new kind of association, but rather as telling us something about how animals react in ambiguous situations.

My first principle throughout this chapter has been that animals learn, and remember, virtually everything that happens to them in a conditioning experiment. If this is true, then it is clear what happens in an excitatory conditioning session: The animal learns that B follows A. It remembers that and comes to expect B whenever A appears in the same context. What does it learn if we introduce a new stimulus, C, and follow it with nothing? I would suggest that it learns that nothing follows C, and that this learning is structurally just like the learning that B follows A. The animal remembers it and comes to expect nothing whenever C appears in the same context. I believe that is all we need to explain simple, on-baseline stimulus control. Note, though, that if the context is changed, we may or may not see evidence of the appropriate expectancy; that depends on a variety of circumstances, including what experimental design and procedures we are using. What I suggest, however, is that the associative structure itself is very simple, and rather uninteresting (this is the sense in which my view is nonassociative). Both sorts of input, that is, both the A–B and C–O associations, are simply kept in long-term storage.

Two questions remain. One is the little question of just what it means to expect "nothing" to happen. Does it mean literally nothing, or is it reinforcer specific so that it means B will not happen. The latent inhibition phenomenon suggests that animals can learn about literally nothing. But the phenomenon of frustration suggests that the nonoccurrence of an otherwise expected event can be a very important psychological reality. I am not prepared at this point to pursue the little question of what nothing is.

Nor do I have much to say about the big question, which is how all the associative contents, which I assume to be permanently stored in memory, are selected at any one moment and acted upon. By what process is memory searched and representations found and then transformed into expectancies that will generate behavior? I can imagine a number of factors being involved including what I have called secondary stimuli, what (in this volume) Holland calls occasion-setting stimuli, and what Rescorla calls facilitators. It is operators like these that transcend the associative structure and operate on it that we have to find out more about.

REFERENCES

Amsel, A. (1950). The combination of a primary appetitional need with primary and secondary emotionally derived needs. *Journal of Experimental Psychology, 40,* 1–14.

Bolles, R. C., Grossen, N. E., Hargrave, G. E., & Duncan, P. M. (1970). Effects of conditioned appetitive stimulation on the acquisition and extinction of a runway response. *Journal of Experimental Psychology, 85,* 138–140.

Bolles, R. C., & Moot, S. A. (1972). Derived motives. *Annual Review of Psychology, 23,* 51–72.
Bouton, M. E. (1984). Differential control by context in the inflation and reinstatement paradigms. *Journal of Experimental Psychology: Animal Behavior Processes. 10,* 56 – 74 .
Bouton, M. E., & Bolles, R. C. (1979). Role of conditioned contextual stimuli in reinstatement of extinguished fear. *Journal of Experimental Psychology: Animal Behavior Processes, 5,* 368–378.
Bouton, M. E., & Bolles, R. C. (in press). Contexts, event-memories, and extinction. In P. Balsam & A. Tomie (Eds.), *Context and learning.* Hillsdale NJ: Lawrence Erlbaum Associates.
Bull, J. A. (1970). An interaction between appetitive Pavlovian CSs and instrumental avoidance responding. *Learning and Motivation, 1,* 18–26.
Davis, H., & Kreuter, C. (1972). Conditioned suppression of an avoidance response by a stimulus paired with food. *Journal of the Experimental Analysis of Behavior, 17,* 277–285.
Estes, W. K., (1943). Discriminative conditioning: I. A discriminative property of conditioned anticipation. *Journal of Experimental Psychology, 32,* 150–153.
Grossen, N. E. (1971). Effect of aversive discriminative stimuli on appetitive behavior. *Journal of Experimental Psychology, 88,* 90–94.
Grossen, N. E., & Bolles, R. C. (1968). Effects of a classical conditioned "fear signal" and "safety signal" on nondiscriminated avoidance behavior. *Psychonomic Science, 11,* 321–322.
Grossen, M. E., Kostansek, D. J., & Bolles, R. C. (1969). Effect of appetitive discriminative stimuli on avoidance behavior. *Journal of Experimental Psychology, 81,* 340–343.
Hammond, L. J. (1966). Increased responding to CS– in differential CER. *Psychonomic Science, 5,* 337–338.
Hendersen, R. W. (1973). Conditioned and unconditioned fear inhibition in rats. *Journal of Comparative and Physiological Psychology, 84,* 554–561.
Hurwitz, H. M. B., & Roberts, A. E. (1969). Suppressing an avoidance response by a pre-aversive stimulus. *Psychonomic Science, 17,* 305–306.
Hyde, T. S. (1976). The effects of Pavlovian CSs on two food-reinforced baselines with and without noncontingent shock. *Animal Learning and Behavior, 4,* 293–298.
Libby, M. E. (1976). The effects of aversive conditioned stimuli on the timing of unsignalled avoidance responding in rats. *Learning and Motivation, 7,* 117–131.
Mackintosh, N. J. (1974). *The psychology of animal learning.* New York: Academic Press.
Marx, M. H., & Murphy, W. W. (1961). Resistance to extinction as a function of the presentation of a motivating cue in the start box. *Journal of Comparative and Physiological Psychology, 54,* 207–210.
Morris, R. G. M. (1974). Pavlovian conditioned inhibition of fear during shuttle box avoidance behavior. *Learning and Motivation, 5,* 424–447.
Mowrer, O. H. (1960). *Learning theory and behavior.* New York: Wiley.
Rescorla, R. A. (1974). Effect of inflation of the unconditioned stimulus value following conditioning. *Journal of Comparative and Physiological Psychology, 86,* 101–106.
Rescorla, R. A., & Heth, C. D. (1975). Reinstatement of fear to an extinguished conditioned stimulus. *Journal of Experimental Psychology: Animal Behavior Processes, 1,* 88–96.
Rescorla, R. A., & LoLordo, V. M. (1965). Inhibition of avoidance behavior. *Journal of Comparative and Physiological Psychology, 59,* 406–412.
Rescorla, R. A., & Solomon, R. L. (1967). Two-process learning theory:Relationships between Pavlovian conditioning and instrumental learning. *Psychological review, 74,* 151–182.
Trapold, M. A. (1966). Reversal of an operant discimination by noncontingent discrimination reversal training. *Psychonomic Science, 4,* 247–248.
Weisman, R. G., & Litner, J. S. (1969). Positive conditioned reinforcement of Sidman avoidance behavior in rats. *Journal of Comparative and Physiological Psychology, 68,* 597–603.

13 Protection From Extinction: New Data and A Hypothesis of Several Varieties of Conditioned Inhibition

S. Stefan Soltysik
University of California, Los Angeles

INTRODUCTION

This chapter is devoted to a theory of "protection from extinction" (PFE). Also a new experimental evidence in support of PFE is presented. Basically, the protection from extinction is explained in terms of blocking of inhibitory conditioning. Such explanation tacitly assumes that extinction consists, or at least initially includes, acquisition of the "inhibitory associative strength" that offsets the positive associative strength producing thus a gradual decrement in the elicitation of a conditioned response (CR).

Blocking of excitatory conditioning is now a well-established phenomenon that has been demonstrated in numerous studies, starting with the seminal papers of Kamin (1968, 1969). It consists of a reduction if not a complete prevention of acquisition of a conditioned response to a novel stimulus paired with an unconditioned stimulus (US) if the novel stimulus is accompanied by a potent conditioned stimulus (CS) that elicits an asymptotic CR, i.e., accurately predicts the occurrence of the US. Whether the crucial variable in the blocking phenomenon is the elicitation of the CR or prediction of the UR, or both, remains to be solved. The issue of blocking and its mechanism certainly continues to be of prime interest to theories of learning.

Blocking of inhibitory conditioning was demonstrated originally by Suiter and LoLordo (1971) and was incorporated into Rescorla and Wagner's (1972) theory of conditioning. It consists in a reduction, if not complete prevention, of the acquisition of inhibitory associative strength by a stimulus paired with the *absence* of the US if this stimulus is accompanied by a potent inhibitory conditioned stimulus (conditioned inhibitor, CI) that fully predicts the absence of the US.

Blocking of inhibitory conditioning may be of limited practical significance when applied to indifferent stimuli, but it certainly is of enormous theoretical and practical interest when considered in the context of the extinction of previously trained CSs. If the extinction procedure, i.e., repetitive presentation of the unreinforced CS, is assumed to result in an increment of the inhibitory associative strength offsetting the excitatory associative strength thus causing the CR to diminish and eventually disappear, any blocking of such acquisition of inhibitory associative strength should provide *protection from extinction* (PFE). In one specific paradigm, where the CI is presented *after* the onset of the CS (sequential CS–CI compound), the CR could be, at least partially, elicited and thus occur without reinforcement. If the blocking of extinction occurs despite the elicitation of the CR, as demonstrated in Soltysik, Wolfe, Nicholas, Wilson, and Garcia-Sanchez (1983) and in this chapter, the PFE is augmented into the phenomenon of the *maintenance of acquired behavior without reinforcement.*

In the past the PFE was demonstrated in a few studies. Although the experimental prototype of testing the PFE could be traced back to some experiments of Chebotareva (1912a,b), who after four CI-CS trials observed no reduction in the number of CS-alone trials required for the extinction of a CR, the first conclusive evidence was not obtained until the study of Chorążyna (1957, 1962). Her experiments were designed by Konorski, who a few years earlier had suggested that the phenomenon of protection from extinction could be involved in avoidance learning. In his comments on the resistance of extinction observed in avoidance responses, Konorski (1948a) explicitly spoke of the "preservation" of the aversive character of the CS when it is accompanied by a "movement constituting a conditioned inhibitor" (p. 231). He admitted that no explanation of this "protection against extinction," using his words, was offered by his theory of conditioning.

In the study of Chorążyna (1957, 1962) an almost perfect protection from extinction was observed for the CSs presented unreinforced but in a successive compound in which the CI preceded (for 10 sec) the onset of the CS. With both classical and instrumental appetitive CRs in well-trained dogs, a nonreinforced presentation of the CI–CS compound was carried on for 40 days, whereas another CS was being normally paired with the food US. No decrement in the CR to the CS "protected" by the CI was found. Also in one recent study (Holland & Rescorla, 1975), the results obtained with such CI-CS compounds were suggestive of PFE. Because the US is not totally withdrawn, such a procedure is more comparable to discrimination training than extinction. However, the difference between extinction and discrimination training is probably irrelevant in respect to the issue of inhibitory conditioning, and there is an advantage of having the general situational context (density of US presentations, the level of tonic, background expectancy, anxiety, etc.) unchanged over several stages of the experiment. At any rate, in our recent study (Soltysik et al., 1983) no difference was found when the PFE was studied with or without intercalated CS-US trials with a control CS. Therefore we use the term PFE for both procedures, extinction sensu strictiori and differential conditioning (discrimination training).

There is an obvious benefit for the subject possessing a mechanism for preservation of CSs, when they are not reinforced in the presence of preceding or fully overlapping CIs. It allows the organism to maintain the unreinforced CSs as potential elicitors of CRs for the occasion when the CI-free situation reoccurs. In the presence of a CI (the changed situational context may become such a CI), the CSs are suppressed, dormant, and do not elicit unnecessary CRs. It should be noted, however, that, because of this absence of CRs in the serial CI–CS compound or in a simultaneous presentation of the CI and the CS the phenomenon of PFE seems to be explainable in terms of configural conditioning or discrimination unless the need for a specifically trained CI in the CI-CS compound is determined.

Already in Chorążyna's study a control group, in which a novel stimulus was used instead of a CI, provided evidence that a specifically trained CI is necessary to the PFE phenomenon. This conclusion is strengthened by the results of a study, where the CI was improperly trained (it was introduced only *after* withdrawal of the US, i.e., during the extinction stage), and no PFE was observed (Johnston, Clayton, & Seligman, 1972, unpublished, reported in Seligman & Johnston, 1973).

Much more valuable and theoretically interesting, however, would be a mechanism that protects the CSs from extinction when the CS precedes the CI so that at least part of the CR is elicited on the nonreinforced CS-CI trial. Such a paradigm for a serial CS–CI compound could be deduced from Miller and Konorski's (1928) notion that an avoidance response serves as a CI, and Konorski's (1948a) suggestion that this CI "somehow preserves" the aversiveness of the preceding CS (p. 231). A few studies were performed to verify this idea. In the pilot study on one dog trained with an appetitive reinforcer (Soltysik, 1960b), some evidence was obtained for the PFE. In a recent study on second-order conditioning by Holland (1980), in which a CS-CI serial compound was used as a reinforcer in second-order conditioning procedure, data were obtained that suggested the involvement of the PFE.

With an aversive reinforcement, however, two other studies in which CS–CI serial compounds were used failed to detect any PFE effect. In these studies (Johnston et al., 1972; LoLordo & Rescorla, 1966) no independent measures of inhibitory strength were used and, judging from a rather brief CI training, the alleged CIs might have been undertrained. In addition to that, there is a possibility that the selection of stimuli used as CIs and the type of CI procedure in these studies might have been less than optimal. Early studies on the CIs of different modalities in dogs found the visual stimuli to yield only weak CIs (Mishtovt, 1906, 1907; Vasil'ev, 1906). Thus, using a visual stimulus as a CI in dogs of LoLordo and Rescorla's experiment might have prevented establishment of a strong CI. Thirdly, there is a possibility that the CIs used in the unsuccessful attempts to obtain the PFE effect were of the incorrect (for PFE) type; this is explained later. Only in our recent study (Soltysik et al., 1983) was a strong PFE effect obtained with a serial CI-CS compound.

In this chapter we present some of the unpublished results obtained with an improved behavioral technique of aversive conditioning in cats (Wolfe & Soltysik, 1981) and with an experimental design that avoids some of the weaknesses of the earlier studies. In our technique the defensive arousal elicited by an aversive CS was monitored directly in the form of changes in heart rate and respiration, in addition to collecting the standard data on more specific motor CRs, such as vocal responses and leg flexions.

Two important changes in procedure distinguish our experiments from the earlier unsuccessful attempts to obtain the PFE. One is a choice of a trace conditioning procedure, and the other is a selection of a type 2 (Konorskian) CI, rather than a "safety signal" type CI or a type 1 (Pavlovian) CI (cf. Table 13.1 following). The choice of a trace conditioning procedure was dictated by an observation described later, that a delay conditioning procedure introduces an additional signal, namely a *termination of the CS,* on nonreinforced trials. Such a signal becomes an unintended CI after some training and may be responsible for the PFE effect in the control subjects.

The choice of a procedure of inhibitory conditioning with the CI following the CS was dictated by a suspicion that, although the CIs obtained by various inhibitory procedures may all possess a CR-suppressing capacity (as judged by summation or retardation tests), they may be, nevertheless, not equivalent in their informational role and in their effectiveness to promote the PFE. This issue is discussed later in the theoretical section.

NEW DATA IN SUPPORT OF PFE

Because the method was presented in detail by Wolfe and Soltysik (1981) and also in Soltysik et al. (1983) or Nicholas, Wolfe, Soltysik, Garcia, Wilson, and Abraham (1983), I only briefly describe our apparatus and procedures.

First of all, our method, although strictly of the classical conditioning variety, allows for considerable freedom of motor behavior and provides data from different parts of the subject's body: (1) somatic, consummatory (US-directed and localized), such as leg flexion, when a footshock US is used; (2) somatic "preparatory" (less specific in the sense of coping with the US), occurring prior to the consummatory response, such as locomotion; (3) vocalizations, which, although also motor and overt, are much less of a reflexive type of behavior than the leg flexion. In addition to these overt behaviors, intraorganismic indices of arousal are recorded: (4) changes in heart rate, and (5) changes in respiration.

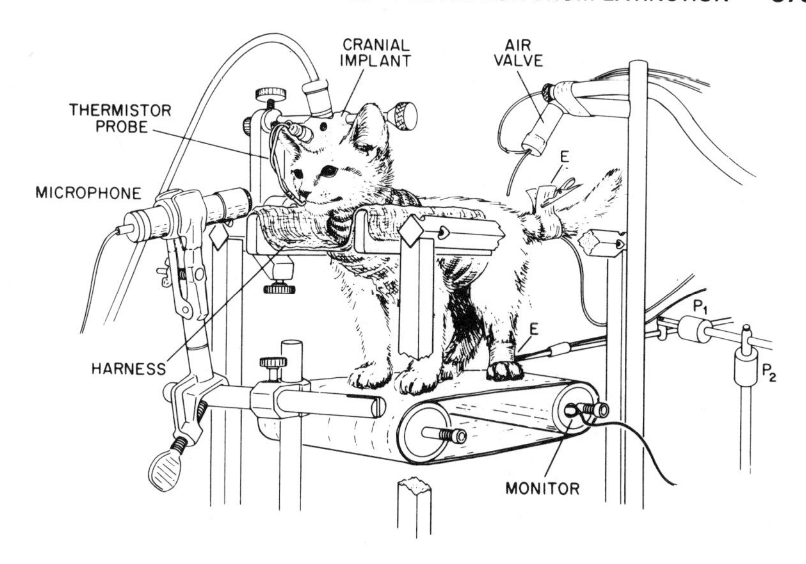

FIG. 13.1. Schematic view of the treadbelt stand. From Soltysik et al. (1983; Courtesy of Learning and Motivation). P = potentiometers at pivot points; E = US-shock electrodes.

Figure 13.1 illustrates the apparatus for restraining the animal and the attachments for collecting the data. Although the subject's head is firmly fixed to the apparatus' frame by means of a cranial acrylic implant, the movement of legs, including ambulation on the treadbelt, is unhampered, and the spacious enclosure makes the subjects relatively uninhibited in respect to overt movements. Strict restraint and small enclosures tend to promote a passive attitude (tonic immobility), which prevents or retards acquisition of active conditioned responses (Carli, Coltelli, & Sabourin, 1974; Tyler, 1971). In spite of such a permissive experimental setup, recording from single neurons in the brain is possible during sessions (the head is immobilized), and recording of the previously mentioned behavioral and organismic responses is relatively easy. Leg flexions are recorded by a precision potentiometer (P_1) activated by a light lever attached to the left hind leg; the same attachment serves also as an electrode (E) for the US-shock delivery. Treadmill monitor records the locomotor activity. A microphone registers vocal responses. A thermistor placed in front of one naris detects respiratory movements: Both respiration rate and respiratory amplitude are thus monitored. The heart rate is recorded either from implanted or externally attached EKG electrodes. The data are recorded on magnetic tape or on the strip chart recorder and are consequently analyzed by a computer. Some improvements in heart rate and respiration data processing consist in computer-assisted fine-grain analysis of the patterns of changes during the CS-US interval (cf. Wolfe & Soltysik, 1981). The increased sensitivity of such analysis was necessary to detect individual and-age related differences in learned responses in our study on the ontogeny of learning in kittens. As it happens, this method proved to be of considerable value in adult studies on excitatory and inhibitory conditioning.

The PFE study was performed on adult cats and the results in respiratory indices of learning were recently described (Soltysik et al., 1983). Here some unpublished data is presented from these same subjects, to illustrate the robustness of the phenomenon of PFE. Before describing the experimental design for PFE, let me make a few general comments about the outcome of this study. First of all, it is significant that the PFE effect was obtained in all conditionable types of responses, namely conditioned leg flexion, conditioned vocalization, conditioned changes in heart rate, and conditioned changes in respiration. Locomotor activity does not seem to be a classically conditionable behavior with this preparation; although ambulation was observed occasionally in the intertrial intervals, it seldom occurred during the trials, except during the extinction and test phases of the study, and then only in one or two subjects. Even when it occurred, it was ''superimposed'' on the leg flexion CR (cf. Fig. 13.2, rightmost top records) rather than occurring as an alternative response.

No less significant is the fact that the PFE occurred in all subjects used in this study, and that this particular type of CI (type 2 in Table 13.1), selected for its

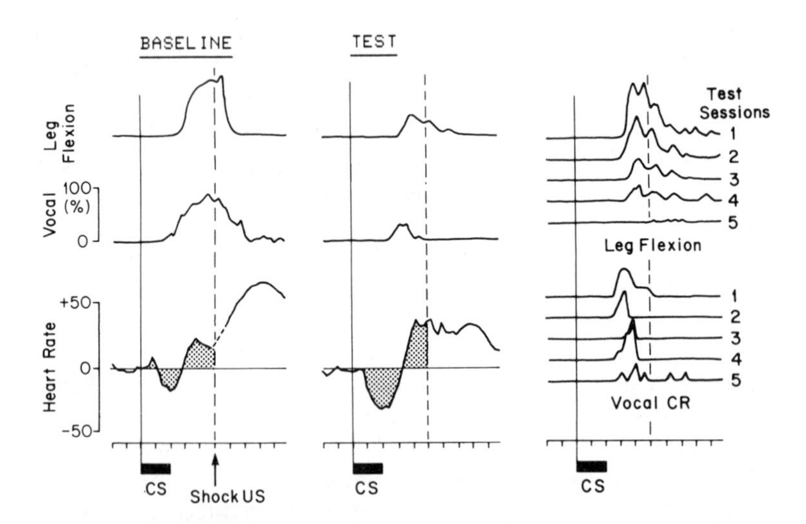

FIG. 13.2. Average leg flexion CRs (upper graphs), vocal CRs (middle graphs), and heart rate responses (lower graphs) to the CS_1 in one subject that received the protection from extinction treatment. The leftmost graphs (Baseline) show the responses at the conclusion of the baseline training, just before the extinction/protection stage began. The middle graphs (Test) show the average responses from the five test sessions; note the presence of all types of CRs. The rightmost graphs illustrate the leg flexion and vocal CRs in each of the five test sessions.

formal similarities to the putative CI in avoidance learning, was easily trained in all subjects. To continue presenting new data it is necessary now to describe the general design for a PFE experiment.

Experimental Design for the PFE

The experimental design for demonstrating the phenomenon of the PFE consists of (1) a baseline training of CSs and a CI, (2) the treatment stage during which two CSs are differentially treated, one extinguished and the other "protected from extinction", and, finally (3) a test stage during which the two CSs are compared as to their CR-eliciting strength. When a within-subject procedure is used, two CSs are needed in each subject. This creates a problem of possible inherent differences in the resistance to extinction between the CSs. To control for such a possibility, the entire experiment should be run twice. After the test stage of the first run, the CSs should be retrained and the entire sequence of treatment stage and test stage should be repeated with the reversal of the treatment for both CSs: The previously extinguished CS should be protected and vice versa. In such design each subject serves as a double control: (1) for the comparison of the outcome of the extinction procedure with one CS and the protection procedure with another CS, and (2) for the comparison of the differential rate of extinction (or resistance to extinction) inherent to the CSs, which usually are of different modalities (to avoid generalization) and could have different properties as CSs.

Thus, the following three types of trials are used in the study of the PFE:

(1) CS–US: [CS, US] = ACQUISITION TRIAL

(2) CS–CI: [CS, CI] = PROTECTION TRIAL

(3) CS: [CS] = EXTINCTION TRIAL

SECONDS

Three stages of the experiment have the following composition of trials, randomly mixed:

I. BASELINE TRAINING:

(1) CS_1—US (2) CS_1-CI

(3) CS_2—US (4) CS_2-CI

and optionally (5) CS_3—US

The CS_3—US trials provide an excitatory background and prevent the complete loss of the US expectation in an experimental situation; in preparations with weaker CRs a floor effect might obscure the PFE. In our preparation it was not a problem and the PFE was observed with and without the interspaced CS_3–US trials.

II. PROTECTION/EXTINCTION: protection of one CS and extinction of the other CS

(1) CS_1-CI (2) CS_2

and optionally (3) CS_3—US

III. TEST: extinction of both CSs

(1) CS_1 (2) CS_2

and optionally (3) CS_3—US

After the completion of the Test, both CSs are retrained (Baseline training) and the Protection/Extinction and Test stages are repeated with the CS_1 subjected to extinction and the CS_2 subjected to the protection procedure.

Here we illustrate the effectiveness of the described design of PFE in a subject that was deliberately overtrained and thus had nearly perfect scores on both positive and inhibitory trials. Also the extinction and protection stages in this subject were of long duration in order to test the reliability of the PFE under long-term conditions.

Thus, Fig. 13.2 shows a composite graph showing several response measures in two stages of our PFE experiment. The leftmost set of graphs shows the average responses from the last 20 trials of the baseline training; these are responses to one of the CSs (flashing light) that in the next stage, was "protected," i.e., presented only in a serial compound with a CI. Note a well-established leg flexion CR (this is an averaged amplitude curve), the amplitude of which approaches the amplitude of the UR and which is well timed so as to smoothly merge the CR and UR components. The same is true for the vocal CR (this is a probability measure, expressed as percentage of occurrence). An interesting feature of the heart rate "CR" is its complicated pattern, with an initial brief acceleration, followed immediately by deceleration and reversing back to acceleration for the last 2 seconds of the CS-US interval. The heart rate (HR) patterns in cats are highly individualized, not only between subjects, but even for the CSs of different modalities within the same subject. The patterns are stable and statistically reliable. The subject's specificity of HR patterns often prevents the option of averaging across the subjects.

The second column of graphs represents the data from the Test stage of the experiment, after this CS was subjected to a prolonged Treatment stage of 30 daily sessions on which this CS was presented only in a CS-CI compound and never reinforced. With 6 such trials per day, this amounts to a total of 180 unreinforced but "protected-by-CI" trials. At the same time the other CS was presented alone for the same number of 180 trials, i.e., unreinforced and unprotected. A third, control CS was regularly reinforced on randomly intercalated trials (6 per session) to assure steady situational background (one third of the trials reinforced) throughout the entire experiment.

When on the 5 test sessions (total of 30 extinction trials) this CS was presented alone, it reliably elicited CRs. Because the average response from 5 days of extinction does not convey the information of how well the CR was preserved, Fig. 13.2 presents in the rightmost column of graphs the leg flexion and vocal CRs from all 5 consecutive test sessions. Now not only the full size CR on the first test day (average of 6 trials) can be appreciated, but also progressive extinction over the 5 days of test stage is demonstrated. Interestingly, the HR response pattern is anything but reduced in the entire 5-day-long test stage as compared to the baseline, preprotection response.

Figure 13.3 shows what happened to the other CS (air puff directed to the cat's back, cf. Fig. 13.1: Air valve) that was subjected to extinction without protection. The CRs during the test stage are completely extinguished. Note that leg flexion and vocal CRs from baseline training are quite comparable for both CSs, but the HR patterns differ. The air puff CS has much more pronounced initial acceleratory component and considerably reduced late acceleratory component. This difference was stable over many months of training, and the HR was the only CR measure that had such stimulus-specific differential "topography." This last observation that the CSs tend to be unequal under even prolonged training forced us to repeat on each subject the entire experiment with the reversed treatment of the CSs. In all subjects this was a successful procedure, showing that the differences between CSs could not account for the PFE effect.

Fig. 13.4 is included here to illustrate how powerful the effect of reversal of the treatment could be. The response measure here is the reduction of amplitude

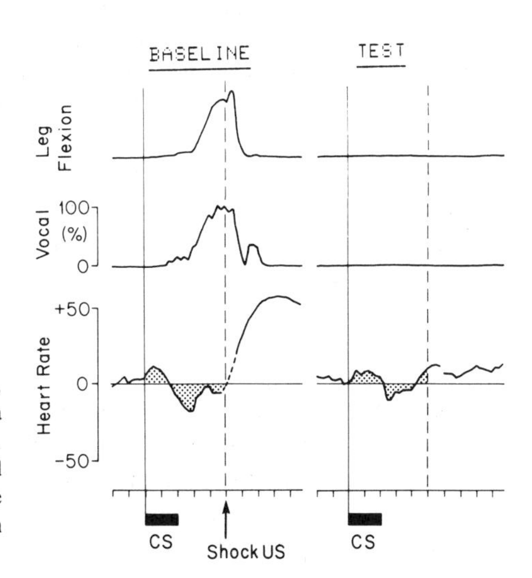

FIG. 13.3. Average leg flexion, vocal, and heart rate responses to a CS_2 in the same subject as in Fig. 13.2. This CS was extinguished and not protected (CS-alone trials). Note complete extinction of the leg flexion and vocalization CRs.

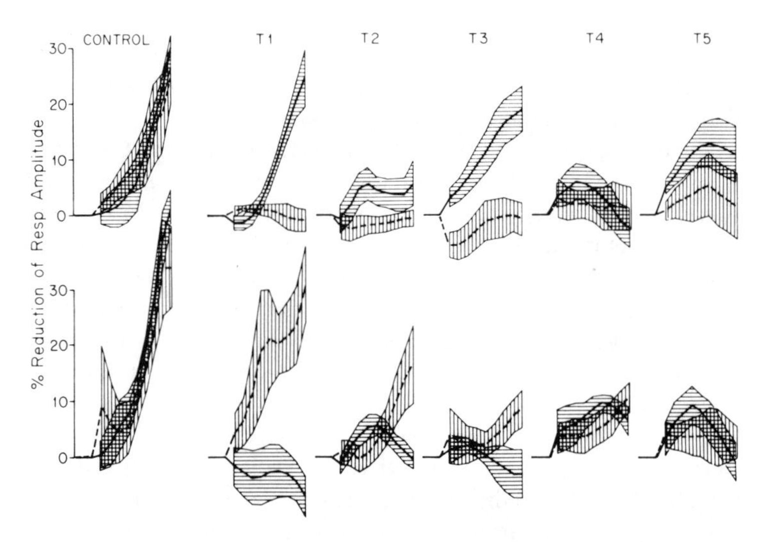

FIG. 13.4. Respiratory CR (reduction of respiration amplitude) in one subject. Solid (for CS_1) and broken (for CS_2) lines show average values of respiratory amplitude during 5 sec of the CS-US interval in percentage of the mean pre-CS respiratory amplitude. Horizontal and vertical hatching illustrates the standard errors of consecutive means (the data were sampled every .5 sec) for the CS_1 and CS_2, respectively. The data from the baseline training (control) and from five test sessions (T1 - T5) are shown. Upper graphs show the results from the first experiment in which the CS_1 was protected and the CS_2 not protected. Lower graphs show the results when the treatment of the CSs was reversed.

of respiratory movements. As demonstrated elsewhere (Soltysik et al., 1983) this measure is the most reliable index of aversive classical conditioning, possessing several useful features. It gives large numerical data, because the reduction could attain 50% or more of the pre-CS average value. It occurs in all subjects and practically on all trials and has much less individual variability and background noise than any other measure we have used. Finally, it has short latency and could be observed at early periods of the CS-US interval prior to the occurrence of overt behavioral responses. The leftmost graphs show the baseline (control) respiratory CRs to both CSs; top graph is from the first original training and the bottom graph is from the retraining after the first treatment and test stages were completed. There is full overlap of the responses to both CSs in both cases. The response to CS_1, protected in the first series (top graphs), is represented as a solid line (means) with horizontal hatching (standard errors); only the data from 5 seconds after the CS onset are shown. The five graphs labeled as T1 through T5 are averages from 5 test sessions, on which the CSs were extinguished. Top graphs are from the first PFE experiment and the bottom graphs are from the second PFE experiment. The response to a CS_2, extinguished in the first experiment, is represented by a broken line (means) and vertical hatching (standard errors). As in the case of motor and HR responses, there is almost perfect "saving" for the protected CS with a complete extinction for the unprotected one on the first day of the test. A complete reversal is seen in the bottom set of graphs on day 1 of the test. The responses to protected CSs become extinguished over the 5 test session, but even on the third day the difference is clear.

By averaging the results from both experiments, and subtracting the CRs to a protected CS from the CRs to an unprotected CS during the test trials, we arrive at a simple set of data that represents the time course and magnitude of the PFE. Fig. 13.5 is such a graph from seven cats. It summarizes the results presented in more detail elsewhere (Soltysik et al., 1983). It also illustrates the occurrence of the PFE effect in the early part of the CS-US interval, i.e., prior to the onset of the CI on the protection trials. It is this early part of the protected from extinction CR that has a practical and theoretical significance in the PFE account of the avoidance responses. If the fear CR is a motivating force for the avoidance response and if the motor avoidance response serves as a CI, the fear CR could be retrogradely (Soltysik & Wolfe, 1980) protected from extinction, thereby solving the puzzle over the maintenance of the avoidance behavior without an apparent motivator (the primary motivator, the US, is absent, and a secondary one, the CS, should be extinguished).

A CI Added to an Extinguished CS does not Promote a Reacquisition of the CR

The same theory (Rescorla & Wagner, 1972; Wagner & Rescorla, 1972), which predicts blocking, by a well-trained CI, of conditioned inhibition, or extinction, also predicts acquisition of a CR to a novel stimulus (or an extinguished CS) when such a stimulus is paired with the CI. Such a prediction was tested for a novel stimulus by Rescorla (1971) and Baker (1974). The first study provided

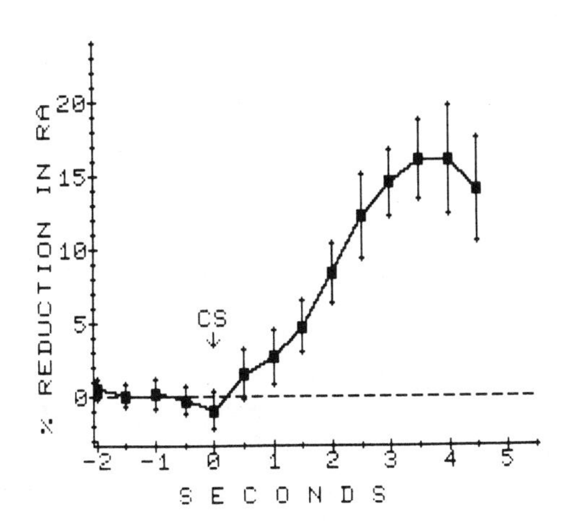

FIG. 13.5. PFE of respiratory responses to CSs in seven cats. This graph represents differential scores: respiratory responses to protected CS minus responses to not protected CS. Note that the PFE effect is evident already in the first two sec after the onset of the CS.

some evidence supporting the prediction, but the second one seemed to disprove it.

In four cats after extinction of the CS during the TEST stage, an additional training and test were carried out. Namely, in the next 5 days, we paired this CS with a CI (Soltysik & Wilson, 1982; unpublished data). The idea behind this treatment of an extinguished CS was as follows. If the extinguished CS is endowed with approximately equal excitatory and inhibitory associative strength (relative to the US), the addition of the CI causes the compound to possess an excessive inhibitory associative strength (overprediction of the omission of the US), which results in a positive value of the discrepancy in the "$\lambda - V_{tot}$" expression of the conditioning model of Rescorla and Wagner (1972) in which:

$$\Delta V_{CS} = \alpha\beta\,(\lambda_o - V_{tot})$$

where λ_o for nonreinforcement equals zero. Assuming that the associative strength for an extinguished CS (V_{CS}) equals zero, the subtracted negative value of V_{CI} gives the positive discrepancy and therefore an increment of positive associative strength for the extinguished CS should occur. Thus, the reacquisition of the CR to the extinguished CS is predicted. The data from one subject are shown in Fig. 13.6. The leftmost graphs show the average CRs before extinc-

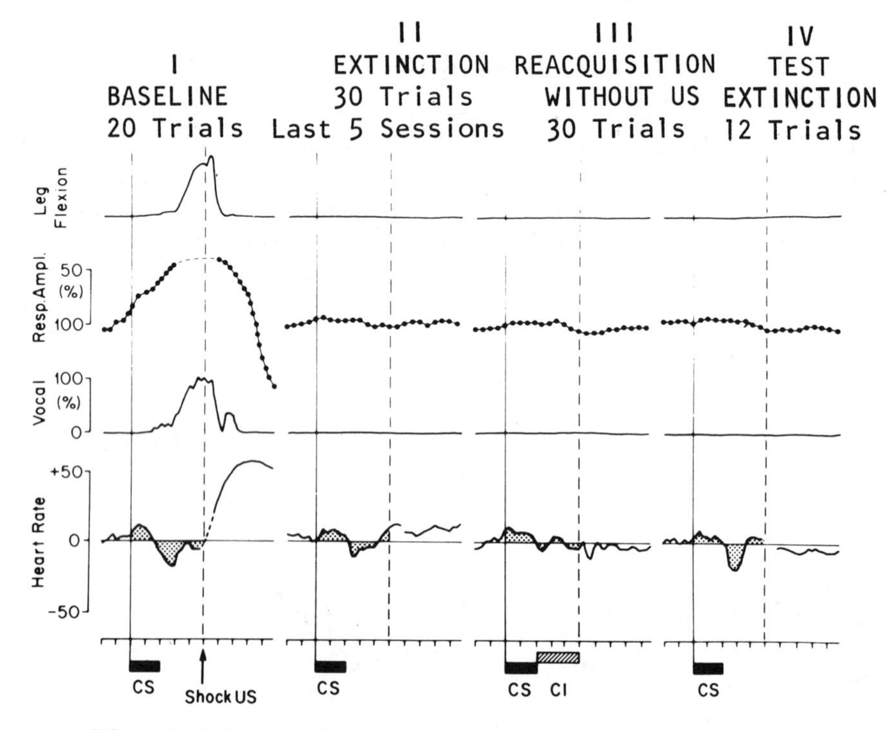

FIG. 13.6. Failure to retrain a CR to an extinguished CS by pairing it with a CI.

tion. The next set of graphs is from the last five sessions of extinction and shows no detectable CRs. When after 5 days of pairing this CS with a CI, again the CS alone was presented for 2 days (rightmost graphs) and no recovery of the CRs was observed. Similar failure of the *reacquisition without reinforcement* was observed in three other cats.

Thus, the results of this experiment failed to support the prediction that the CI may serve as a substitute reinforcer promoting the accruement of the positive associative strength and confirmed the negative results of Baker (1974).

Protection from Extinction without an Explicit CI

The following relevant observation was obtained unintentionally in one subject. Its extremely reliable (repeated over days) and measurable effect and its potentially great procedural importance justify a brief summary here. In one well-trained animal a comparison of protection and extinction was made with two CSs; both were trained with a 5-sec CS-US interval, and both had a duration of 5.2 sec, so that there was a .2-sec overlap with a .3-sec-long shock US. During the extinction/protection stage, the protected stimulus (a light) was presented for 5 sec and the CI was added 2 sec after the onset of this CS. The unprotected CS (an air puff) was presented for only 2 sec, to make it more comparable to the protected one, which also had only 2 sec of "unmasked" duration. After considerable (25 sessions) training a clear protection from extinction was found for the CS compounded with a CI, whereas complete extinction was observed for the unprotected CS. When, after retraining of both CSs in a new baseline series of sessions, the same treatment was repeated with the reversal of the CSs, a strange observation was made. Not only did the protected CS show during the test stage a fair evidence of retaining the ability to elicit the leg flexion CR, but also the "extinguished" CS seemed to have fully retained its excitatory properties.

This result is shown in Fig. 13.7 and 13.8. The leftmost set of records shows the last 24 trials during the baseline training stage. Note how reliable is this leg flexion CR: these are the *raw records of single trials*. The middle set of records shows the first 24 trials during the stage when this stimulus was protected from extinction, i.e., presented only in CS-CI compound. Note the perfect inhibition of the CR on CS-CI trials (Fig. 13.7), but also the same degree of suppression on trials with the shortened CS (Fig. 13.8). The third (rightmost) set of records represents the data from the test stage when these CSs were presented alone. Note that the "unprotected" CS seems to be an even stronger elicitor of the leg flexion CR than the protected CS. The middle set of records, which should show extinction when the unprotected CS is presented over many trials, strongly suggests the solution for this puzzling result. There was no extinction to the CS alone and it looked as if the subject learned to "discriminate" the long (5 sec) CS from the short (2 sec) CS. That this "discrimination" was in fact based on the acquisition of a CI to the termination of the CS was suggested by the fact that

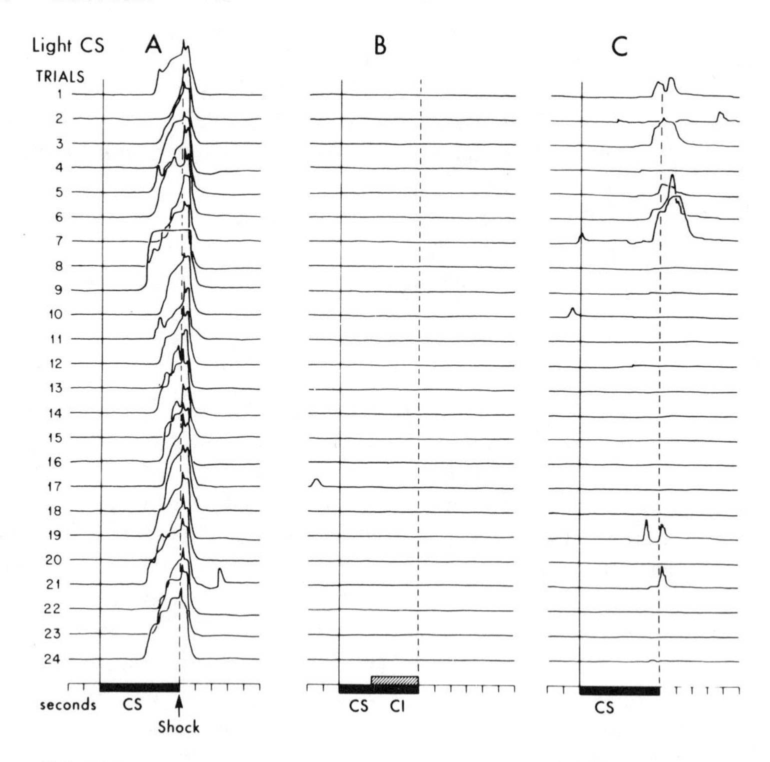

FIG. 13.7. Leg flexion CRs to CS_1 (light) during acquisition training (leftmost graphs), at the end of the extinction/protection stage (middle graphs), and on the first four test sessions. Individual records for each trial are shown. This CS had 5-sec duration and overlapped with the US.

the "discriminated" CS did elicit heart rate and respiratory responses, which after its termination were suppressed in a way as they were suppressed on the CS-CI trials.

Thus both, CS compounded with a CI, and CS whose duration was shortened to 2 seconds, were in fact protected from extinction. Learning not to respond to the shortened CS did not occur immediately, as evidenced by the first experiment where the short CS was easily extinguished. But that experience must have facilitated the learning to treat the "free termination of the CS" as a CI in the consecutive experiment, and it resulted in perfect PFE effect for a CS presented without a reinforcer and without an explicit CI.

This finding is important for two reasons. Firstly, it helped to avoid a mistake of introducing into the experimental design of PFE a stimulus event that could help to discriminate reinforced from unreinforced presentations of the CS. If, during acquisition training, the CS overlaps with a US, the unreinforced CS will have a new component, namely a free termination (an offset not masked by the

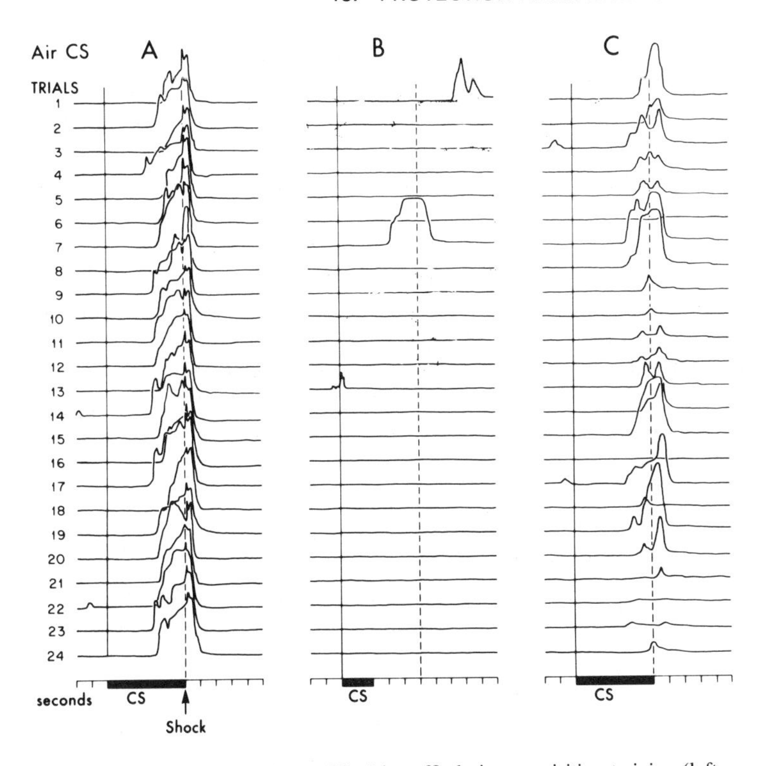

FIG. 13.8. Leg flexion CRs to CS_2 (air puff) during acquisition training (leftmost graphs), at the end of extinction treatment (middle graphs), and on the first four test sessions. Note remarkable PFE (responding on test sessions); this CS was shortened to 2 sec during extinction but no explicit CI was used to protect it from extinction.

US) that could acquire the functional role of a CI. Secondly, an acquisition of a CI to such a free termination of the CS should occur whenever the CS-US (with overlap) trials are intermixed with CS-alone trials. This is exactly a procedure of partial reinforcement and it is known to increase the resistence to extinction of the partially reinforced CS. Could it be that the phenomenon labeled by Konorski as PFE, and postulated to explain the persistence of well-trained avoidance responses, has also played a role in at least some of the studies on partial reinforcement that used overlapped CS-US trials and had shown the so-called partial reinforcement effect (PRE)? If so, both phenomena might best be explained as instances of blocking of inhibitory conditioning by a CI, explicit in the PFE studies, but unintended and unnoticed in the PRE studies.

As explained in Soltysik et al. (1983) no such unintended CI could be acquired by the subjects during extinction or discrimination training if the trace conditioning procedure is used, because the termination of the CS precedes the

US and therefore also serves as a CS. It occurs both on reinforced and nonreinforced trials and therefore cannot serve as a discriminative stimulus. With the trace conditioning procedure we never observed any failure of extinction of the CS presented alone.

In summary, the data published elsewhere (Soltysik & Wolfe, 1980; Soltysik et al., 1983) and presented here provide strong evidence for the phenomenon of protection from extinction. The CI that served as a stimulus blocking the inhibitory conditioning to the CS (i.e., interfering with the extinction of the CR to this CS) is of the variety that we called Konorskian, namely a signal of the omission of the US. Such a CI is acquired when a stimulus is presented after the onset of the CS but before the expected onset of the signaled, but omitted, US. An important new observation was also made when the CS/US overlapped during acquisition and a short CS was used during extinction. The termination of the CS that occurs only on nonreinforced trials acquired the role of the CI and caused the PFE effect to occur.

Finally, using a CI of Konorskian type as a reinforcer of the extinguished CS failed to reinstate the positive associative strength to the CS. This result is contrary to the prediction of the theory of Rescorla and Wagner (1972) and disagrees with Rescorla's (1971) data but supports similar results obtained by Baker (1974). It does not conflict, however, with more recent formulations of Rescorla (1975), particularly those (Rescorla, 1979, this volume) that ascribe to a CI the role of modulating of the threshold for elicitation of the CR.

THEORY OF THE PROTECTION FROM EXTINCTION

Since its first formulation the concept of protection from extinction (PFE) was closely tied to the notions of conditioned inhibition. Thus, Konorski (1948a) suggested a connection between two functions of an avoidance response: (1) playing the role of a conditioned inhibitor (CI); and (2) protecting the CS from extinction. However, he did not try to explain the relationship between CI and PFE. Soltysik (1960a,b, 1963) elaborated on this theme and produced relevant experimental data. The first powerful theoretical justification for such a liaison between conditioned inhibition and the PFE came from the theory of blocking. Here the logic is clear. If the extinction consists of, or at least includes in its initial stage, inhibitory conditioning to the unreinforced CS, compounding this CS with a CI during nonreinforcement should block such inhibitory conditioning to the CS, preventing or retarding the extinction. Although extinction is certainly more complex than just adding inhibitory associative strength to the existing excitatory associative strength of the CS (Pearce & Hall, 1980; Rescorla, 1979), even *some contribution* of inhibitory conditioning would suffice for the PFE effect to occur.

Before continuing on the blocking account of the PFE, three other possible explanations of the PFE should be mentioned.

1. "Protective Inhibition" Hypothesis. (Asratian, 1969; Pavlov, 1928, lecture 22). According to this view, the activation of the central representation of the CS by a CS-alone presentation continues for some time after the termination of the CS, and this prolonged active state within the CS "center" leads to the exhaustion and loss of "excitatory reactivity" on the part of neurons representing the CS. This is particularly important on extinction trials when the activation of the CS "center" lingers long after the termination of the unreinforced CS. Exhausted and depleted neurons, according to this view, respond to the sensory input (the CS) not with excitation but with an inhibitory process, which: (1) prevents transmission of excitation via the "temporary connection" (associative pathway) to the "center" of US; hence no CR, and (2) eventually spreads to the US "center," suppressing thus the CRs elicited by other homogeneous CSs.

However, when the CS is followed by the US that elicits strong concurrent activation of the US-representing neurons, the activation of the CS neurons is terminated. Pavlov explained this "inhibition" of the CS "center" by the activation of the US center in terms of negative induction; a strong focus of excitation within the US center produces an inhibitory surround. The CS "center" becomes, therefore, inhibited. Because in Pavlov's neurophysiological views the excitation is paramount to exertion and depletion, whereas the inhibition is metabolically connected with restoration and repletion, such negative-inductive termination of the excitation within the CS center amounts to the preservation of the excitability and "capacity to work" of the CS neurons. By extrapolating this view to the CS-CI compound it is possible to predict the PFE. Namely, the CI following an unreinforced presentation of the CS inhibits the CS representation and protects its neural elements from exhaustion. In a sense the CI emulates the effect of the US. If an established CI could be envisioned as exerting an inhibitory effect upon the center of the CS, the result would be similar to that of reinforcement; the CS neurons would be protected from exhaustion and therefore prevented from altering their reactivity toward inhibition. Thus, protection from extinction would be predicted by combining the theory of protective role of inhibition with the assumption of the CI terminating the excitation of the CS representation.

There is at least one problem with the "Protective Inhibition" hypothesis. Because the localization of conditioned inhibition within the CS "center" conflicted with the evidence from binary conditioning, where the same stimulus could be simultaneously a CS for one US and an extinguished CS for another US, Asratian (1969) proposed that the locus for the inhibitory "protective" effect from the US is exerted upon the neural elements of the associative pathway rather than the representation of the CS. Interestingly, Rescorla (1975, p. 25) also considered such a location for conditioned inhibition. However, no theoretical or experimental data exist to verify such a mechanism of conditioned inhibition and its involvement in PFE.

The preceding comments should not be interpreted to imply that the "protective" hypothesis was Pavlov's only view on conditioned inhibition, from which the theory of the PFE could be construed. Another position, according to which conditioned inhibition originates from the interaction or competition between two behavioral acts, was expressed by Pavlov on many occasions and was developed more recently by Anokhin (1958). This approach anticipated the "alternative response" theories of conditioned inhibition. In particular, emotional states elicited by the omission of the expected reinforcer, such as "biologically negative reaction" of Anokhin (1958), frustrative nonreward of Amsel (1962), antidrive (Konorski, 1967), relaxation response (Denny, 1971), or the opponent process of Solomon and Corbit (1974), are the best-known examples of hypothetical organismic (covert) responses that could play a role in conditioned inhibition and extinction.

2. Consolidation Hypothesis. Another explanation of the PFE effect is possible in terms of interference with the consolidation of association between two memory processes. Soltysik and Zieliński (1963) suggested that the CI could interfere with the "short-term memory traces" (active representations) of paired events. If so, a retardation of acquisition should occur when the CS is followed by a US and then by the CI (CS-US-CI). By extrapolating from this view to the extinction procedure, and assuming that the interacting central events are (1) a CS representation, and (2) the representation of no-US, or more precisely, the representation of the omission of the expected US, the following hypothesis is available.

If extinction depends on making an association between the CS (or, strictly speaking, its slowly decaying memory trace, or representation) and the similarly slowly decaying trace of the omission of the US, the specific CI, which may suppress one or both of these traces, would prevent or retard the extinction. This hypothesis may have several variants and enriched terminology. One could apply the Konorskian concepts of no-US or "antidrive" (Konorski, 1967) to describe the central perceptual and emotional effects of the omission of the expected US. One could also replace the term *trace* with *rehearsal within the STM* (Wagner, 1978), backward scanning, etc. The gist of this hypothesis remains as follows: Extinction is a learning process that requires, for establishment of inhibitory association, some posttrial processing of information and the CI interferes with it. In terms of performance such interference would be observed as the PFE.

3. Amsel's Theory of Persistence. Amsel (1972a,b) proposed a theory that should be classified as a theory of acquired resistance to extinction; as such it is related to the concept of protection from extinction. Amsel rejected the notion of "inhibition," and his alternative response elicited by nonreinforcement (R_X, e.g., "frustration" in appetitive, and "relief" in aversive conditioning), although initially conflicting with the original conditioned response, is not totally incompatible with it and eventually starts contributing to its strength and elicitation. This theory explains "persistence," i.e., increased resistance to extinction not by postulating any interference with the extinction process, as the first two hypotheses did, but by assuming that the response elicited by nonreinforcement becomes a mediator (r_X--s_X type in Hullian-Spenceian terms) associatively strengthening the conditioned response in S-R fashion. Note that this explanation of persistence does not include any "inhibitory" stimuli. Instead, a conflicting response resulting from nonreinforcement (in its conditioned form: r_X) is postulated to function as an elicitor of the CR. Consequently it is difficult to apply this theory to any study that uses a specifically trained CI for obtaining the persistence effect, particularly when the CI is independent of the subject's behavior and retains a strong suppressive effect upon the CRs.

Theory of Blocking and Its Application to the PFE

In relation to the PFE, the first of the three preceding hypotheses, i.e., the notion of protective inhibition, was never seriously considered. The concept of interference with the "consolidation of extinction," however, has reappeared in the form of the theory of blocking and was applied directly to the phenomenon of PFE. The theory of blocking is best described as a consolidation theory, cleansed of its structural and neurophysiological "impurities" and recast in more behavioral terms. The equations for acquisition of positive (excitatory) and negative (inhibitory) associative strength proposed by Rescorla and Wagner (1972) directly and unequivocally predict the PFE. Mackintosh (1974, p. 336) made explicitly such a suggestion when commenting on the PFE account of avoidance learning.

Briefly, when a CS is presented unreinforced but in the presence of a CI, and both CS and CI are well trained so that they possess asymptotic positive and negative associative strength ($V_{CS} = \lambda$; and $V_{CI} = -\lambda$), the outcome of such a trial should be no change in the positive associative strength of the CS because the total combined associative strength of the CS and the CI equals zero (see Equation 1). In other words, when there is no discrepancy between the limit of learning with no-US (which is zero) and the combined associative strength of the CI/CS compound (which is also zero), no learning, or in this case, no extinction should occur.

$$\Delta V_{CS} = \alpha\beta\ [0 - (V_{CS} = V_{CI})] \qquad (1)$$

An interesting variant of blocking theory is obtained if the behavioral terms λ and *associative strength* are changed into informational ones: *limit of signalling value* or *full predictability* for λ and *predictability of US* and *predictability of no-US* for V_{CS} and V_{CI}, respectively (cf. Moore and Stickney, this volume). Such a Kaminian theory of conditioning and blocking would also predict no extinction on nonreinforced trials if the omission of the US is made not surprising by the presence of the CI, a predictor of no-US. Rescorla and Wagner's theory of conditioning, which makes the associative strength a variable of learning related to behavior (responding) rather than acquiring information (anticipation, belief), predicts that *any* CI should produce PFE, when compounded with the nonreinforced CS. In contrast, the Kaminian variant of the theory of conditioning seems to be more partial to the exact informational content of the CIs. If it is the "surprisingness" of the reinforcing event, e.g., omission of the US on extinction trial, that promotes learning, then different inhibitory stimuli may affect this aspect of nonreinforcement differently, even if they all pass the criterion of being CIs on summation and retardation tests. To explain this statement, it is necessary, and heuristically useful, to review and compare the procedures that are or could be used for training inhibitory stimuli.

Procedures of Conditioned Inhibition

So far we generally referred to a conditioned inhibitor (CI) as a stimulus that, after one or another form of inhibitory training, has acquired capacity to suppress the CR. There are, however, at least five different procedures, based on distinctly different theoretical premises, which are, or could be in theory, effective in yielding the CIs. Not included are, of course, the procedures of extinction and differential conditioning (CS+ and CS−), because they yield stimuli possessing mixed, excitatory and inhibitory, properties. Rather, this list restricts itself to stimuli that at the onset of inhibitory procedure could be considered associatively neutral in respect to the US:

1. Historically the first procedure of CI was identical with the second-order conditioning: A novel stimulus was followed by an established CS. Conditioned inhibition, instead of second-order conditioning, was serendipitously obtained by this procedure (Mishtovt, 1906, 1907, Vasilev, 1906); hence the sequential CI-CS compounding has been used in most studies until recently. This procedure for obtaining conditioned inhibition we call Pavlovian.
2. The procedure in which the CI follows the CS, and which was found effective in our PFE study, we call Konorskian (Soltysik et al., 1983). It corresponds to Miller and Konorski's (1928) original recognition that the avoidance response, which by necessity follows the CS and precedes the expected US, plays the role of a CI.
3. Based on Konorski's (1948b) idea that the termination of the US is a critical event in inhibitory conditioning, a procedure was proposed in which a novel stimulus is presented after the onset of the US but prior to its termination.

A few studies with appetitive (Zbrożyna, 1958) or aversive (Segundo, Galeano, Sommer-Smith, and Roig, 1961) reinforcers, confirmed the effectiveness of this procedure.

4. Based on Rescorla and Wagner's (1971) concept of CS/US contingency, i.e., a probabilistic treatment of the CS–US pairings and situational cues-US pairing, the defining procedure for CI was identitied as presentation of a stimulus explicitly unpaired with the US. Whether presented immediately after the US or in the middle of the intertrial interval, such a stimulus, if followed by a US-free period of time, becomes a predictor of the absence of the US, i.e., a CI. Although some recent data cast doubt on the correctness of the contingency approach (cf. Fowler, this volume) to inhibitory conditioning, it is almost certain that such a stimulus, consistently followed by the absence of the US, must acquire new significance that could be only opposite in some sense to the CS. This type of CI I propose to call Rescorlan, to acknowledge the contribution of this author to the contingency theory and inhibitory conditioning, in particular. In the context of aversive conditioning the Rescorlan CI was aptly dubbed a "safety signal" (Mowrer, 1960).

5. The last and still untried procedure is derived from the concepts of counterresponses or counterstates occurring after the termination (or earlier) of the unconditioned response (UR), or after the omission of the expected reinforcer. Konorski's (1967) antidrive, or no-US, and Solomon and Corbit's (1974) opponent process are the well-known terms for such counterresponses. They could be of either positive or negative hedonic value. For a counterresponse occurring after the omission of an appetitive reinforcer Anokhin (1958, p. 129) proposed a "biologically negative reaction," and Amsel (1958), primarily for instrumental learning theory, spoke of "frustrative nonreward." Within the context of aversive conditioning, a "relief" (Mowrer, 1960, p. 419; Konorski, 1967, p. 34) and a "relaxation response" (Denny, 1971) were suggested. Assuming conditionability of such an anti-UR (all the aforementioned authors suggested that these responses or states are conditionable and more recently Schull (1979) also proposed a similar theory) the CI could be envisioned as a CS, which after being paired with such an anti-UR (S-R learning?) elicits an anti-CR and acts as a CI.

Training of such a CI would require monitoring the occurrence of the anti-UR and presenting a novel stimulus before its onset. One would have to compare this CI procedure with the similar ones, such as Rescorlan and the CI type 3.

The following table describes the basic features of the five types of CI. Not included is a procedure of simultaneous presentation of the CS and X, because its status is uncertain. Either it is a subvariety of CI type 2, or it is different from any serial case (Holland, this volume). Also the problems with masking, configuring, and a "unique stimulus" hypothesis seem to be more serious with simultaneous than with serial compounds.

TABLE 13.1
Training Procedures and Behavioral Outcomes of Inhibitory Conditioning

Cond. Inhibitor Type	*Procedure (Types of Trials)*	*Behavioral Effect*	*Informational Content*
1. (Vasilev 1906; Mishtovt 1906)	1. X--CS 2. CS--US	X prevents elicitation of CR	X predicts unreinforced CS
2. (Miller & Konorski 1928)	1. CS--X 2. CS--US	X inhibits elicited CR	X predicts an omission of US
3. (Konorski 1948)	1. US . . . . . . US X ↑	X inhibits CR & UR	X predicts US' termination
4. (Rescorla 1969)	1. CS--US---X 2. CS--US	X inhibits CR	X predicts US-free period
5. (Konorski 1967; Denny 1971; Solomon & Corbit 1974)	1. X--(-UR) 2a. US alone or 2b. CS-US	X inhibits CR (conjectural)	X predicts on set of "antiresponse"

What all these CIs have in common is their inhibitory potential that could be revealed on summation and retardation tests; they suppress CRs and resist being transformed into positive CSs by pairing with the US. But are they equivalent as CIs in other respects? Would they easily transfer from one type CI to another? Not much work has been done in this respect, particularly with types 3, 4, and 5. But the available evidence suggests that, whereas type 1 readily transfers into type 2, the opposite transfer is difficult (Soltysik & Berg, in preparation). This observation of asymmetry in transfer corroborates with a recent study by Holland (this volume), which shows that a CI type 1 (Pavlovian, or serial CI-CS) and simultaneous CS/CI compound differ in respect to such features as specificity of inhibitory effect, retardation test, and susceptibility to CI-US pairing. Fowler's (this volume) data indicate that CI procedure type 4 (Rescorlan) and type 5 are very different in inhibitory power, specifically type 5 is much more effective. If CIs obtained by serial CI-CS, CS-CI, and simultaneous CS/CI presentations are in fact different, even more profound differences could be expected between types 3, 4, and 5.

How does this possibility reflect upon the theory of PFE? For the lack of experimental data, not much can be concluded, but the review of existing evidence might at least point at the directions for future studies. All previous studies could be classified into two groups. Irrespective of what type of CI was used, an important variable of the PFE study is whether during the presentation of a CS/CI compound the CR is elicited or not. In procedures where the PFE was tested with the simultaneous CI/CS compound or with serial CI-CS compound, the CR was absent and the PFE was obtained (Chorążyna, 1957, 1962). When the CR at least partially occurs before the onset of the CI, as in the CS-CI compound, no PFE was reported by LoLordo and Rescorla (1966) and Johnston et al. (1972, cited by Seligman & Johnston, 1973), but a strong evidence for PFE was observed by Soltysik et al. (1983). Some evidence for such retrograde PFE was given in an early preliminary study of Soltysik (1960), and also Holland (1980) obtained data on second-order conditioning, which could be interpreted as a support for the PFE with a CS-CI serial compound.

Why did the two first mentioned CS-CI studies fail to show PFE? It is not possible to answer this question satisfactorily at this time, but at least the following suggestions might help in future studies. In the introduction we suggested that the CIs were not sufficiently trained and that in LoLordo and Rescorla's (1966) study the visual modality of the CI might have contributed to the weakness of the CI. Now, taking into account a hypothesis that there are *different types of CI,* possessing different "information content," a third possibility emerges: A wrong type of the CI may have been selected in unsuccessful studies. It is possible, that in the LoLordo and Rescorla (1966), as well as in the Johnston et al. (1972) studies, a rather late onset of the CI within the CS-CI compound might have prevented it from becoming a CI type 2 and instead it was a CI type 5 or 4. If our suggestion is correct that information content of the CI is important for the PFE effect, this feature of both unsuccessful studies might be crucial. The effective CI in a PFE study must signal the impending omission of the US rather than merely the immediate absence of US. It is the cancellation of the US expectation that prevents extinction on a nonreinforced trial rather than simple "safety signal," announcing a US-free period of time. Future comparisons of the PFE effects, with CIs equally effective in suppressing the CR but carrying different information, will help to construct better explanation for the PFE mechanism.

In conclusion, our data and theoretical speculations provide a basis for acceptance of the PFE as an important behavioral phenomenon. By inserting a CI between the CS and the expected but not occurring US, the subject can maintain acquired behavior without reinforcement. Because CI may be provided by the subject's action (e.g., avoidance response), the PFE mechanism may intervene in many phenomena characterized by increased resistance to extinction. Avoidance learning and partial reinforcement, motivational persistence and pathological fixation, all may be better understood if the concept of PFE is included in their analyses.

ACKNOWLEDGMENT

The author wishes to express his gratitude to Dr. W. J. Wilson for his proofreading and comments.

REFERENCES

Amsel, A. (1958). The role of frustrative nonreward in non-continuous reward situations. *Psychological Bulletin, 55,* 102–119.

Amsel, A. (1962). Frustrative nonreward in partial reinforcement and discrimination learning: Some recent history and a theoretical extension. *Psychological Review, 69,* 306–328.

Amsel, A. (1972a). Inhibition and mediation in classical, Pavlovian and instrumental conditioning. In R. A. Boakes & M. S. Halliday (Eds.), *Inhibition and learning* (pp. 275–299). London/New York: Academic Press.

Amsel, A. (1972b). Behavioral habituation, counterconditioning, and a general theory of persistence. In A. H. Black & W. F. Prokasy (Eds.), *Classical conditioning II: Current research and theory.* New York: Appleton-Century-Crofts.

Anokhin, P. K. (1958). *Vnutrennee tormozhenie kak problema fiziologii.* (Internal inhibition as a problem of physiology). Moscow: Medgiz.

Asratian, E. A. (1969). Mechanism and localization of conditioned inhibition. *Acta Biologiae Experimentalis, 29,* 271–291.

Baker, A. G. (1974). Conditioned inhibition is not the symmetrical opposite of conditioned excitation: A test of the Rescorla-Wagner Model. *Learning and Motivation, 5,* 369–379.

Carli, G., Coltelli, M., & Sabourin, M. (1974). Effects of animal hypnosis on the performance and the extinction of an avoidance response. *Brain Research, 66,* 365–366.

Chebotareva, O. M. (1912a). K fiziologii uslovnogo tormozhenia. (Physiology of conditional inhibition). *Transaction of the Society of Russian Physicians in St. Petersburg, 79,* 151–152.

Chebotareva, O. M. (1912b). Dal'neishie materialy k fiziologii uslovnogo tormozhenia. (Further data on physiology of conditional inhibition). *Doctoral Dissertation.*

Chorążyna, H. (1957). Some data concerning the mechanism of conditioned inhibition. *Bulletin de L'Academie Polonaise des Sciénces, Série des sciences biologiques, 5,* 387–392.

Chorążyna, H. (1962). Some properties of conditioned inhibition. *Acta Biologiae Experimentalis, 22,* 5–13.

Denny, M. R. (1971). Relaxation theory and experiments. In F. R. Brush (Ed.), *Aversive conditioning and learning* (pp. 235–295). New York: Academic Press.

Holland, P. C. (1980). Second-order conditioning with and without unconditioned stimulus presentation. *Journal of Experimental Psychology: Animal Behavior Processes. 6,* 238–250.

Holland, P. C., & Rescorla, R. A. (1975). Second-order conditioning with food unconditioned stimulus. *Journal of Comparative and Physiological Psychology. 88,* 459–467.

Johnston, J. C., Clayton, K. N., & Seligman, M. E. P. (1972). Unpublished experiment reported in Seligman & Johnston (1973).

Kamin, L. J. (1968). ''Attention-like'' processes in classical conditioning. In M. R. Jones (Ed.), *Miami Symposium on the Prediction of Behavior 1967: Aversive Stimulation* (pp. 9–33). Coral Gables, FL: University of Miami Press.

Kamin, L. J. (1969). Predictability, surprise, attention, and conditioning. In B. A. Campbell & R. M. Church (Eds.), *Punishment and aversive behavior* (pp. 279–296). New York: Appleton-Century-Crofts.

Konorski, J. (1948a). *Conditioned reflexes and neuron organization.* London: Cambridge University Press.

Konorski, J. (1948b). K voprosu o vnutrennem tormozhenii. (On internal inhibition). In L. A. Orbeli, I. P. Razenkov, P. K. Anokhin, & K. Kh. Kekcheev (Eds.), *Ob'edinennaia Sessia Posviashchennaia 10-ti letiu so Dnia Smerti I.P. Pavlova.* Moskva: Izdatel'stvo Akademii Meditsinskikh Nauk SSSR.

Konorski, J. (1967). *Integrative activity of the brain. An interdisciplinary approach.* Chicago/London: The University of Chicago Press.

LoLordo, V. A., & Rescorla, R. A. (1966). Protection of the fear-eliciting capacity of a stimulus from extinction. *Acta Biologiae Experimentalis, 26,* 251–258.

Mackintosh, N. J. (1974). *The psychology of animal learning.* London/New York/San Francisco: Academic Press.

Miller, S., & Konorski, J. (1928). Sur une forme particuliére des reflexes conditionnels. *Comptes Rendus des Séances de la Société de Biologie et de ses Filiales, 99,* 1155–1157.

Mishtovt, G. V. (1906). Opyty tormozheniia iskustvennogo uslovnogo reflexa (zvukovogo) razlichnymi razdrazhiteliami. [Experiments with inhibition of an artifical conditional reflex (to a sound) by various stimulu]. *Transaction of the Society of Russian Physicians in St. Petersburg, 74,* 89–103.

Mishtovt, G. V. (1907). Vyrabotannoe tormozhenie iskustvenoogo uslovnogo reflexa (vukovogo) na sliunnye zhelezy. (Acquired inhibition of the artificial conditional reflex (to a sound) of the salivary gland.). *Doctoral Thesis, St. Petersburg.*

Mowrer, O. H., (1960). *Learning theory and behavior.* New York/London: Wiley.

Nicholas, T., Wolfe, G., Soltysik, S. S., Garcia, J. L., Wilson, W. J., & Abraham, P. (1983). Postnatal development of heart rate patterns elicited by an aversive CS and US in cats. *The Pavlovian Journal of Biological Science, 18,* 144–153.

Pavlov, I. P. (1927). *Conditioned reflexes.* London: Oxford University Press.

Pavlov, I. P. (1928). *Lectures on conditioned reflexes. Twenty-five years of objective study of the higher nervous activity (behavior) of animals.* New York: International Publishers.

Pearce, J. M., & Hall, G. (1980) A model for Pavlovian learning: Variations in the effectiveness of conditioned but not of unconditioned stimuli. *Psychological Review, 87, 532–552.*

Rescorla, R. A. (1971). Variation in the effectiveness of reinforcement and nonreinforcement following prior inhibitory conditioning. Learning and Motivation, 2, 113–123.

Rescorla, R. A. (1975). Pavlovian excitatory and inhibitory conditioning. In W. K. Estes, (Ed.), *Handbook of learning and cognitive processes* (Vol. 2, pp. 7–35). Hillsdale, NJ: Lawrence Erlbaum Associates

Rescorla, R. A. (1979). Conditioned inhibition and extinction. In A. Dickinson & R. A. Boakes (Eds.), *Mechanisms of learning and motivation* (pp. 83–110). Hillsdale, NJ: Lawrence Erlbaum Associates.

Rescorla, R. A., & Wagner, A. R. (1972). A Theory of Pavlovian conditioning: Variations in the effectiveness of reinforcement and nonreinforcement. In A. H. Black & W. F. Prokasy (Eds.), *Classical conditioning II: Current research and theory* (pp. 64–99). New York: Appleton-Century-Crofts.

Schull, J. (1979). A conditioned opponent theory of Pavlovian conditioning and habituation. *The Psychology of Learning and Motivation, 13,* 57–90.

Segundo, J. P., Galeano, C., Sommer-Smith, J. A., & Roig, J. A. (1961). Behavioural and EEG effects of tones 'reinforced' by cessation of painful stimuli. In J. F. Delafresnaye (Eds), *Brain mechanisms and learning.* (pp. 265–291). Oxford: Blackwells Scientific Publications.

Seligman, M. E. P., & Johnston, J. C. (1973). A cognitive theory of avoidance learning. In F. J. McGuigan & D. B. Lumsden (Eds.), *Contemporary approaches to conditioning and learning* (pp. 69–110). Washington, DC: Winston.

Solomon, R. L., & Corbit, J. C. (1974). An opponent-process theory of motivation: I. Temporal dynamics of affect. *Psychological Review, 81,* 119–145.

Soltysik, S. (1960a). Studies on the avoidance conditioning: 2. Differentiation and extinction of the avoidance reflexes. *Acta Biologiae Experimentalis, 20,* 171–182.

Soltysik, S. (1960b). Studies on the avoidance conditioning: 3. Alimentary conditioned reflex model of the avoidance reflex. *Acta Biologiae Experimentalis, 20,* 183–192.

Soltysik, S. (1963). Inhibitory feedback in avoidance conditioning. *Boletín del Instituto de Estudios Médicos y Biológicos, México, 21,* 433–449.

Soltysik, S., & Wolfe, G. (1980). Protection from extinction by a conditioned inhibitor. *Acta Neurobiologiae Experimentalis, 40,* 291–311.

Soltysik, S. S., Wolfe, G. E., Nicholas, T., Wilson, W. J., & Garcia-Sanchez, J. L. (1983). Blocking of inhibitory conditioning within a serial conditioned stimulus-conditioned inhibitor compound: Maintenance of acquired behavior without an unconditioned stimulus. *Learning and Motivation, 14,* 1–29.

Soltysik, S., & Zieliński, K. (1963). The role of afferent feedback in conditioned avoidance reflex. In E. Gutmann & P. Hnik (Eds.), *Central and Peripheral Mechanisms of Motor Functions* (pp. 215–221). Prague: Publishing House of the Czechoslovak Academy of Sciences.

Suiter, R. D., & LoLordo, V. M (1971). Blocking of inhibitory Pavlovian conditioning in the conditioned emotional response procedure. *Journal of Comparative and Physiological Psychology, 76,* 137–144.

Tyler, T. J. (1971). Effects of restraint on heart-rate conditioning in rats as a function of US location. *Journal of Comparative and Physiological Psychology. 77,* 31–37.

Vasil'ev, P. N. (1906). Vliianie postoronnego razdrazhitelia na obrazovavshiisia uslovnyi reflex. (The effect of an extraneous stimulus on the established conditional reflex). *Transactions of the Society of Russian Physicians in St. Petersburg, 73,* 389–392.

Wagner, A. R. (1978). Expectancies and the priming of STM. In S. H. Hulse, H. Fowler & W. K. Honig (Eds.), *Cognitive processes in animal behavior.* Hillsdale, N.J.: Lawrence Erlbaum Associates.

Wagner, A. R., & Rescorla, R. A. (1972). Inhibition in Pavlovian conditioning: Application of a theory. In R. A. Boakes & M. S. Halliday (Eds.), *Inhibition and Learning* (pp. 301–336). London/New York: Academic Press.

Wolfe, G. E., & Soltysik S. S. (1981). An apparatus for behavioral and physiological study of aversive conditioning in cats and kittens. *Behavior Research Methods & Instrumentation, 29,* 637–642.

Zbrożyna, A. (1958). On the conditioned reflex of the cessation of the act of eating. I. Establishment of the conditioned cessation reflex. *Acta Biologiae Experimentalis, 18,* 137–162.

Author Index

Italics denote pages with bibliographic information.

SUBJECT INDEX

D

E

R

S

T U

W

Z